FOOD PRODUCT and PROCESS INNOVATIONS

About the Editor

Professor Hari Niwas Mishra has over 30 years of professional experience in teaching and research. A Professor of Food Technology in the Agricultural & Food Engineering Department Dr. Mishra is former Chairman of the Post Harvest Technology Centre, IIT Kharagpur. Professor Mishra teaches Food Product & Process Technology, Non Thermal Processing of Food, Industrial Processing of Foods & Beverages and Food Chemistry. His research interests include RTE Health Foods & Nutraceuticals, Novel Food Product & Process Development and Extension of Shelf Life of Perishable Foods. A former President of the Association of Food Scientists & Technologists of India (AFSTI), Professor Mishra has many laurels and awards to his credit. To name a few are All India Food Processors' Association (AIFPA) Presidents award for outstanding contribution to growth and development of food processing industry, Dr JS Pruthi award for new product & process development, Best paper award of AFSTI and other professional bodies.

Professor Mishra has published more than 482 research papers & popular articles, 4 books, 4 e-books, 31 book chapters and has 9 Indian patents to his credit. Besides, he is on the editorial boards of several reputed journals. He has supervised more than 245 student research projects including 9 Post-doctoral and 39 Ph D research scholars. Professor Mishra has handled several international and national sponsored research & industrial consultancy projects. He has been an Expert Member of the National Mission Steering Group (NMSG), Integrated Child Development Services (ICDS), Ministry of Women & Child Development, Government of India. He is also a member of the Board of Directors of M/s Parson Solutions, Cochin.

Professor Mishra has worked in different capacities on various academic and administrative committees of IIT Kharagpur and many other institutions within the country and abroad. He has been actively involved in several committees of different ministries & departments of the Government of India such as MoFPI, MWCD, DBT, ICAR, CSIR, ICMR, FSSAI, BIS, etc. He has worked as examiner of several Ph D, M Tech & M Sc theses; paper-setter and evaluator of various PG and UG courses in food technology and engineering of many Universities. Professor Mishra has been on the Research Advisory Committees, Academic Council, Senate and Board of Studies of several Institutions and Universities such as CFTRI, NDRI, NIFTEM, NABI, CIAB, IBSD, BHU, Allahabad University, Tezpur University, Guwahati University, etc. He has visited several countries abroad including Taiwan, Singapore, Thailand, Malaysia, Australia, Turkey, France, Switzerland, Denmark, Germany, Sweden, Norway, Ireland, UK, USA and Canada to name a few.

Professional details including research publications of Professor Mishra can be seen at http://www.iitkgp.ac.in/department/AG/faculty/ag-hnm or http://www.fctliitkgp.in

FOOD PRODUCT and PROCESS INNOVATIONS

Volume-II

Editor

Professor Hari Niwas Mishra

Food Chemistry and Technology Laboratory
Agricultural & Food Engineering Department
Indian Institute of Technology Kharagpur

NEW INDIA PUBLISHING AGENCY
New Delhi, India

NEW INDIA PUBLISHING AGENCY
101, Vikas Surya Plaza, CU Block, LSC Market
Pitam Pura, New Delhi 110 034, India
Phone: + 91 (11) 27 34 17 17 Fax: + 91 (11) 27 34 16 16
Email: info@nipabooks.com
Web: www.nipabooks.com

Feedback at feedbacks@nipabooks.com

ISBN: 978-93-86546-15-9 (Set)

Composed and Designed by NIPA

Preface

The Food Chemistry and Technology Laboratory (FCTL) is devoted to excellence in learning and research in the field of food science and technology. It has been a fundamental part of the Agricultural & Food Engineering Department, Indian Institute of Technology Kharagpur for the last over three decades. The major focus of FCTL is the development of novel and value added RTE/RTC/RTD health food & beverage products and process technologies and shelf life extension of perishable foods.

This book on "Food Product and Process Innovations" is a compilation of the major research and development work carried out in the FCTL in the last one and a half decade. The contributors of the book are the past and present research scholars and scientists, associated with the FCTL, who worked under the supervision and guidance of Professor H N Mishra. Many of these contributors are now leading researchers and teachers in reputed academic institutions and food industry in the country and abroad. The book comprises of two volumes of 11 chapters each. Volume-I includes chapters on products and processes related to the manufacturing, shelf-life extension and quality evaluation of grain based RTC/RTD/RTE health foods and beverages. Chapters on manufacturing technologies and shelf-life extension of health foods and beverages prepared from milk, fruits & vegetables; instant tea, tea based products, natural antioxidants, colours and preservatives are included in Volume-II

There is a large growing demand for novel food products which not only provide health & nutrition benefits and convenience to the consumers but also offer immense economic promise to the food processing industry and farmers.This book, which discusses the recent and emerging engineering and technological innovations for novel and convenience food products manufacturing and shelf life extension, will play a pivotal role in development and commercial manufacture of such products.This book can be used as a comprehensive resource and reference book by the food processing professionals including scientists, technologists, researchers and students working in industry, reasearch institutions and universities.The book will be beneficial to all those interested in finding out the latest innovations in food processing, packaging and storage research and developments which have emerged and can be launched in the near future in the global market setting up new trends in the food industry.

The hard work and untiring efforts of all the contributors in compilation and publication of this book is greatefully acknowledged. Thanks to all those who helped directly or indirectly in the publication of this book. Special thanks are due to FCTL Scholars for providing editorial assistance and preparation of the manuscript. Information taken from other published literature has been duly cited; any missing is inadvertent. Comments and suggestions of the readers of the book are welcome for further improvements in future editions.

Professor H N Mishra

Contents

Contents of Volume-I

List of Contributors

Anupriya Mazumder, Agricultural & Food Engineering Department, Indian Institute of Technology Kharagpur, West Bengal, India

Apurba Giri, Department of Nutrition, Mugberia Gangadhar Mahavidyalaya Bhupatinagar, Purba Medinipur, West Bengal, India

Chandani Sen, Agricultural & Food Engineering Department, Indian Institute of Technology Kharagpur, West Bengal, India

Chitrangada Das Mukhopadhyay, Centre for Healthcare Science and Technology, Indian Institute of Engineering Science and Technology Shibpur, Howrah, West Bengal, India

Danie Shajie A, Agricultural & Food Engineering Department, Indian Institute of Technology Kharagpur, West Bengal, India

Gayatri Mishra, Agricultural & Food Engineering Department, Indian Institute of Technology Kharagpur, West Bengal, India

Hari Niwas Mishra, Agricultural & Food Engineering Department, Indian Institute of Technology Kharagpur, West Bengal, India

Indira Dey Paul, Agricultural & Food Engineering Department, Indian Institute of Technology Kharagpur, West Bengal, India

J Chitra, Agricultural & Food Engineering Department, Indian Institute of Technology Kharagpur, West Bengal, India

Mousumi Ghosh, Agricultural & Food Engineering Department, Indian Institute of Technology Kharagpur, West Bengal, India

Mrittika Bhattacharya, Department of Food Science &Technology, University of California, Davis, CA, USA

P Prabuthas, Edward Food Research and Analysis Centre Limited, Kolkata, India

Ranjana Pande, Agricultural & Food Engineering Department, Indian Institute of Technology Kharagpur, West Bengal, India

Rewa Kumari, Food Processing and Technology Department, Bilaspur University, Bilaspur, Chhattisgarh, India

Rohit Upadhyay, General Mills India Private Limited, Mumbai , India

Sanchita Biswas, Agricultural & Food Engineering Department, Indian Institute of Technology Kharagpur, West Bengal, India

Shiby Varghese K, Defence Food Research Laboratory (DRDO-DFRL), Siddharthanagar, Mysore, Karnataka , India

Shubhangi Srivastava, Agricultural & Food Engineering Department, Indian Institute of Technology Kharagpur, West Bengal, India

Sinija V R, Department of Food Engineering, Indian Institute of Crop Processing Technology, Thanjavur, Tamil Nadu, India

Smitha Tripathi, Department of Botany, Shivaji College, University of Delhi, India

Sneha Sehwag, Amity Institute of Food Technology, Amity University, Noida Campus, India

Snehasis Chakraborty, Department of Food Engineering and Technology, Institute of Chemical Technology, Mumbai, India

Subrata K Bag, Dairy Engineering Department, West Bengal University of Animal & Fishery Sciences, Nadia, West Bengal, India

Sudhanshi Biloria, Agricultural & Food Engineering Department, Indian Institute of Technology Kharagpur, West Bengal, India

Sunil Manohar Behera, Agricultural & Food Engineering Department, Indian Institute of Technology Kharagpur, West Bengal, India

Swarrna Halder, Agricultural & Food Engineering Department, Indian Institute of Technology Kharagpur, West Bengal, India

1

Extraction, Characterization and Food Utilization of Algal Biomass and Bioactives

A Mazumder, P Prabuthas, A Giri, H N Mishra

1. Algal Biomass

Algae is defined as heterogeneous assemblage of organisms that range in size from tiny cells to giant seaweeds, thus algae mostly are photosynthetic species which include eukaryotes and prokaryotic Cyanobacteria (blue-green algae) and live in aquatic habitats. They are ubiquitous in nature and are found in variety of environments ranging from soil, sand, marshes, brackish water, sea and fresh water. Most of them are capable of manufacturing their own food.

Microalgal research is offering wide potential in food biotechnology and is gaining more attention because of presence of wide range of bioactive and nutraceutical properties and added health benefits.The increasing population is one of the major growing concerns of today's world. To meet the nutritional requirement of the rising population, researchers are looking for optional sources for food which are easy to cultivate, cost effective and produce large amount of bioactive compounds useful to prevent major diseases. Chlorella is unicellular green alga that is a potential source of nutrients like carotenoids, minerals, vitamins, and polyunsaturated fatty acids. Especially, omega-3 fatty acid and carotenoids have received considerable interests due to their roles in the prevention of chronic diseases. Lutein and zeaxanthin from Chlorella have been shown to be associated with eye health and function. Beta-carotene is known to reduce the risk of CVDs and certain cancers. Animal studies also demonstrated that Chlorella supplementation reduced the serum cholesterol levels.

Similarly, the blue green microalgae Spirulina, possesses diverse biological activities and nutritional significance. Regular consumption of Spirulina fortified food improves immune system by increasing phagocytic activity of macrophages, stimulating the production of antibodies and cytokines, increase accumulation of NK cells into tissue and activation and mobilization of T and B cells. Spirulina have also shown to perform regulatory role on lipid and carbohydrate metabolism by exhibiting glucose and lipid profile correcting activity in experimental animals and in diabetic patients. They were also found to be active against several enveloped viruses including herpes virus, cytomegalovirus, influenza virus and HIV. They are capable to inhibit carcinogenesis due to antioxidant properties.

Dunaliella, a halotolerant microalga, is exploited for the production of β-carotene on commercial base. One of this alga's most valuable characteristics is its antioxidant effect. Antioxidant activity was found to be increased in rats receiving natural beta carotene from Dunaliella. Feeding of this alga to rats also found to protect the cornea from UVB-induced radiation damage. It does so by decreasing oxidation and increasing antioxidant activity in the cornea. Dunaliella was also shown to protect the liver by increasing antioxidant activity and decreasing lipid peroxidation.

Haematococcus species is well known for its high production of a strong antioxidant, astaxanthin. The high amount of astaxanthin is accumulated when the environmental conditions become unfavorable which includes high salinity and low nutrients. Natural astaxanthin is a unique antioxidant in that it can protect both the fatsoluble and the watersoluble parts of cells. Most antioxidants only protect one or the other, but astaxanthin can span the cell membrane and have one end of the molecule in the fatsoluble part and one end in the water-soluble part, thereby protecting the entire cell.

The nutritional contents of algae are rapidly gaining importance as a renewable source to substitute the conventional ingredients in the aquaculture/animal feed. To date, it seems that microalgae have the potential to play pivotal roles in order to remedy the energy, environment and food crisis prevailing in the world. Algae are the primary producers and are a source of many essential nutrients. Some of the algae are recognized as balance foods which offer sufficient quantity of proteins, carbohydrates, vitamins, and minerals for normal functioning of human body. Blue green algae, red algae, green algae are reported to have higher contents of dietary fiber. Algae are used as one of important medicinal source due to its antioxidant, anticancer, antimicrobial properties.

In last few decades functional foods have emerged rapidly due to high market demands and promising health benefits. The new trend of research is inclined in development of algal fortified functional food items consumed popularly like pasta, salad dressing, mayonnaise and gelled desserts (Batista et al., 2012). Nutritional composition of algae is well documented and it mainly contains proteins, carbohydrates, lipids and trace nutrients, including vitamins, antioxidants, and trace elements. Incorporation of algae to food products serves various purposes like as natural food colour and also as a source of high value bioactive compounds. Different species of microalgae are used for natural compounds for food industry. The bioactive compounds present in algae include carotenoids, sulphated polysaccharides, essential fatty acids, polypeptides, amino acids, vitamins and minerals (Gouveia et al., 2006). Microalgal pigments are an excellent source of natural food colours which can replace synthetic food colours used in industry. European commission has recently committed to replace 46 synthetic food colours as there are allergic and showing hyperactivity in children. Apart from food, algae have various other application in feed, pharmaceutical and production of biofuel.

1.1 Factors affecting bioactives concentration in algal species

A number of studies deal with the effect of different stress conditions on the metabolism of various bioactives in algal species. The commercial exploitation of the microalgae for the production of carotenoids and PUFAs has already started up with few strains of *Chlorella*, *Dunaliella*, *Haematococcus* etc. The impacts of abiotic stresses such as light, UV-radiation, nutrient starvation, temperature and metals on microalgal metabolism and on the production of biological active compounds is discussed briefly here.

Photosynthetic apparatus of most microalgae adjust themselves according to light level and light quality by optimizing its pigment content. Microalgae have developed different photoprotective mechanisms to manage various sunlight intensities. One of these is the xanthophyll cycle which consists in the reversible conversion of violaxanthin, antheraxanthin and zeaxanthin in green algae and in the reversible conversion of diadinoxanthin and diatoxanthin in brown algae. Acclimation to low irradiance intensity or blue enriched light increases carotenoids synthesis such as Fucoxanthin (Guihéneuf et al., 2009). The photo-protection or the low photo-acclimation leads carbon to carotenoids formation whereas in non-stressfull conditions, carbon serves mainly to growth. Similarly, UV radiation also affects microalgal bioactives synthesis. Microalgae experience high levels of UV-radiation in shallow areas and in low turbid habitats. Several authors have shown that UV exposure increases carotenoids content in microalgal species.

Nutrient starvation is another major factor which induces the pigment metabolism. The reorientation of the metabolism induced by nutrient starvation is illustrated by the accumulation of Astaxanthin in H. pluvialis under nitrogen limitation and phosphate starvation. Astaxanthin accumulation is related to a massive increase in carbohydrate content up to 63% of the cell dry weight (Boussiba et al., 1992) and an increase of lipid content in the cytoplasm. Riyahi et al. (2007) have shown that the production of β-carotene in Dunaliella salina was increased with nitrate 1 mM and salinity 30%. Some metals such as Cu, Fe, Zn are essentials for cellular metabolism as they are essential components of electron transporters involved in photosynthesis and respiration. When metals are present in excess, they induce an oxidative stress and antioxidant defense mechanisms. Many microalgae excrete exo-polysaccharides that adsorb metals in the medium and prevent them to enter inside the cells.

Microalgae can synthesize VLC-PUFA as major fatty acid components. Experiments in controlled conditions are necessary in order to select species producing those PUFAs, and to find out the conditions of their growth and class of lipid. Tonon et al. (2002) have shown that fatty acids accumulate during the stationary phase of growth when nitrate concentration in the growth medium was low. In the marine diatom Nitzschia leavis cultivated at 15, 19 and 23°C, growth is enhanced at the highest temperature but the lowest temperature favours the distribution of PUFAs in phospholipids.

2. Principles and Practices

The demand for algae biomass is increasing worldwide, yet the ability for producers to meet that demand requires development of technologies to reduce the cost of production while maintaining and improving product quality. Producers, processors and breeders are in need of systems that successfully identify, promote and harness algae biomass to maximize profits, secure supply, reduce environmental impacts, increase market competitiveness, sustain small and mid-sized scale producers and earn consumer confidence. The overall concepts and steps involved in the processing of biomass for direct food utilization or for bioactives extraction is presented Figure 1.

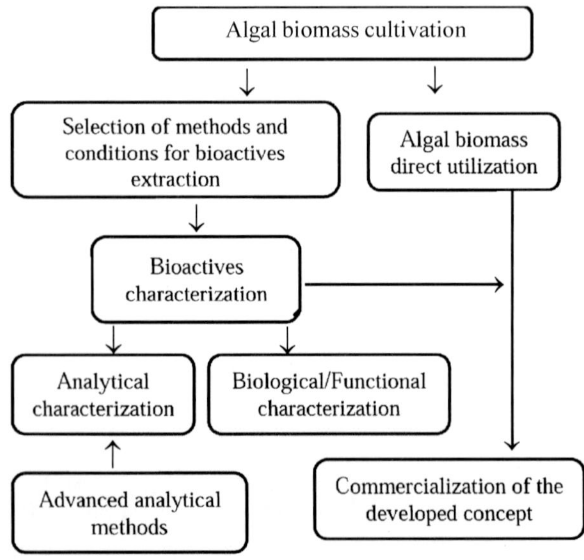

Fig. 1: Steps involved in algal biomass processing

The initial step of processing of algae for food or feed is the concentration of its biomass. Concentrated algae paste can be formed by concentrating algae from mass cultures and preserving the resultant paste through refrigeration, freezing or drying. The harvesting procedures to concentrate microalgae can be obtained by centrifugation, flocculation, filtration, foam fractionation or photo bioreactors. Another alternative to artificial diets that may overcome the costly and unpredictable production of fresh microalgae is the use of algae that has been preserved by drying.

Other than food or feed purposes, the obtained biomass can also be used for extraction of bioactives or natural colours. Understanding the nature of targeted functional ingredients is foremost and essential step before its extraction from the matrices (micro and macro algae). Increasing the permeability of cell walls plays a very important role to increase extraction yield and/or extraction rate. Pretreatment methods to modify the cell structure and increase the permeability of cell walls or membranes include physical, biological, and chemical treatments. In physical methods, mechanical disruption, high-pressure (HP) process, pulsed electric field (PEF) application, ultrasonic treatment, and freeze–thaw, etc are available. In biological methods, different cell wall-degrading enzymes are available commercially to break-down cell walls or membranes and to diminish the overall internal resistance for transporting bioactive compounds from internal matrix to the external solution. In chemical methods, various chemicals for increasing the cell wall permeability are used.

The extraction method may be single or combination of various extraction methods for obtaining better yield of desired bioactive (s). However, the method must also be economical, selective, and environment friendly. The chosen method should follow the legal requirements like use of food-grade solvents and maximum residual limits in the final product. Traditional methods of extraction techniques like solid-liquid extraction, liquid-liquid extraction involves high usage of solvents and longer extraction time. The extraction yield, selectivity and reproducibility of these methods are usually low. More selective methods like supercritical fluid extraction, ultrasound assisted extraction method, microwave assisted extraction, enzyme assisted extraction and pressurized liquid extraction may however, give better results than the conventional extraction procedures.

While large-scale commercialization of algae is still to be realized as a source of biofuels, algaculture is already producing food for humans and animals as well as fertilizers, dyes and colorants for a wide range of industries. However, the commercialization of algae presents a number of challenges, from processing capacity limitations to costs. Significant challenges confront the algae cultivation, including costs. Since the most common method for growing algae is by fermentation, costs are relatively fixed. Tanks and other equipment are required as well as sugar or yeast, a high value feedstock, are necessary. But many are finding ways to address this challenge. Using renewable resources like sunlight and seawater and CO_2 , a renewable feedstock, to cut down on the fermentation costs. Providing separators to be a cost-effective way to extract algae cells from the liquid growth medium. A major challenge is producing enough feedstock to test and improve technologies that allow for large-scale production along the entire production stream. To increase dewatering capacity in the fermentation stage, for example, production facilities would require an equivalent boost in separation capacity. In the end, the success of algae-based products, such as biofuels, depend on the cost per unit. To compete with fossil fuels, the production of biofuels would have to be so efficient that the cost per gallon is economically competitive. The technical problems are solved by introducing new technologies. However, it is difficult to manage the cost at various stages of processing which pose a major challenge.

3. Extraction of Bioactives from Algal Biomass

Algal biomass are rich sources of protein, carotenoids, vitamins, minerals and neutraceutical compounds which are having immense potential for anti microbial, anticancer, antioxidant, antidiebetic and other useful health benefits. The potential and commercial strains of microalgae are *Spirulina* sp., *Chlorella* sp., *Heamatococcus* sp, *Dunaliella salina*. To gain the benefits from algae the bioactive components need to be separated by extraction. Extraction is a technique to withdraw or separate target component from matrix by either using solvent or otherwise by advanced technology like high pressure processing, super critical fluids, pulse electric field, microwave, enzymes and ultrasound assisted extractions. These technologies are useful to separate valuable biomolecules from various algae resources. Extraction of biomolecules from algae can be carried out in two ways, (i) conventional technology and (ii) advanced technology. Conventional technology includes soxhlet extraction, solvent extraction using mechanical shaker or orbiter, freezing and thawing and hydro distillation. Advanced technologies are ultrasound assisted extraction, microwave assisted extraction, enzyme assisted extraction, pressurized liquid extraction and supercritical fluid extraction. Traditionally, extraction of various bioactive compounds from natural products has been performed in many industries using different solvents. Recently, investigators have focused on novel extraction techniques like ultrasound assisted extraction (UAE), microwave assisted extraction, supercritical fluid extraction and pressurized fluid extraction to overcome these drawbacks.

3.1 Methods of extraction

3.1.1 Conventional methods

The natural bioactive compounds present in algae are traditionally extracted with solvents. Lipid soluble or non-polar bioactives are extracted with hexane, dichloromethane, chloroform, acetonitrile and polar compounds are extracted with water, ethanol or methanol. After extraction the solvents are evaporated either using heat or vacuum evaporator. The main advantages of solvent extraction are its low cost and ease of operation in industrial scale. Maceration is an age old technique which was used to prepare homemade tonic in earlier days. It follows few simple steps. Firstly, the algal material is grinded in small particle size to expose the surface area more to the extracting solvent. Then in a closed vessel appropriate solvent or menstruum is added. Lastly, the liquid is strained off and the marc or the residual solid part pressed tightly to recover occluded extract. There is occasional shaking in maceration to increase diffusion and remove concentrated solutions from the sample surface and expose to new solvent to increase extraction yield.

Hydro-distillation is other type of conventional extraction technique where generally water, direct steam and water along with steam is used for extraction of volatile components instead of solvent. This technique is not used in algae extraction due to its low selectivity of target components.

The conventional extraction methods such as maceration and soxhlet extraction have many drawbacks like large amount of solvent utilization, long extraction time, lower extraction yield, large volume of solvent, low extraction selectivity and degradation of heat sensitive compounds specially polyphenols and antioxidants (Azmir et al., 2013).

Chlorella vulgaris were extracted for lutein, β-carotene using maceration and soxhlet technology. The yield was highest in soxhlet extraction rather than maceration. The yields of extraction in maceration were 2.97± 0.31 and 0.08 ± 0.01mg/g in lutein and β-carotene respectively. Whereas, using soxhlet extraction the yields were 3.42 ± 0.11 and 0.26 ± 0.09 mg/g in lutein and β-carotene respectively. In this process soxhlet has shown better extraction efficiency than maceration. In all the extraction 90% ethanol has shown highest extraction yield compared to acetone, hexane, water and various concentration of aqueous ethanol (Cha et al., 2010). In another study the yield of lutein extraction was highest in soxhlet extraction compare to super critical extraction of *Chlorella vulgaris* (Kitada et al., 2009).

3.1.2 Ultrasound assisted extraction (UAE)

Among all the novel techniques, UAE is an inexpensive and simple substitute to traditional extraction techniques. Applications of UAE for the extraction of lipids, proteins, flavonoids, carotenoids from algae are well reported in literature (Ma et al., 2009). UAE uses acoustic mode for producing cavitation bubbles which implodes resulting into high shear forces. This helps in disrupting the cell wall allowing the solvent to penetrate into the plant material and increases the contact surface area between the solvent and compound of interest, resulting into increased mass transfer along with good mixing. Therefore, UAE provides increased extraction yield, increased rate of extraction, reduced extraction time and higher processing throughput along with the advantage of usage of reduced temperature and solvent volume which is very useful for the extraction of heat labile compounds (Vinatoru, 2001).

The application of ultrasonic assisted extraction (UAE) in food processing technology is of interest for enhancing extraction of components from algal biomass. The use of ultrasonication for extraction purposes in high-cost raw materials is an economical alternative to traditional extraction processes, which is an industry demand for a sustainable development. Ultrasound is an oscillating sound pressure wave with a frequency greater than the upper limit of the human hearing range. Ultrasound is thus not separated from 'normal' (audible) sound based on differences in physical properties, only the fact that humans cannot hear it. Although this limit varies from person to person, it is approximately 20 kilohertz (20 kHz) in healthy, young adults. Ultrasound devices operate with frequencies from 20 kHz up to several gigahertz as shown in Figure 2.

The application of ultrasound as a laboratory based technique for assisting extraction from plant material is widely published. Several reviews have been published in the past to extract plant origin metabolites, flavonoids from algae

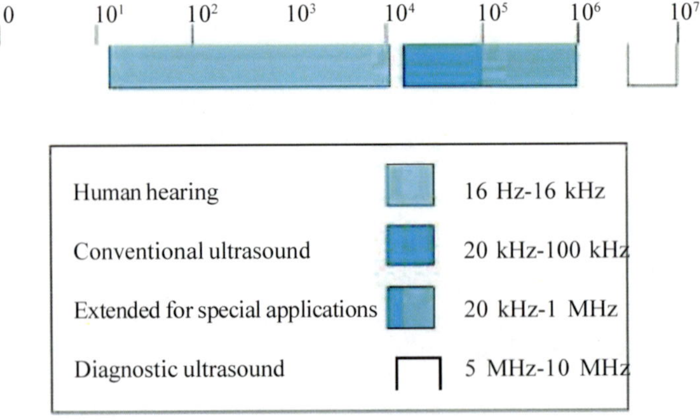

Fig. 2: Frequency ranges of sound

using a range of solvents and bioactives from chlorophycean algae like *Chlorella, Dunaliella* and other species of commercial value.

Using conventional stirred extraction ethanol was significantly less effective than ethyl acetate and butanone. The application of ultrasound improved the relative performance of ethanol such that it was comparable to butanone and ethyl acetate alone. Ultrasound assisted extraction has high influence on selectivity of lutein extraction which has shown better results than pressurized liquid extraction in *Chlorella vulgaris* (Cha et al., 2010). Polysaccharide extraction was highest in *Chlorella pyrenoidosa* with 400 w of ultrasound with 800 s. The total yield of polysaccharide was 44.8 g/ kg of dried algae (Shi et al., 2007).

3.1.3 Microwave assisted extraction (MAE)

The principle mechanism behind microwave assisted extraction is electromagnetic wave which changes the cell structure. In microwave extraction generally two common phenomena takes place simultaneously. First one is dipole rotation and second one is ionic conduction trough reversals of dipoles. The charged ion of solute and solvent get displaced and heat up the medium. The MAE delivers high extraction yield compared other extraction techniques. There are 3 major steps involved in MAE from plant or algal biomass. In the beginning the solutes get separated from the active site of the sample matrix with increased temperature and pressure. In the second step, solvent get diffused across the sample matrix. At last step, the solutes get released into the solvent. MAE is also considered as a green technology as it reduces the solvent consumption rather than other conventional extraction technology (Alupului et al., 2012). Microwave assisted extraction has lot of advantages over conventional extraction processes. MAE provides quicker heating time for bioactive extraction, simple equipment design, and reduced thermal gradient and high extraction yield. This technique is highly selective to organic and organo-metallic components.

Astaxanthin from *Haematococcus pluvialis* was extracted using microwave assisted extraction technology and has shown good yield (594 μg/ 100 mg dry algae powder) with 141 w power, 83 s extraction time and 9.8 mL solvent (Zhao et al., 2009). This extraction technology has given best results for sulphate polysaccharide (fucoidin) from brown algae *Fucus vesiculous*. The yield was highest at 120 psi, 1 min and 1g algae/25 mL of water (Rodriguez-Jasso et al., 2011)

3.1.4 Enzyme assisted extraction (EAE)

The phytochemicals present in algal biomass are sometimes bound with polysaccharide-lignin material by hydrogen or hydrophobic bonding. The bioactive compounds of algae are usually dispersed in cytoplasm and are inaccessible to extraction solvent. Enzyme assisted pretreatment is a novel and useful technique to unbind the material from the sample matrix (Rosenthal et al., 1996). There are some specific enzyme like α-amylase, cellulase and pectinase which significantly increases extraction yield and reduces extraction duration by disrupting the cell wall and hydrolyzing the polysaccharide and solubilizing the lipids (Singh et al., 1999). Various process parameters affect the enzymatic extraction efficiency from sample matrix. Mainly, particle size of sample, concentration and composition of enzyme, solid to water ratio, hydrolyzing time and moisture content are the key factors for EAE (Niranjan et al., 2004; Dominguez et al., 1995).

The main principle behind enzyme enhanced extraction is that it degrades cell wall polysaccharide present as complex sulphated and branched polysaccharides. The heterogeneous polysaccharides are associated with protein and metal ions like calcium, cadmium and iron. The red algal cell wall is made up with carrageenan, agar and cellulose. Enzyme assisted extraction is very useful in antioxidant extraction of seaweeds from *Palmaria palmate* using umamizyme. There are specific enzymes acting on particular cell wall composition like cullulase and carrageanase increases the extraction yield of *Chondrus crispus.* To digest cell wall of *Ulva pertusa* and remove polysaccharide barriers cullolase and macerozyme mixture act as best enzyme compostions. Agarase and cullolase enzymes are used in *Gracilaria verucosa* seaweeds for protein extraction (Fleurence ety al., 1995).

3.1.5 Pressurized liquid extraction (PLE)

The main principle behind the PLE is to keep the solvent in high pressure to extend its boiling point and extract the target component with the help of high pressure and temperature. Higher extraction temperature enhances higher solubility and mass transfer rate of analyte and reduces viscosity and surface tension of the solvent (Ibañez et al., 2012). Pressurized liquid extraction is generally known by few other name like accelerated solvent extraction (ASE), enhanced solvent extraction (ESE) and high pressure solvent extraction (HPSE) (Nieto et al., 2010). In comparison to soxhlet extraction PLE takes less time and gives high yield and reduces solvent consumption.

Pressurized liquid extraction has shown higher extraction efficiency than maceration, soxhlet and ultrasound assisted extraction from *Chlorella vulgaris* with 90 % ethanol as solvent. Temperature (110 °C) was the key parameter for getting highest extraction of carotenoid and chlorophylls. In addition high temperature inhibit the formation of pheophorbide a harful derivate of chlorophyll (Cha et al., 2010).

3.1.6 Super critical fluid extraction

Super critical fluid extraction (SCFE) is a safe technology, consumes less time, highly selective, produce no toxic by-products. Hence, the technology is termed as green technology without any harmful residuals. All the material exists in three basic state namely, solid, liquid or gaseous state. In super critical stage substances behave like an intermediate between liquid and gas and posses both of the properties. The range of pressure and temperature beyond which substances act like super critical material are known as critical points (Inczedy et al., 1998). In super critical fluids surface tension, diffusion and viscosity follows gas like properties and solvation power and density are the liquid like properties. All of these properties makes super critical fluids an excellent solvent with high yield and short time.

Super critical extraction technology is widely used in recovery of novel bioactive compounds from algal biomass. In *Scenedesmus almeriansis* strain 50 % of pigment recovery is possible when 60 °C temperature and 400 bar pressure is applied for 5 h. the extraction efficiency is higher in SCFE compare to the reference method (Macías-Sánchez et al., 2010). Addition of modifiers or entrainers in Supercritical extraction increases extraction yiled. The yield of zeaxanthin is increased when ethanol is added as modifier (16.7% ethanol). The total carotenoid yield is increased to 15.5 mg from 7.61 mg (Liau et al., 2010).

4. Quality Analysis and Characterization of Bioactives

Algae are gaining more attention from medical scientists as a nutraceutical and source of potential pharmaceuticals. Phycocyanin, blue coloured pigment, is a water-soluble protein found to exhibit anti-inflammatory and antioxidizing effects. The essential fatty acid like ϒ-linolenic acid (GLA) which stimulates growth in some animals and makes skin and hair shiny soft and more durable. GLA also acts as an antinflammatory, sometimes alleviating symptoms of arthritic conditions. Supercritical fluid (SF) extracts from various microalgae showed antioxidant and antimicrobial activity and proved as a safe alternative to synthetic antioxidants and antimicrobials (Mendiola et al., 2007). The antioxidant activity of the extracts was lower when compared with BHT and trolox, but significantly higher than α-tocopherol. Organic acids like propionic, benzoic and mandelic acids were also found to exhibit antimicrobial activity. Studies at the Harvard

University School of Dental Medicine found a reduction in mouth cancer when β-carotene extracts, obtained from microalgae, were consumed. The β-carotene solution applied to oral cancer tumors in hamsters reduced the tumor number and size and in some cases these disappeared (Schwartz & Shklar, 1987).

A study was conducted in Food Chemistry & Technology Laboratory, IIT Kharagpur, India to find out the different functional properties in-vitro conditions using the extracts obtained from fresh *Spirulina* biomass. The microalga *Spirulina platensis* var *lonor,* used in this study, was procured from *Spirulina* Production Research & Training Centre (SPRTC), Madurai, Tamilnadu, India. The organism was grown and maintained in 500 mL Erlenmeyer's flasks containing CFTRI (Central Food Technological Research Institute) medium. Extraction of biochemicals from the fresh *Spirulina* biomass was carried out using different solvent systems (Hexane: Acetone and Distilled water : Ethanol), with ultrasound assistance for the maximum yield of β-carotene and phycocyanin. Two sets of optimization experiments were carried out. During the first set of experiments, extraction using non-polar solvent system (Hexane : Acetone, ratio 3 : 2) was employed, and in the second set of experiments, extraction using polar solvent system (Water : Ethanol, ratio 2 : 1) was employed. Ultrasound sound assistance at 30 kHz frequency was provided during extraction. The independent variables used for optimization of extraction process were solvent volume (ml), extraction temperature (°C), extraction time (min) and sonicator amplitude (%). The dependent variable for first set of experiments was β-carotene and for second set of experiments was phycocyanin. The maximum yield of β-carotene (0.179 mg/g) and phycocyanin (30.72 mg/g) was achieved at solvent (hexane : acetone and water: ethanol) volume 20 mL and extraction time 4 min. The level of extraction temperature and sonicator amplitude for β-carotene and phycocyanin was 35°C and 15°C, and 48% and 74%, respectively. The extracts of *Spirulina* obtained from solvent extraction were concentrated in a rotary vacuum evaporator at 40°C and 110 mm Hg for 60 min. The concentrated extract was subjected to initial pre freezing and followed by freeze drying at -55 ± 5 °C at 0.001–1000 mbar (0.0001–100 kPa) vacuum range.

The protein digestibility index; antioxidant, anti-diabetic, anti-microbial and anti-cancer activity of the freeze dried solvent extract of *Spirulina platensis* biomass was performed under in-vitro conditions. The fresh *Spirulina* biomass and its solvent extract counterpart showed the protein digestibility index of 79.40% and 64.74%, respectively. The trend of protein digestibility index is presented in Figure 3.

The antioxidant studies using DPPH method showed IC_{50} value of *Spirulina* extracts and the standard, butylated hydroxyl toluene (BHT), as 155.80 µg/mL and 173.44 µg/mL, respectively. The anti-diabetic activity of extract was

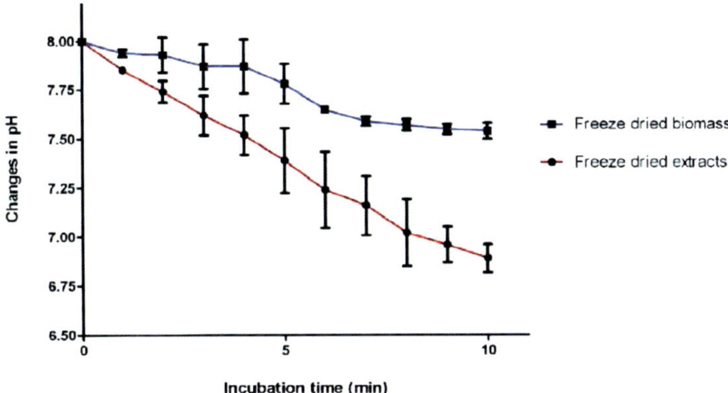

Fig. 3: Changes in the mean pH values with time during digestion with multi-enzyme system for freeze dried *Spirulina* biomass and extracts

demonstrated by in-vitro α-amylase and α-glucosidase inhibitory assay. The results showed that the IC_{50} value of standard acarbose drug was 74.54 µg/mL whereas the extract exhibited an appreciable amount of inhibition on α-amylase and α-glucosidase activity with IC_{50} value of 58.47 and 51.28 µg/mL, respectively. With respect to anti-microbial activity, the strains of *E. coli* showed its sensitivity to the extract with reference to streptomycin as the standard drug. *K. aerogenes* is highly susceptible to the extracts. The detailed results obtained are presented in Table 1.

Table 1: Antimicrobial activity of Streptomycin and *Spirulina* extract against selected bacterial species by disk diffusion method

	Spirulina extract MID[a] (0.05 mg/disk)	*Spirulina* extract MID[a] (0.10 mg/disk)	StreptomycinMID[a] (0.01 mg/disk)
E. coli	12±0.00	21.5±0.50	16.5±0.05
K. aerogenes	45±1.04	76.0±1.00	18.0±0.00
A. faecalis	7±0.50	10.5±0.50	0[b]

[a]MID = Mean Inhibition in diameter; [b]0 = No zone of inhibition

The anti-cancer activity against HCT-15 (human colon cancer) cell lines showed that at 400 µg/mL concentration almost 90% of the cell growth was inhibited with IC_{50} value of 91.43 µg/mL. These results showed that the freeze dried extracts from microalgal could retain its functional properties during solvent extraction process and provide a positive way for improving health and biomedical research.

Similarly, chemical characterization studies of the *Spirulina* extract was also conducted. Various chromatographic techniques (TLC & HPTLC) were performed to confirm the presence of β-carotene and to understand the number of other carotenoids present in the *Spirulina* extract. Thin layer chromatography

(TLC) was carried out using acetone hexane ethyl acetate (ratio 20 : 79.5 : 0.5) solvent systems. The β-carotene separated by TLC was further confirmed by HPLC using reverse phase C_{18} column. The extracts were analyzed for different biochemical constituents using HPTLC. The TLC studies showed the presence of β-carotene (R_f 0.55) with acetone : hexane (8:2) mobile solvent system (Figure 4).

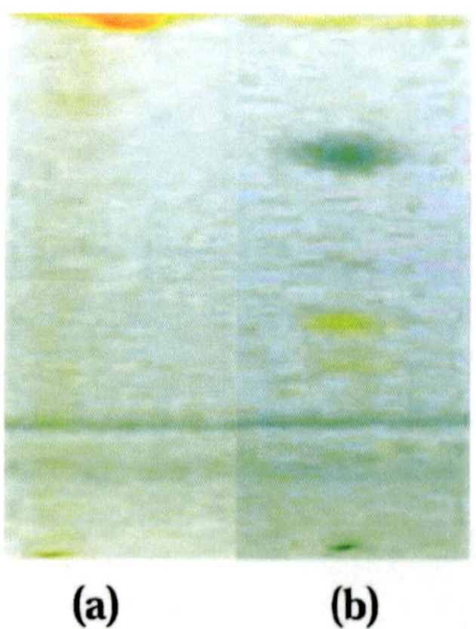

Fig. 4: Pigments separated with *Spirulina* extract sample by TLC using acetone and hexane (8:2) solvent system (a) standard β-carotene, (b) extract sample

Preparative HPLC studies also confirmed the presence of β-carotene (R_t 27.80) in the extract. HPTLC analysis of extract showed the presence of zeaxanthin (R_f 0.13), violaxanthin (R_f 0.20), chlorophyll a (R_f 0.32), and β-carotene (R_f 0.94) (Figure 5).

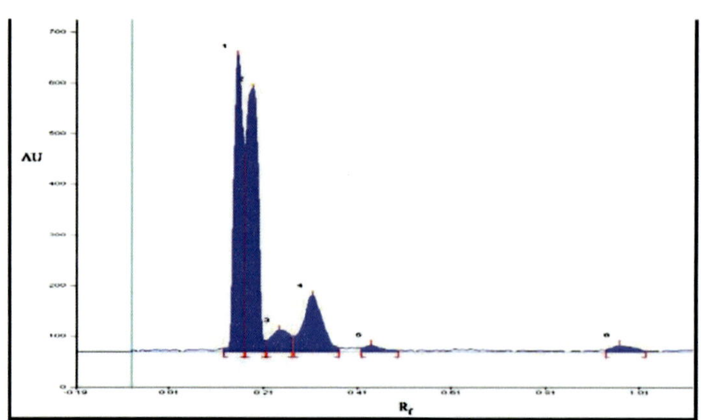

Fig. 5: HPTLC chromatogram showing different peaks obtained with *Spirulina* extract sample

5. Food Utilization of Algal Biomass and Extracts

Now-a-days several combination of algae or mixers with other health foods are available in the market in the form of liquids, pastilles, capsules, tablets, powders as nutritional supplements. These different forms of algal biomass can be added into different food products (e.g. soft drinks, biscuits, breads, yoghurt, candies, pastas, *Shrikhand* and other fermented products). Recently demand of functional food is increasing. Algal biomass may be used as a functional ingredient to develop functional food. In the markets of countries like USA, Japan, France, Germany, Thailand, China functional foods added with algae are available.

5.1 Novel health foods

The compatibility of algal biomass in food system depends on the processing condition and the nature of food matrix or interaction with other food components. The addition of algae in the products changes products' colour, microstructure, rheology and nutritive value. Some food preparations with algae are discussed as follows.

5.1.1 Food gel

Food gel is a suitable media for incorporation of microalgal biomass. Nunes and others (2003, 2006) incorporated micro algal biomass in protein polysaccharide based vegetable dessert which is made of pea protein isolate, κ-carrageenan and starch. Different microalgal biomass such as *Spirulina maxima*, *Chlorella vulgaris*, *Haematococcus pluvialis* and *Diacronema* were used and compared with the products where commercial pigments (phycocyanin, astaxanthin, β-carotene, canthaxanthine and lutin) were used. It was found that gel incorporated with micro algae had lesser intense tonalities and texture modification as compared to commercial pigment addition. Batista et al., (2007) prepared gel system where pea protein, κ-carrageenan and starch were mixed with 0.75% (w/w) micro algae. During cooling its rheological behaviour such as maturation kinetics and visco-elastic function have been monitored. The *Spirulina* addition decreased the gels rheological parameters because of thermodynamic incompatibility of micro algal protein and other component of the gel (Chronakis, 2001).

5.1.2 Food emulsion

Production of oil-in-water emulsion using microalgae is an innovative technique. The colours formed from this colourant are good appealing and stable.

Coloured emulsion with algal pigments: Batista and others (2006a,b) successfully incorporated microalgal pigment in oil- In-water emulsion made with 3% (w/w) pea protein isolate and 65% (w/w) vegetable oil. Phycocyanin extracted from *Spirulina maxima* were used. Phycocyanin is hydrophilic proteinaceous pigment so, it is added at continuous aqueous phase before emulsification process. It has been reported that this pigment have significant effect on emulsion's structure and rheological properties. Phycocyanin protein molecules interact with the surface protein molecule of the oil droplet (Chronakis et al., 2000).

Coloured emulsion with algal biomass: Gouveia et al., (2006) incorporated microalgae *Haemato coccuspluvialis* and *Chlorella vulgaris* to colour oil-in-water pea protein-stabilized emulsion. It was found that there was increase resistance to oxidation for emulsion containing microalgae. The *Haemato coccuspluvialis* showed higher oxidation stability as compared to *Chlorella vulgaris.* The microalgal biomass addition increased the emulsion textural parameters. Addition of 75% (w/w) biomass showed optimal concentration level. At higher level addition of algae, the emulsion became excessively firm as a result the viscosity increased. Raymundo and others (2005) used *Chlorella vulgaris* biomass as a fat mimetic and its emulsifier ability in a pea protein emulsion. It was observed that an emulsion with 2% micro algae is more structural as compared to those without microalgal biomass.

5.1.3 Functional biscuit

Gouveia and others (2007) incorporated *Chlorella vulgaris* in butter biscuit as colouring agent. Upto 1% (w/w) level addition the green colour was stable during storage period. Beyond that the green tonality (-a*) difference in CIELAB system was not significant and the colour was not stable during storage period. The texture of the biscuits significantly changed as algal biomass has been added. Replacement of little quantity of flour with microalgae made a complex matrix which was rich in protein and polysaccharide. The water absorption ability of matrix improved as a result the firmness of biscuit increased. In another study Gouviea et al., (2008) used *Isochrysis galbana* biomass in biscuits. They observed an increase in the textural properties and stability of colour and texture during three months storage period. They also noticed that when biomass addition increased from 1% to 3% the colour of biscuits became green to a brownish. The addition of *Isochrysis galbana* in biscuits gave higher concentration of ω-3 fatty acids (EPA and DHA). Although during manufacturing of biscuits there was drastic thermal treatment. The ω-3 fatty acids remain stable during storage time; the thermal resistance of ω-3 fatty acids was due to the encapsulation of ω-3 fatty acids inside microalgae.

5.2 Chlorella biomass fortified pasta

Mazumder and others (2014) prepared algal pasta with commercial durum semolina flour, 2% Chlorella biomass (w/w) and water. The mixture was extruded using cold extrusion technology to preserve the bioactive component present in algae (Figure 6 a & b). Cooking time, cooking loss, swelling index and water absorption was higher in Chlorella fortified pasta as compared to control (without *Chlorella* biomass). After preparation and cooking pasta samples were analyzed for xanthophyll and β-carotene. The xanthophylls (0.65 mg/ 100g pasta) and β-carotene (0.163 mg/100g pasta) were lesser than the powder Chlorella biomass. This loss happened during processing and cooking of pasta.

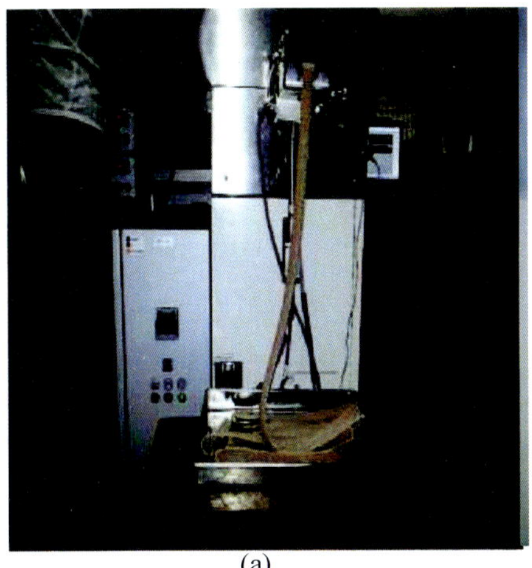

(a) (b)

Fig. 6: (a) Manufacture of extruded pasta, and (b) Chlorella fortified pasta

Fradique et al., (2013) prepared pasta where *Isochrysis galbana* and *Diacronema vlkianum* biomass were incorporated (0.5-2g/100g DW) separately. They noticed that the prepared products showed high resistance to the thermal treatment during cooking. The product containing high amount to omega-3 fatty acid (EPA and DHA). It has been reported that at higher level (2g/100g DW) addition of these algae produce fishy flavour so, it is recommended to use this product in fish based culinary preparation.

5.3 *Spirulina* biomass fortified Shrikhand

Shrikhand is an indigenous fermented and sweetened milk product of Indian origin and regularly consumed in Gujarat, Maharashtra and certain parts of Karnataka, Madhya Pradesh and Rajasthan. In addition to these places, because of its typical sweet-sour taste, it is becoming popular in other parts of the country. Shrikhand is prepared by blending *Chakka* with sugar, cream and other ingredients like fruit pulp, nut, flavor, spices, color, etc. to achieve the finished product of desired composition, consistency and sensory attributes. Shrikhand has a typical semi-solid consistency with a characteristic smoothness, firmness and pliability that make it suitable for consumption directly after meal or with puree or bread.

Giri and et al., (2014) developed a functional shrikhand where *Spirulina* biomass was incorporated (Figure 7). The Shrikhand was prepared by partially removing water from curd or *dahi* (*Chakka)* and it was mixed with sugar (@ 76.4% w/w of *Chakka*), Cardamom (@ 1g / kg *Chakka*) and *Spirulina* powder (@ 0, 0.5, 1 & 1.5% w/w of *Chakka* in different batches) and blended to smooth and homogenous consistency (Figure 8). As the addition of *Spirulina* increased from 0 to 1.5%, protein acid ash percentage in Shrikhan increased whereas moisture, fat, and carbohydrate percentage gradually decreased. Addition of *Spirulina* into Shrikhand diluted the acidity and as a result pH increased. The strong diacetyl flavour of the Shrikhand masked this undesirable fishy flavour of *spiruline* powder. Though, the addition of 1.5% *Spirulina* made the product quite thicker but that was not significant. However, in *Spirulina* added sample colour and appearance scored significantly ($p<0.05$) lower value as compared to the control. *Spirulina* powder did not properly dissolve in the Shrikhand matrix. So, sensory panelist distinguishe dark green spots on the product.

Shrikhand (contains 1.0% *Spirulina* biomass) contained β-carotene - 1.34 mg/100 g, phycocyanin 121.08 mg/100 g, chlorophyll a 4.16 mg/100 g and chlorophyll b 3.37 mg/100 g.

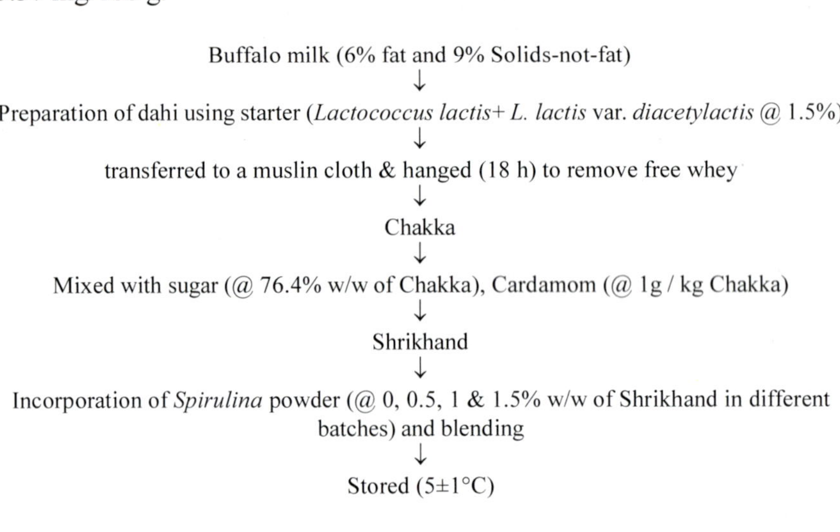

Fig. 7: Flow diagram for manufacture of *Spirulina* fortified Shrikhand

Fig. 8: *Spirulina* fortified *Shrikhand* along with control *Shrikhand;* (a) Control; (b) *Shrikhand* with 0.5% *Spirulina*; (c) *Shrikhand* with 1% *Spirulina*; (d) *Shrikhand* with 1.5% *Spirulina*
(See colour version on page 333)

5.4 Algal biomass fortified jam

Singh and others (2014) reported development of functional jams. Two kinds of fortified jams were prepared, one with the addition of the Chlorella powder in varying percentages (1%, 2% and 3%, w/w) and the other with the addition of the extract (5mg/100g jam). Upon sensory evaluation it was noticed that the jams with 2% and 3% chlorella concentrations were not satisfactory and hence the jam with 1% chlorella powder concentration was good. The jam which was fortified with the extract showed very promising result on the sensory score card. Texture analysis showed that samples with higher concentration of chlorella powder showed a decrease in hardness along with an increase in deformation at hardness. The sample fortified with Chlorella extract showed more than 10 times the higher vitamin A content as compared to control and considerable loss in the sensory characteristics.

5.5 *Spirulina* based RTS beverage

For preparation of any standard soft beverages the British Soft Drinks Association (BSDA, 2009) recommended certain common ingredients like acids, colourings, flavorings, sugars or sweeteners, preservatives, health promoting substances like vitamins and minerals, etc. In this study, extracts from the fresh biomass of *Spirulina* was used for the preparation of RTS beverage. Prepared extract (0.1 g) was mixed with 200 mL of purified, pre-sterilized and cooled drinking water under sterile environment. The other ingredients which included 0.1% (w/v) table salt, 20% (w/v) sugar, 0.25% (w/v) citric acid, 0.05% (w/v) sodium benzoate and 0.02 % (v/v) of pineapple flavor were added previously into the purified water (Figure 9). Four different samples were prepared which differed only in terms of pectin (0 % and 1 %) and aloe vera juice with TSS 0.9 °Bx (0 % and 10 %). Samples 1, 2, 3 and 4 contains pectin and aloevera with the concentration of 1 & 0 %, 0 & 10 %, 0 & 0 % and 1&10 %, respectively. The prepared beverages (Figure 10) were sealed in the pre-sterilized glass bottles under sterile environment and stored under refrigerated conditions.

Studies conducted on sensory evaluation of RTS beverage with different combinations of ingredients showed that there was no much variation in the other components except the pectin used which showed difference in the viscosity of the beverage at different concentrations affecting the mouth feel. The aloe vera juice with TSS 1.5 °Bx was also tested to improve the thickness of beverages in place of pectin. The aloe vera, apart from improving thickness is also expected to improve the digestion and cleaning effect in stomach. The extract of *Spirulina* was considered as the important raw material for preparation of this RTS beverage. Even though there is no specific RDA for intake of phycocyanin, there is a patent which recommend the quantity of

Spirulina platensis* var *lonor
20 % inoculum (0.1 - 0.2 OD at 56 nm)
↓
Inoculated into CFTRI medium (g/L)
$NaNO_3$ 2.25, $MgSO_4$ 0.5, $CaCl_2$ 0.02, pH 10
↓
Growth for 16 days
33 °C, 67.5 µm/m^2/s light intensity
↓
Harvesting
Centrifugation at 5000 rpm for 10 min
↓
Extraction
Non-polar (hexane:acetone, 3:2) and polar (H_2O:ethanol, 2:1) solvents
Fresh biomass 10 g; solvent volume 20 mL; time 4 min; temperature 35 °C and 15 °C for β-carotene and phycocyanin, respectively
↓
Centrifugation
10000 rpm for 10 min
↓
Vacuum concentration
40 °C at 110 mm of Hg for 60 min, final concentration 28 °Bx
↓
Pre-freezing and freeze drying
-55.5 °C, 0.001-1000 mbar for 4 to 6 h
↓
***Spirulina* extract powder**
0.68 g, 4.7 % moisture (wb)
↓
Preparation of RTS beverage
200 mL of beverage contains - 0.2 g of extract; 1 % pectin; 0.1 % NaCl; 20 % sugar; 0.25 % citric acid; 0.05 % sodium benzoate
↓
Storage
Transparent glass bottles, at 5 °C for 2 months or at 28 °C to 45 °C for 30 days

Fig. 9: Process and material flow-chart for the preparation of RTS beverage from fresh *Spirulina* biomass

phycocyanin intake in a range of 0.25 to 2.5 g/day which promotes normal cell activities and at the same time controls cell function that prevents generation of malignancy such as cancer or inhibits its growth or recurrence. The RTS beverage was prepared by using 0.05 % w/v (0.1g /200 mL) of the extract which contains about 15 % of RDA for phycocyanin and about 30 % of RDA for β-carotene, along with other ingredients. Finally, four beverage samples were prepared for final sensory evaluation by ranking using fuzzy logic method. Sensory evaluation and ranking of *Spirulina* beverages by fuzzy logic method showed that Sample 1 which contained 1 % (w/v) pectin along with other common ingredients (0.1 % w/v table salt, 20 % w/v table sugar, 0.25 % w/v citric acid, 0.05 % w/v sodium benzoate and 0.02 % w/v of pineapple flavor) was ranked first, followed by Sample 2 containing 10 % (v/v) aloe vera juice. The third and fourth ranks were for Sample 3 (no pectin and aloe vera juice) and Sample 4 (1 % w/v pectin and 10 % v/v Aloe vera), respectively. Taste had the highest value (0.7824) and was considered to be 'highly important' for beverage in general, followed by aroma (0.9840) and mouth feel (0.6619).

The sample showing best rank was taken for storage studies in transparent glass bottles under 3 temperature conditions viz 5±2°C, 30±2°C and 45±2°C maintained thermostatically in refrigerator/incubator. Samples were taken out at 0th, 5th, 15th, 30th and 60th day interval and analyzed for colour change (ΔE), total microbial count, quantity of phycocyanin and β-carotene, pH, acidity, and viscosity using standard analytical techniques. The rate of change in colour (ΔE) values was found to be steadily increasing (0 to 49) with storage days irrespective of the storage conditions. The increase in ΔE value indicates that storage days have negative effect on the product quality in terms of colour. The TSS (21.5 to 21.8 °Bx) and acidity (0.42 to 0.47 % w/v) of the beverage was found to be increasing with increase in storage period. During refrigerated condition (5±2 °C) there was no variation in the total microbial count (2 to 4 CFU/mL) upto 60 days of storage. In all the cases, the refrigerated storage (5±2 °C) showed no appreciable variation in phycocyanin (36 to 30.4 mg/100 mL), β-carotene (0.22 to 0.19 mg/100 mL) and chlorophyll content (1.12 to 1.11 mg/100 mL) upto 60 days of storage.

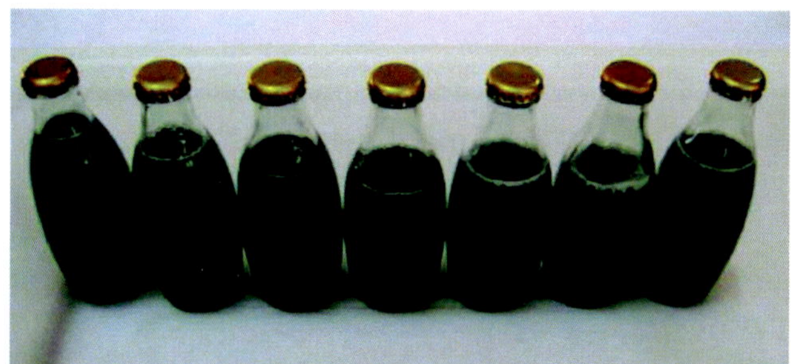

Fig. 10: The developed RTS *Spirulina* beverage from the extract of fresh *Spirulina* biomass (See colour version on page 333)

6. Quality & Safety Issues and Challenges

Some marine algae contain toxic substance and they are eaten by the marine shellfish (clams, mussels, oysters), finfish etc. which create adverse effect to human. Different poisoning syndrome created by the algal toxins are- neurotoxic shellfish poisoning, paralytic shellfish poisoning, amnesic shellfish poisoning, diarrhoeic shellfish poisoning, ciguatera fish poisoning etc. For food supplements algae such as *Isochrysis*, *Tetraselmis suecica*, *Skeletonema costatum*, *Nannochloropsis* sp., *Chaetoceros gracilis*, *Pavlova lutheri*, *Dunaliella tertiolecta*, *Phaeodactylum tricornutum*, *Chlorella* sp. can be used because they do not produce any toxins. *Chlorella* and *Spirulina* are commonly sold as food supplements both in the USA and in EU. *Dunaliella* derived β-carotene and *Crypthecodinium cohnii* derived decosahexaenoic acid are permitted as food ingredient by European Food Safety Authority.

For food and feed application of algae, knowledge of the safety of algal strain is very important. The Department of Botany of the Smithsonian National Museum of Natural History has prepaered an website of an overview of harmful Dinoflagellates and diatoms. In the USA *Spirulina* has been GRAS status for use as ingredient in foods at small level (0.5-3.0 g/serving). However, precautionary measures would be necessary on the consumption of *Spirulina* and *Spirulina* related products for people in dialysis, pregnant women and nursing mothers.

On microalgae based food products, two USA laws are applicable when they are traded on the consumer market: (1) the Federal Food, Drug and Cosmetic Act (1938) for all foods and food additives, (2) the Dietary Supplement Health and Education Act (1994), which amended the Food, Drug and Cosmetic Act to cover dietary ingredients and supplements.FDA reviewed several algae based GRAS food ingredients such as microalgal oil derived from *Ulkenia* sp. SAM2179, *Spirulina*: the dried biomass of *Arthrospira platensis*, "Calcified seaweed" derived from *Phymatolithon calcareum*or *Lithothamnium corallioides*, algal oil (*Schizochytrium* sp.) etc.

There are two aspects for biosafety of GM algae potential adverse environmental effect and potential harm to animal or human health for food, feed or pharmaceutical applications. In Europe, the European Food Safety Authority and in the USA the FDA, are responsible for this biosafety evaluation. The risk evaluation is done by identification of harmful characteristics of microalgae due to properties of the vector, the insert, recipient organism and the resulting characteristics of GM micro-algae and their derived products with respect to the environment and human health (Enzing et al., 2014).

7. Summary

Algae are potential purveyors of various natural substances of commercial value. Natural products derived from algae include protein, pigments, sulphated polysaccharides, alginate, carotenoids, polyphenols, flavanoids, ω-3 and ω-6 fatty acids, minerals and other essential nutrients. Microalgae derived additives like luteins, zeaxanthin, and β-carotene have huge application in pharmaceutical and food industry as natural colorant and antioxidant. Alginates produced from macroalgae are widely used in bakery, ice-cream and pharmaceutical industry. In the era of functional food, incorporation of different algae and algal extract in the food products is an innovative idea. In this regard some research has been carried out by incorporating it into food products like food gel, food emulsion, biscuit, pasta, shrikhand, jam, beverage, etc. In future, algae or algal extract will be tried to fortify in more products. But, during algal fortification some points should be keep in mind such as- addition of algae or algal extract should be in widely or regular consumed food, during addition of algae or algal extract the level of incorporation should be such a way that the product's unique sensory characteristics should not be compromised, the technology of fortification or the fortified product should be economical, packing of the algal food should be very attractive to the consumer and should contain details of the health benefit of the algal food beside its other general information. Till now availability of algal food product in the market is very limited. In this regard industry should come forward to explore it to the consumer. In future algal biomass can represent as a strong agent of potential source of neutraceuticals, antimicrobials and antioxidant if industry commercializes the extraction technologies.

Reference

Alupului, A., Călinescu, I., Lavric, V. 2012. Microwave extraction of active principles from medicinal plants. *UPB Science Bulletin, Series B*, 74(2).

Azmir, J., Zaidul, I.S.M., Rahman, M.M., Sharif, K.M., Mohamed, A., Sahena, F., Omar, A.K.M. 2013. Techniques for extraction of bioactive compounds from plant materials: a review. *Journal of Food Engineering*, 117(4): 426-436.

Batista, A.P., Gouveia, L., Nunes, M.C., Franco, J.M., Raymundo, A. 2007. Microalgae biomass as a novel functional ingredient in mixed gel systems. In: Williams PA, Phillips GO, editors. Gums and Stabilisers for the Food Industry, 14th Edition. Cambridge: Royal Society of Chemistry p 487-494.

Batista, A.P., Nunes, M.C., Fradinho, P., Gouveia, L., Sousa, I., Raymundo, A., Franco, J.M. 2012. Novel foods with microalgal ingredients–Effect of gel setting conditions on the linear viscoelasticity of *Spirulina* and *Haematococcus* gels. *Journal of Food Engineering*, 110(2): 182-189.

Batista, A.P., Raymundo, A., Sousa, I., Empis, J., Franco, J.M. 2006a. Colored food emulsions – implications of pigment addition on the rheological behaviour and microstructure. *Food Biophysics*, 1:216-227.

Batista, A.P., Raymundo, A., Sousa, I., Empis, J. 2006b. Rheological characterization of coloured oil-in-water food emulsions with lutein and phycocyanin added to the oil and aqueous phases. *Food Hydrocolloids*, 20:44-52.

Boussiba, S., Fan, L., Vonshak, A. 1992. Enhancement and determination of astaxanthin accumulation in green alga Haematococcus pluvialis. *Meth Enzymol*, 213:386-391.

BSDA 2009. http://www.britishsoftdrinks.com/default.aspx?page=432.

Cha, K.H., Lee, H.J., Koo, S.Y., Song, D.G., Lee, D.U., Pan, C.H. 2010. Optimization of pressurized liquid extraction of carotenoids and chlorophylls from Chlorella vulgaris. *Journal of Agricultural and Food Chemistry*, 58(2):793-797.

Chronakis, I.S., Galatanu, A.N., Nylander, T., Lindman, B. 2000. The behaviour of protein preparations from blue-green algae (*Spirulina platensis* strain Pacifica) at the air/water interface.Colloids and Surfaces A: *Physicochemical and Engineering Aspects*, A173:181-

192.

Chronakis, I.S. 2001. Gelation of edible blue-green algae protein isolate (*Spirulina* platensisstrain Pacifica): thermal transitions, rheological properties, and molecular forces involved. *Journal of Agricultural and Food Chemistry,* 49:888-898.

Dominguez, H., Nunez, M.J., Lema, J.M. 1995. Enzyme-assisted hexane extraction of soya bean oil. *Food Chemistry,* 54(2):223-231.

Enzin, C., Ploeg, M., Barbosa, M., Sijtsma, L. 2014.Microalgae-based products for the food and feed sector: an outlook for Europe. In: Vigani M, Parisi C, Cerezo ER, editors. JRC Scientific and Policy Reports. Luxembourg: Publications Office of the European Union.

Fleurence, J., Massiani, L., Guyader, O., Mabeau, S. 1995. Use of enzymatic cell wall degradation for improvement of protein extraction from Chondrus crispus, *Gracilaria verrucosa* and Palmaria palmata. *Journal of Applied Phycology,* 7(4):393-397.

Fradique, M., Batista, A.P., Nunes, M.C., Gouveia, L., Bandarra, N.M., Raymundo, A. 2013. Isochrysisgalbana and *Diacronema vlkianum* biomass incorporation in pasta products as PUFA's source. *LWT-Food Science and Technology,* 50(1):312-319.

Giri, A., Mazumder, A., Mishra, H.N. 2014. Development of *Spirulina* Biomass Incorporated Functional Shrikhand. National Seminar on 'Non Thermal Processing Techniques: Emerging Innovations for Sustainable, Safe & Healthy Foods', organized by Dept. of Food Technology, HIT, Haldia, West Bengal, on 21-22nd March, Published in Proceedings. ISBN: 978-81-927768-2-8. Page: 74.

Gouveia, L., Batista, A.P., Miranda, A., Empis, J., Raymundo, A. 2007. Chlorella vulgaris biomass used as colouring source in traditional butter cookies. *Innovative Food Science & Emerging Technologies*, 8:433-436.

Gouveia, L., Batista, A.P., Raymundo, A., Sousa, I., Empis, J. 2006. Chlorella vulgaris and Haematococcuspluvialisbiomass as colouring and antioxidant in food emulsions. *European Food Research and Technology*, 222, 362-367.

Gouveia, L., Coutinho, C., Mendonça, E., Batista, A.P., Sousa, I., Bandarra, N.M., Raymundo, A. 2008 Sweet biscuits with *Isochrysis galbana* microalga biomass as a functional ingredient. *Journal of the Science of Food and Agriculture,* 88(5):891-896.

Guihéneuf, F., Mimouni, V., Ulmann, L., Tremblin, G. 2009. Combined effects of irradiance level and carbon source on fatty acid and lipid class composition in the microalga Pavlova lutheri commonly used in mariculture. *Journal of Experimental Marine Biology and Ecology,* 369:136-143.

Ibañez, E., Herrero, M., Mendiola, J.A., Castro-Puyana, M. 2012. Extraction and characterization of bioactive compounds with health benefits from marine resources: macro and micro algae, cyanobacteria, and invertebrates. In *Marine Bioactive Compounds* (pp. 55-98). Springer US.

Inczedy, J., Lengyel, T., Ure, A.M. 1998. Supercritical Fluid Chromatography andExtraction. Compendium of Analytical Nomenclature (Definitive Rules 1997), Third ed. Blackwell Science.

Kitada, K., Machmudah, S., Sasaki, M., Goto, M., Nakashima, Y., Kumamoto, S., Hasegawa, T. 2009. Supercritical CO2 extraction of pigment components with pharmaceutical importance from Chlorella vulgaris. *Journal of Chemical Technology and Biotechnology,* 84(5):657-661.

Liau, B.C., Shen, C.T., Liang, F.P., Hong, S.E., Hsu, S.L., Jong, T.T., Chang, C.M.J. 2010. Supercritical fluids extraction and anti-solvent purification of carotenoids from microalgae and associated bioactivity. *The Journal of Supercritical Fluids,* 55(1):169-175.

Ma, Y.Q., Chen, J.C. Liu, D.H., Ye, X.Q. 2009. Simultaneous extraction of phenolic compounds of citrus peel extracts: Effect of ultrasound. *Ultrasonics Sonochemistry,* 16(1):57-62.

Macías-Sánchez, M.D., Fernandez-Sevilla, J.M., Fernández, F.A., García, M.C., Grima, E.M. 2010. Supercritical fluid extraction of carotenoids from Scenedesmus almeriensis. *Food Chemistry*, 123(3): 928-93

Mazumder, A., Giri, A., Mishra, H.N. 2014. Development of process technology of microalgae enriched pasta products. Proceduing of the National Seminar on 'Non Thermal Processing Techniques: Emerging Innovations for Sustainable, Safe & Healthy Foods', Haldia, West Bengal, 21-22nd March.

Mendiola, J.A., Jaime, L., Santoyo, S., Reglero, G., Cifuentes, A., Ibanez, E., Senorans, F.J. 2007. Screening of functional compounds in supercritical fluid extracts from *Spirulina platensis, Food Chemistry,* 102: 1357-1367.

Nieto, A., Borrull, F., Pocurull, E., Marcé, R.M. 2010. Pressurized liquid extraction: a useful technique to extract pharmaceuticals and personal-care products from sewage sludge. *TrAC Trends in Analytical Chemistry,* 29(7):752-764.

Niranjan, K., Hanmoungjai, P., Dunford, N.T., Dunford, H.B. 2004. Enzyme-aided aqueous extraction. *Nutritionally Enhanced edible Oil and Oilseed Processing*, 89-90.

Nunes, M.C., Batista, P., Raymundo, A., Alves, M.M., Sousa, I. 2003. Vegetable proteins and milk puddings.*Colloids and Surfaces B: Biointerfaces,* 31:21-29.

Nunes, M.C., Raymundo, A., Sousa, I. 2006. Rheological behaviour and microstructure of pea protein / k"carrageenan / starch gels with different setting conditions. *Food Hydrocolloids,* 20:106-113.

Raymundo, A., Gouveia, L., Batista, A.P., Empis, J., Sousa, I. 2005. Fat mimetic capacity of Chlorella vulgaris biomass in oil-in-water food emulsions stabilised by pea protein. *Food Research International,* 38:961-965.

Riyahi, J., Haouazine, Y., Akallal, R., Mouradi, A., Creach, A., Givernaud, R., Mouradi, A. 2007. Influence des nitrates, de la salinité et du stress lumineux sur la teneur en acides gras et en β-carotène de Dunaliella salina. *Bull Soc Pharm Bordeaux,* 146:235-250.

Rodriguez-Jasso, R.M., Mussatto, S.I., Pastrana, L., Aguilar, C.N., Teixeira, JA. 2011. Microwave-assisted extraction of sulfated polysaccharides (fucoidan) from brown seaweed. *Carbohydrate Polymers,* 86(3):1137-1144.

Rosenthal, A., Pyle, D.L., Niranjan, K. 1996. Aqueous and enzymatic processes for edible oil extraction. *Enzyme and Microbial Technology,* 19(6):402-420.

Schwartz, J., Shklar, G. 1987. Regression of experimental hamster cancer by beta-carotene and algae extracts. *Journal of Oral and Maxillofacial Surgery,* 45: 510-515.

Shi, Y., Sheng, J., Yang, F., Hu, Q. 2007. Purification and identification of polysaccharide derived from *Chlorella pyrenoidosa. Food Chemistry,* 103(1): 101-105.

Singh, R.K., Sarker, B.C., Kumbhar, B.K., Agrawal, Y.C., Kulshreshtha, M.K. 1999. Response surface analysis of enzyme assisted oil extraction factors for sesame, groundnut and sunflower seeds. *Journal of Food Science and Technology,* 36(6):511-514.

Singh, Y., Giri, A., Mazumder, A., Mishra, H.N. 2014. Optimization of ultrasound assisted solvent extraction of β-carotene from Chlorella, and its utilization in fortification of apple jam. 2nd International Conference on 'Algal Biorefinery: A potential source of food, feed, biochemicals, biofuels and biofertlizers' at Denmark Technical University, Denmark on 27th-29th August.

Tonon, T., Harvey, D., Larson, T.R., Graham, I.A. 2002. Long chain polyunsaturated fatty acid production and partitioning to triacylglycerols in four microalgae. *Phytochemistry*, 2002, 61: 15-24.

Vinatoru, M. 2001. An overview of the ultrasonically assisted extraction of bioactive principles from herbs. *Ultrasonics Sonochemistry,* 8(3):303-313.

Zhao, L., Chen, G., Zhao, G., Hu, X. 2009. Optimization of microwave-assisted extraction of astaxanthin from *Haematococcus pluvialis* by response surface methodology and antioxidant activities of the extracts. *Sep Sci Technol,* (1):243-262.

2

Instant Soluble Tea Powder and Ready-To-Use Tea Products

V R Sinija, S M Behera, H N Mishra

1. Introduction

Each day, millions of people take a small bag, drop it into a cup, pour in boiling water, and add a dash of milk plus a spoonful of sugar. People start the day with it, end the day with it, and also serve it socially or in times of distress (Herath and De Silva, 2011). The reason behind its popularity is based on the fact that the combination of theanine amino acid contained in tea and melatonin and B-complex vitamins contained in milk offers superior relaxing effects. Furthermore, the harmony of tea's sweet aroma and milk's natural sweetness will create a healthy low calorie drink.

Tea is one of the most pleasant drinks consumed by two third of the world's population. It is not only the most popular and one of the oldest known beverages in the world but also an antioxidative agent available in everyday life which may help to prevent a wide variety of diseases such as cancers and heart diseases (Yang and Landau, 2000; Luczaj and Skrzydlewska, 2005). Tea provides a significant source of phenolic compounds in the diet. While polyphenols have long been of interest for their contribution to the astringenttaste of teas (Ding et al., 1992), research into the health benefits of tea has been actively pursued only in more recent times (Dufresne and Farnworth, 2001; Yamamoto et al., 1997). Even though the results are mixed for the relationship between tea and health in epidemiological and clinical studies, in vitro studies point to a range of mechanisms by which tea phenolics may help to offset chronic diseases such as cancer and cardiovascular diseases (Kris-Etherton and Keen, 2002).

More than 482 million metric tonnes of tea was produced globally in the year 2013 out of which above 1.86 million metric tonnes or 39 % of total production found place in the international export market (Bordoloi, 2012). Though tea is produced by a large number of countries, the production is dominated by four countries, viz. China, India, Kenya and Sri Lanka. These four countries account for more than three quarters of global tea production. The tea industry in India occupies an important place and plays a useful part in the national economy. India is the second largest tea producer in the world with production of 1205.40 million kg in 2013-14. Interestingly, India is also the world's largest consumer of tea with the domestic market consuming 890 million kg of tea during 2012-13 (Figure 1).

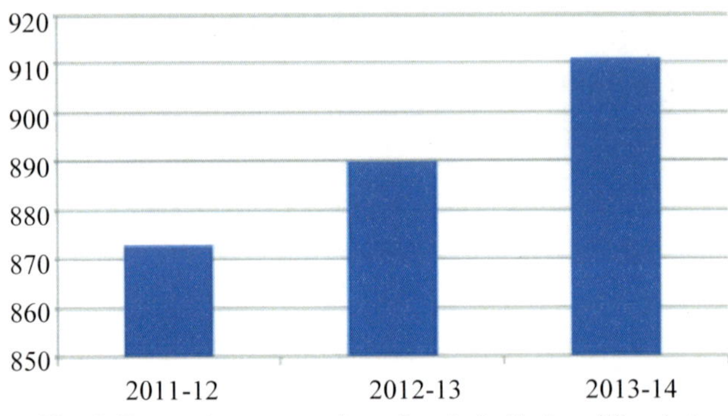

Fig. 1: Domestic consumption of tea in India (in million kg)

India is ranked fourth in terms of tea exports, which reached 218.12 million kg during 2013-14 and were valued at US$ 727.04 million. The top importing markets were Commonwealth of Independent States (CIS, 51.58 million kg), UAE (23.19 million kg) and Iran (22.42 million kg). All varieties of tea are produced by India. While CTC (Crushing, Tearing, Curling) accounts for around 89 % of the production, orthodox/green and instant tea account for the remaining 11%. India has around 579.35 thousand hectares of area under tea production. Tea production is led by Assam, West Bengal, Tamil Nadu and Kerala as shown in Figure 2. According to estimates, the tea industry is India's second largest employer. It employs over 3.5 million people across some 1,686 estates and 157,504 small

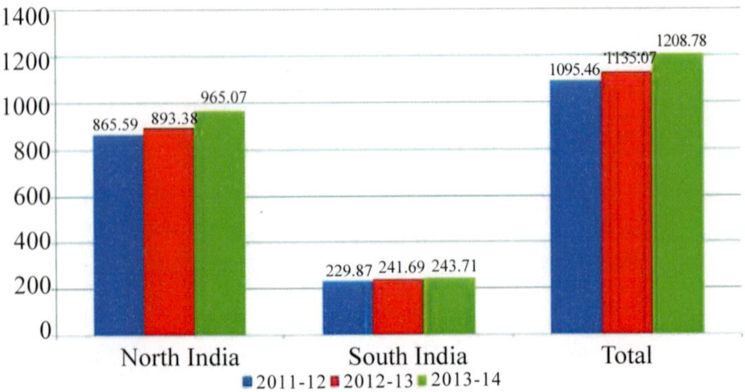

Fig. 2: Tea production in different parts of India

holdings; most of them women. Production reached 1208.78 million kg in 2013-14. Around 965.07 million kg was produced in North India and 243.71 million kg was produced in South India. (*http://www.teacoffeespiceofindia.com/tea/tea-statistics*)

Tea processing had many incarnations over the last one hundred years from loose tea to blended, to packet teas, to tea bags and finally to instant teas, ready-to-drink teas and flavoured teas. Instant tea, a product dried from tea infusion was first produced from black tea in England during 1940's. Although its production has been under study for a long time, instant tea is the major problem for the tea industry now- a- days, with regard to both its production and acceptance (Hart, 2008).

Due to its aroma, stimulatory effect and health benefits, green tea is now becoming more popular and one of the most widely consumed beverage worldwide. Green tea ploymphenols are considered beneficial to human health, especially as chemo preventive agents (Yamamoto et al., 1997). The cardio protective effect of flavonoids from green tea can be attributed to not only antioxidant, antithrombogenic and anti-inflammatory properties but also improvement of coronary flow velocity reserve (Sinija and Mishra, 2008).

Only in the last few years has there been any serious attempt to market instant tea, outside the catering and vending sector. The market is still in an early stage of development with several freeze dried products of reasonable quality consisting of 100% tea solids retailing alongside mixes containing whiteners or sugar and other flavouring. The manufacturing process is essentially similar to that for instant coffee; however, control of the processing condition is much more critical in case of tea to avoid product degradation. After extraction, the liquors require very careful handling as it deteriorates quickly at normal temperature. So, it must be processed to powder immediately by spray drying or freeze drying or has to be stored under very low temperature. Soluble/instant tea has recently become of urgent practical importance largely because of the success of instant coffee and the sale of beverages from vending machines. Instant tea is now used for iced drinks which have greatly increased its demand. The growing popularity of instants takes place at the expense of leaf tea. Convenient products have the advantage of having little or no waste in the household. Improvements not only of the product itself, but also of its suitability for providing an attractive drink from vending machines are constantly under study (Hart, 2008).

1.1 Health benefits of tea

The secret of green tea lies in the fact that it is rich in catechin, polyphenols, particularly EGCG. The EGCG is a powerful antioxidant; besides inhibiting the growth of cancer cells, it kills cancer cells without harming healthy tissue. It has also been effective in lowering LDL cholesterol levels, inhibiting the abnormal formation of blood clots, reduction of platelet aggregation, lipid regulation, and inhibition of proliferation and migration of smooth muscle cells. Understanding the molecular mechanisms of these effects of green tea is a subject of investigation in many laboratories (Katiyar and Elmets, 2001; Fujiki et al., 2003; Hirasawa and Takada, 2004).

In general, green tea has been found to be superior to black tea in terms of antioxidant activity owing to the higher content of EGCG (Cheng, 2000). The processes used in the manufacture of black tea are known to decrease levels of the monometric catechins to a much greater extent than the less severe conditions applied to other teas (Cheng, 2004). Treatment of green tea polyphenols to skin has been shown to modulate the biochemical pathways involved in inflammatory responses, cell proliferation and responses of chemical tumor promoters as well as ultraviolet light-induced inflammatory markers of skin inflammation. Cells that migrate towards the surface of the skin normally live about 28 days, and by day 20, they sit on the epidermis getting ready to die and slough off.

Current research seems to show that EGCG reactivates epidermis cells (Lee et al., 2002). Green tea also exhibits antifungal activity (Hirasawa and Takada, 2004). EGCG enhances the antifungal effect of amphotericin B or fluconazole against antimycotic-susceptible and resistant *Candida albicans*. Combined treatment with catechin allows the use of lower doses of antimycotics and induces multiple antifungal effects. It is hoped that this lower doses may help to avoid the side effects of antimycotics.

EGCG and ECG were found to be potent inhibitors of influenza virus replication in cell culture. Quantitative analysis revealed that, at high concentration, EGCG and ECG also suppressed viral RNA synthesis in cells, whereas EGC failed to show similar effect. Similarly, EGCG and ECG inhibited the neuraminidase activity more effectively than the EGC. Neuraminidase is an antigenic glycoprotein enzyme found on the surface of the influenza virus (Song et al., 2005). EGCG found in green tea can help to boost one's immune system and thereby helping to prevent HIV. The EGCG prevents the binding of HIV to human T cells, the first step in HIV infection. EGCG inhibited the binding of human immunodeficiency virus (HIV) to human CD4 (+) lymphocytes, which is a crucial step in HIV infection. EGCG showed a strong affinity for CD4, and by binding them, could effectively inhibit the binding of the HIV envelope (gp120). This data opens new perspectives for the treatment of this life threatening disease (Nance and Shearer, 2003).

Green tea has almost become synonymous with weight loss and diet now a day. Green tea diet is an excellent source of polycatechin polyphenols, a group of antioxidants that act on free radicals. With green tea diet's polycatechin polyphenols, a person has a better chance of avoiding ailments and keeping healthy for a much longer period of time. The interaction of caffeine and EGCG causes green tea diet to promote thermogenesis in the body. Green tea diet helps increase the body's metabolic rates. Because green tea diet has an inhibiting effect on insulin, it helps keep sugar from being stored as fats and instead, send them directly into the muscles for immediate use. Green tea can even help prevent tooth decay. Just as its bacteria-destroying abilities can help prevent food poisoning, it can also kill the bacteria that cause dental plaque (Sinija and Mishra, 2008).

There is also epidemiological evidence that drinking green tea (but not black tea or oolong tea) may help prevent diabetes (Takatoshi et al., 2006) although it is worth noting that this is evidence of an association and that future studies are needed to confirm the effect. EGCG has been found to increase insulin sensitivity and may repair damaged beta cells (Anderson and Polansky, 2002; McKay and Blumberg, 2002). Certain cognitive benefits associated with caffeine content in tea are, such as a reduction in the likelihood of Parkinson's disease and a temporary increase in short term memory. Green tea could be relevant for management of iron overload and oxidative stress. Green tea may also help to reduce inflammation associated with Crohn's disease and ulcerative colitis (Dryden et al., 2006).

To date, the only negative side effect reported from drinking green tea is "insomnia" due to the fact that it contains caffeine. Green tea contains vitamin K and may interfere with warfarin (Taylor and Wilt, 1999). Based on current literature, there does not appear to be any significant side effects or toxicity associated with regular tea consumption. Patients sensitive to caffeine should use caffeine-free tea or a caffeine-free extract. Also the fluoride content in tea reduces the anticancer properties of tea, or even causes cancer as fluoride is considered a cancer promotor. The high fluoride content could also cause neurological and renal damage, especially in the presence of aluminum. Additionally, the high fluoride content could cause osteoporosis, arthritis, and other bone disorders (http://www.bruha.com, accessed on 14th December 2008).

2. Technology of Tea Manufacturing

Tea can be categorized into three types, depending on the level of fermentation, i.e. green (unfermented), oolong (partially fermented) and black (fermented) tea. The term fermentation is often used incorrectly in tea processing. The more correct term should be oxidation, which means exposure to air without any additives during the process. It is mainly the oxidative polymerization and condensation of catechins catalyzed by endogenous polyphenol oxidase and peroxidase. The oxidation products such as the aflavins and the arubigins contribute to tea color and taste of the black tea. Another form of tea is white tea which is made from new growth buds and young leaves that have been steamed to inactivate polyphenol oxidation and then dried. Alternatively, with the combination of the ways of processing and the characteristic quality of manufactured tea, it is classified into six types viz. green tea, yellow tea, dark tea (containing brick tea and pu-erh tea), white tea, oolong tea and black tea. The so called fermentation in tea processing is not the anaerobic breakdown of energy-rich compound.

2.1 Unit operations in tea processing

Tea production technology is essentially a biochemical technology since the transformation of the green tea leaf, which has a bitter taste and grassy odour into an aromatic and tasty manufactured tea is based on biochemical processes. The major unit operations involved in tea processing are briefly described in the following paragraphs.

2.1.1 Plucking

Since the objective of plucking is the commercial production of high quality tea at the highest possible level throughout the life time of a tea plant, plucking and other bush management should be carried out efficiently. Not only adequate leaf of good standard should be plucked, but sufficient mature leaf should be left on the bush. Fresh shoots of two leaves and a bud obtained by fine plucking are the best basic material, because of the high contents of polyphenols and caffeine. Generally plucking interval is 7–10 days (Wherkoven, 1974; Owuor and Obanda, 2001).

2.1.2 Withering

Withering of fresh leaf is the first and indispensable stage in orthodox tea processing. It is carried out to biochemically and physically prepare freshly plucked tea leaves. Withering involved the phenomenon of loss of moisture in tea leaves from 78–80% to about 55–68%, carried out under conditions that prevent the temperature of the respiring tea leaf from rising above 24 °C (Keegal,

1958; Child, 1960). Both physical and chemical changes take place during withering. Physical changes include the loss of moisture which makes it flaccid or rubbery, a most desirable condition to help in its rolling. Chemical wither starts immediately after the leaf is detached from the plant. The various chemical changes includes breakdown of larger molecules to smaller ones that results in increase of contents of amino acids and flavor compounds, increase in caffeine content and increase in permeability of cell membranes which has a great effect on the mixing of polyphenols, enzymes and oxygen for even fermentation (Harler, 1963).

2.1.3 Rolling

Rolling is actually the mechanical procedure for the original hand rolling process and the primary objective is to wring out the juice from the leaf. In rolling, the leaf is damaged in such a way that while twist is produced the semi permeable membrane in the leaf is distorted allowing the contents to be mixed in the presence of air. This will start off the chemical changes necessary for black tea production through the fermentation process. Adequate bruising and damaging of the leaf as well as the spreading of the expressed juice, is obtained by subjecting the leaf to twisting, unrolling, re-rolling and breaking up. The expressed juice spreads on the surface of the leaf making it wet and sticky. The conventional rolling process is now replaced by the continuous machines like CTC machines, rotorvanes etc. which are more efficient and less time consuming (Wherkoven, 1974).

2.1.4 Fermentation

The most important stage in black tea manufacture is fermentation. The biochemical changes developed during the withering stage will be followed by the most rapid changes during fermentation. These charges results in profound qualitative and quantitative changes in tea leaf constituents, contributing to the formation of new taste and aroma products responsible for the character of black tea. The fermentation process starts with the rolling, but does not end there. The central reaction in the fermentation is the oxidation of catechins to othoquinones which, in turn, combine with one another to form theaflavins having bright red colour and a fair solubility (Sanderson, 1972). When the aflavins or epitheaflavic acid condense with oxidized catechins, polymeric thearubigins are formed, having dark brown colour and fair to poor solubility. Many researchers showed that these oxidation reactions of catechins determine the colour and taste of the finished black tea.

2.1.5 Drying

During drying, the tea moisture is reduced to low levels in order to obtain a transportable product which is less prone to deterioration. Although the oxidation of the catechins has practically ceased when the leaf goes to the dryer, the condensation of the theaflavins and thearubigins continues, so that the briskness, quality and strength are steadily decreasing. These changes are brought to a standstill only when the leaves has reached a certain degree of dryness. Some of the unchanged caetchins in the leaf are probably altered in structure, by epimerization, during firing. During the early stages of drying the pectase enzyme produce a varnish which in due course improves the keeping quality of made tea. As the leaf temperature rises during the drying process the pectase is destroyed. Fermented tea leaf can be dried over a considerable range of temperature and time (70–105 °C for 45–190 min) (Harler, 1963).

In tea processing, drying is the major energy consuming process, and may have adverse effects on quality. In order to design and control the drying stage in an accurate way, drying rates have to be established. The final target moisture content of dried tea samples is around 4-6% wb. The dried tea is then going for the sorting and grading operations and the sorted product is packed in different packaging materials as per the requirement.

Drying of liquid extracts

For drying of liquid extracts to prepare instant tea samples either freeze drying or spray drying techniques are used.

Freeze drying

It is an important drying process for heat sensitive products which have to conserve their flavor, bioactivities, and other properties (Liapis and Sadikoglu, 1997). The freeze drying experiments are time consuming and are usually expensive to be carried out in all possible operating ranges. Freeze-drying is based on the dehydration by sublimation of a frozen product. Due to the absence of liquid water and the low temperatures during the process, most of deterioration and microbiological reactions are stopped which gives a final product of excellent quality (Genin and Rene) 1995; Irzyniec et al. (1995). The solid state of water during freeze-drying protects the primary structure and the shape of the products with minimal reduction of volume.

Spray drying

In spray drying finely divided drops of liquid food are dehydrated by direct contact with heated air in extremely short residence time of 3–30 s, which results minimum heat degradation of dried product. Liquid is transferred to fine drops by using either rotary/pressure or pneumatic two-fluid atomizing device. Drops are formed inside a chamber where they move down along with the flow of heated air. Depending on type of atomizer and the liquid to be dried, size of atomized drops and air temperature vary between 20–180 micron and 150 – 250 °C respectively. Surface area created by the atomizer range from 310 m^2 down to 35 m^2/L of feed liquid (Dittman and Cook, 1977; Das, 2005).

2.1.6 Packaging and storage

Since a considerable lapse of time passes before the tea arrives at the blender, packer and consumer, they have to be stored and dispatched under conditions that leave their keeping qualities unimpaired. The use of aluminium lamination is generally adopted now for packaging of tea. It is light, durable, easy to handle, relatively cheap and making a good air tight package when folding over the edges of the sheet. Rather a lot of work has been done on the suitability of lining materials other than aluminium foil like various plastic materials, cardboard packages, etc.

2.2 Instant tea

The convenience-oriented lifestyle of modern society has influenced the tea industry. The consumption of convenient canned tea products has increased in parts of world compared to traditional old-fashioned tea leaves. The canned teas are sold either hot or cold, depending on consumer needs (Hara et al., 1995). Instant tea is of great interest to the general public as it is easy to handle and convenient to use. Little research has been reported on the preparation of instant soluble tea powders from whole green tea leaves. For the preparation of high-quality tea powder, characteristic sensory attributes should be maintained in the powder, and good functional properties such as dispersibility and solubility are desirable.

There are several methods for the preparation of instant tea. The commercial method of production of instant tea is given in Figure 3. The details of the technique involved are mostly kept secret or protected by patents. Hindustan Lever Ltd. (1975) has developed and patented a process for preparing instant green tea by heating the fresh leaf to a temperature sufficient to inactivate the enzymes and then the leaf is comminuted, extracted with hot water and dried by conventional means like spray-drying or freeze drying. Another method was developed by Tea Research Institute of Ceylon (Wickremasinghe, 1977) for the production of cold water soluble tea concentrates and powders. In this method, tea leaves are extracted with hot water and the extract is subjected to gel filtration to effect the separation of non-phenolic compounds like chlorophylls, proteins, polypeptides and polysaccharides, while retaining polyphenolic compounds. Then resultant filtrate is concentrated to obtain a cold water soluble powder.

There are certain difficulties to be solved in producing instant tea, which would give a beverage comparable with an ordinary tea infusion. The problems are connected with the liquoring characteristics, flavour or aroma and with tea quality itself. In the general method of extraction by boiling water, the flavor of this product is generally poor. When hot infusion cools down, it becomes turbid and particles settle down on prolonged cooling. This is called tea cream and decreaming is thus necessary, since it affects the clarity and appearance of

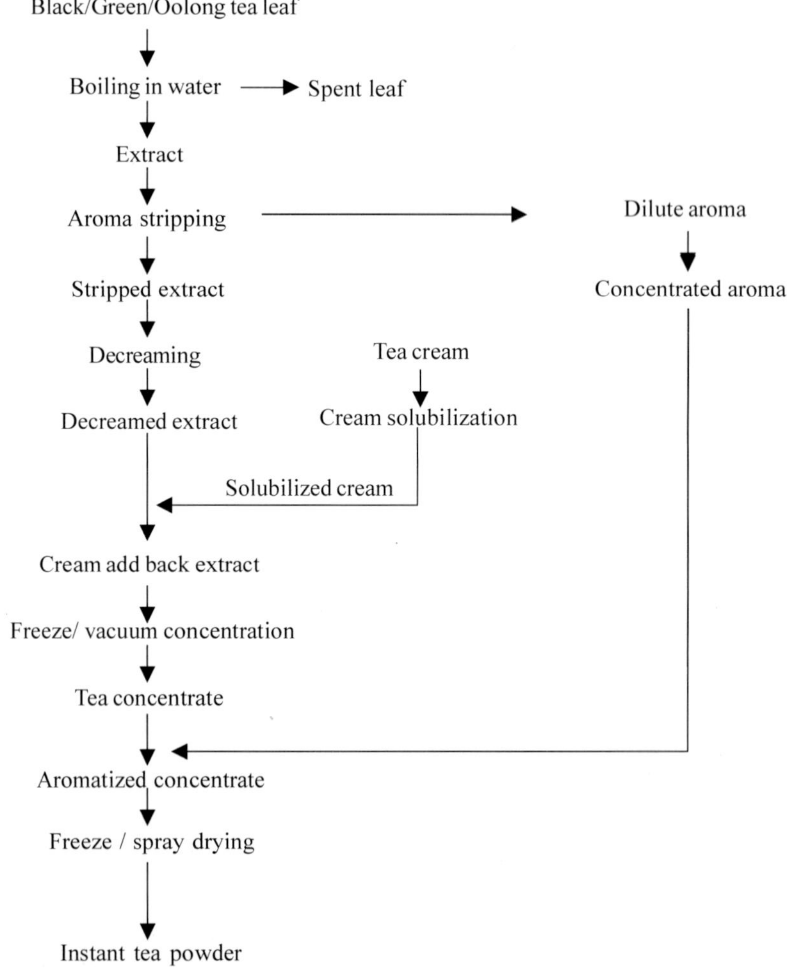

Fig. 3: Flowchart for commercial production of instant tea powder

cold water soluble instant tea. In addition to the quality problems of instant tea like low aroma, poor taste and insufficient cold water solubility, low productivity is also one of the important problems remaining (Pintaro 1977). Moreover when instant tea is made from the prepared black tea, the process is highly expensive and at the same time energy intensive also. Tea contains about 300–450 g/kg of extractable solids, but the yield of instant tea is only 200 g/kg tea in production scale (Schott, 1988).

3. Process Technology for Preparation of Instant Tea Powder and Granules from Fresh Tea Leaves

3.1 Instant green tea powder

Considering the drawbacks of the existing methods for production of Instant tea powder, a novel technique has been developed by Sinija et al. (2007) to produce instant tea powder from the fresh tea leaves itself. The process flow chart for the preparation of Instant black and green tea powder is shown in Figures 4 (a & b). The various unit operations involved are detailed below.

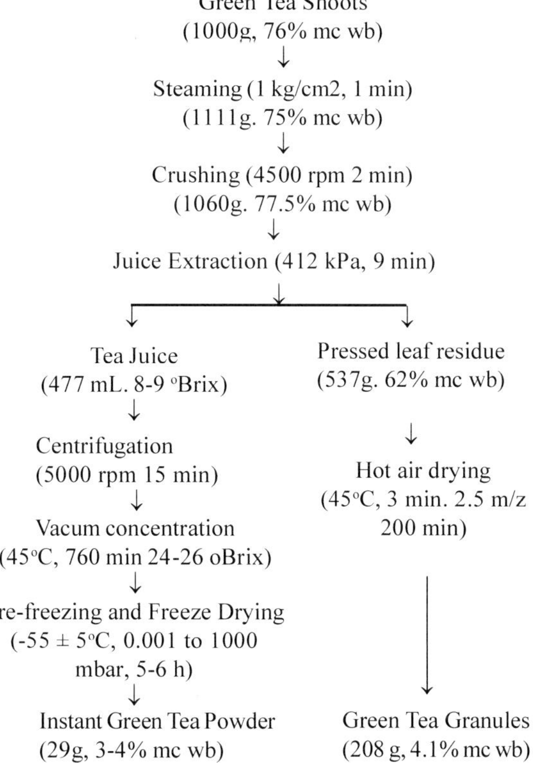

Fig. 4a: Flowcharts of instant green tea production

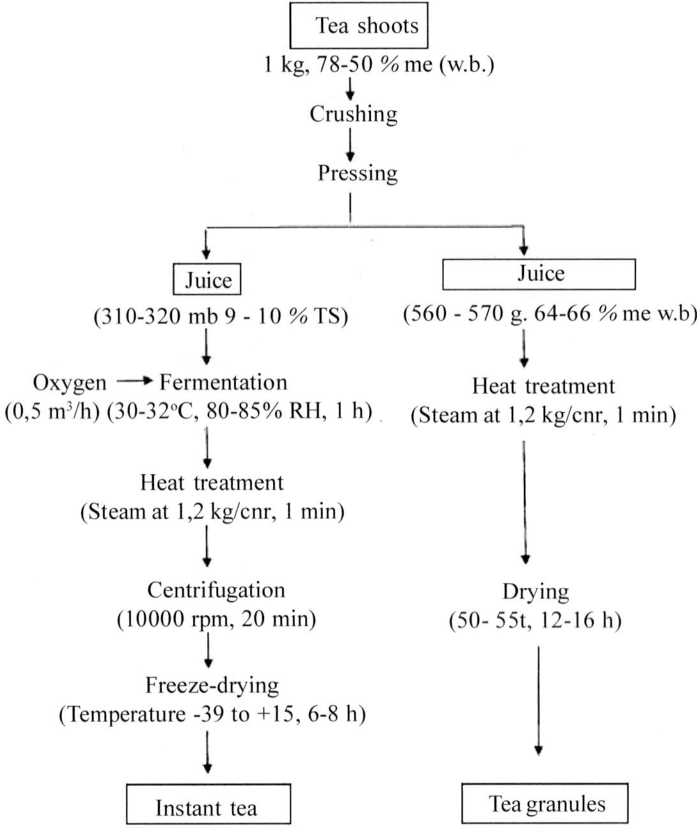

Fig. 4b: Flowcharts of instant black tea production

3.1.1 Plucking of tea shoots

For good quality tea, proper plucking standard and plucking intervals are essential. Tea shoots at the proper stage of development should be selected with minimum physical damage. The weight of fresh tea leaves was determined by a laboratory balance and the moisture content was measured by moisture analyzer with an accuracy of 0.01%. Tea shoots (2 leaves and the unopened terminal bud) were plucked from the experimental gardens of IIT Kharagpur.

3.1.2 Processing of tea leaves

Immediately after plucking, the leaves were subjected to steaming in order to inactivate the enzymes responsible for the fermentation reaction, for green tea production. Figure 5 shows the pictorial view of the set up used for steaming. Steam (1 kg/cm^2) generated in the autoclave is directed through a rubber hose to the fresh tea shoots spread over a wire mesh tray in thin (single) layer for 1 min. The temperature of the leaves rose to 70–80 °C which is sufficient to inactivate the enzyme polyphenol oxidase responsible for the oxidation of various polyphenols to thearubigins and theaflavins, which will give the specific liquor characteristics of black tea (fermented tea) (Wherkoven, 1974; Muthumani and Kumar, 2007). After steaming, the leaves become flaccid.

For the production of instant black tea, the above said steaming step is not required. After plucking the shoots, directly it will go for crushing.

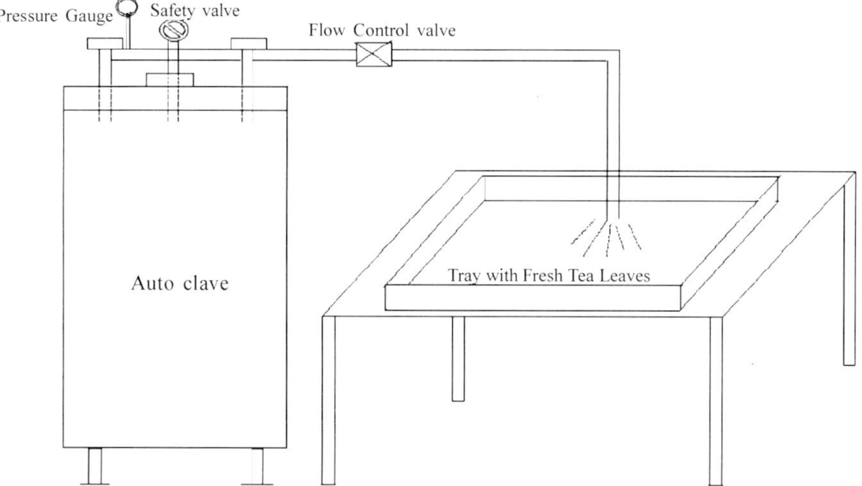

Fig. 5: Pictorial view of steaming set up

3.1.3 Crushing of leaves

The steamed leaves were then crushed to a fine paste in a mixer grinder (Kenstar Excellence). Experiments were carried out with equal amount of raw material for different crushing periods (1 – 4 min) at medium speed (4500 rpm) in order to find out the conditions for maximum juice recovery, which contains maximum amount of polyphenols and total solids. After crushing the paste was taken out and weighed. Rolling process in conventional method of tea processing was replaced by crushing in this method. During crushing, the polyphenols diffuse into the cytoplasm by rupturing the membranes and there is a wringing action on the leaf tissues by the mechanical disruption of its cells. The optimum condition for crushing is 2 min and 4500 rpm.

3.1.4 Pressing

Extraction of juice from tea leaves was carried out with the help of an extraction unit designed and fabricated in the workshop of Agricultural and Food Engineering Department, IIT Kharagpur and a hydraulic press. The cross sectional view of the extraction unit is given in Figure 6.

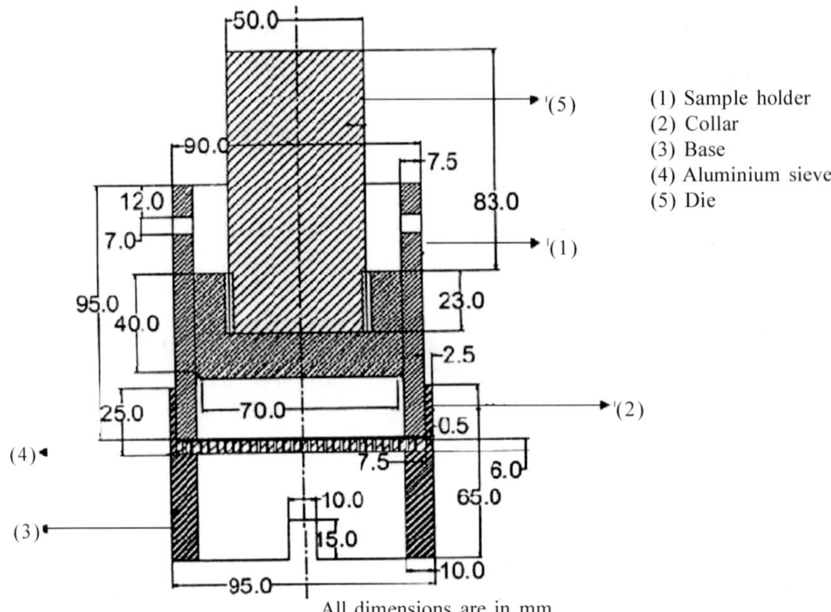

Fig. 6: The expression cell used for juice expression from crushed leaves

For the extraction of juice, all the parts of the extraction unit were assembled. Crushed leaves were placed on the sample holder. The die was placed on top of the sample and tea juice was extracted by applying pressure on the die by using the hydraulic press. Optimized conditions for juice expression are 412 kPa pressure and 9 min extraction time.

3.1.5 Centrifugation

The juice obtained (Total solids 9.3%, Total polyphenols 273.1 mg/g) was centrifuged (5000 rpm, 15 min) in a laboratory centrifuge (R-24 type) at 5,000 rpm for 15 min, to remove the colloidal and other suspended solid particles, which cause turbidity in the tea brew which is not desired for instant tea.

3.1.6 Freeze drying of juice

The centrifuged juice was then concentrated under vacuum before freeze drying. The juice with total soluble solids of 8–9 °Brix was concentrated in a rotary vacuum flash evaporator (Model–PBU–6D, Figure 7) for 30 min (Temperature 45 °C, vacuum 76 mm Hg) to a final concentration of 24-26 °Brix . Concentrated juice was then poured in round bottom flask and subjected to pre-freezing in a pre-freezing bath (Figure 8). The motor within the pre-freezing bath helps the sample holders to rotate slowly within the glycol solution so that thin layers of the frozen juice will be formed in the round bottom flask which is easy to dry.

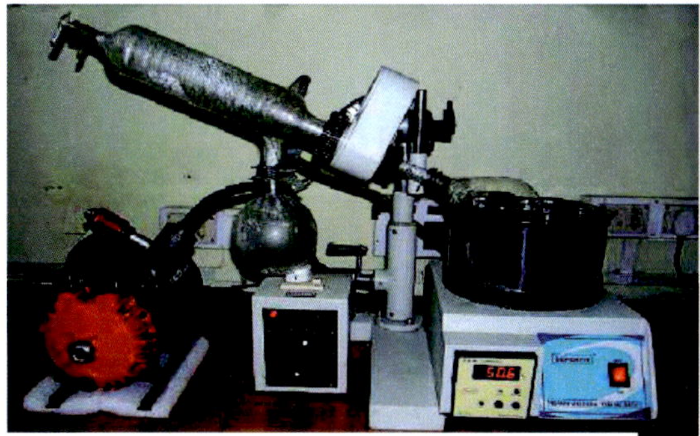

Fig. 7: Vacuum flash evaporator

Fig. 8: Pre-freezing bath

Fig. 9: Freeze dryer with samples loaded for drying

Once the sample gets frozen, the flasks were loaded in the freeze dryer (Lyodel Freeze Dryer) with acrylic drum manifold attachment as shown in Figure 9. The drum manifold accessory is of stainless steel construction with eight ports available for attaching the round bottom flasks. The temperature of freeze dryer condenser is -55 ± 5 °C and the vacuum range is 0.001–1000 mbar (0.0001–100 kPa). At the end of drying the flasks were removed from the dryer and powder was collected in self sealable pouches. The various physicochemical characteristics of the powder were analyzed by standard procedures.

3.2 Convective air drying of the pressed leaf residue for making green tea granules

The leaf residue after juice expression was converted into green tea granules by convective air drying (3–3.5 h). Drying was optimized (response surface methodology) and parameters selected were 45 °C air temperature, 2.5 m/s air

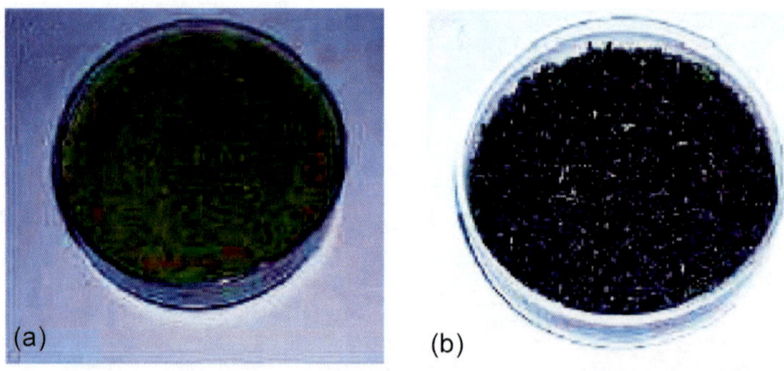

Fig. 10: Instant green tea powder (a) and green tea granules (b)
(See colour version on page 334)

velocity and 3 mm sample thickness. Quality of dried green tea granules samples were evaluated in terms of overall acceptability (combined effect of colour/ appearance, flavour and taste) from sensory evaluation. The L*, a*, b*, values for the tea brew prepared from the samples obtained by different experimental conditions showed that at higher temperature the colour of the brew changed from yellowish green to brownish tinge. A decrease in L* value and increase in'a*' and 'b*' values indicated the development of a brown color. It took nearly 200 min to bring down the moisture content of pressed leaf residue to 4.1% to produce tea granules.

The beauty of this technology lies in the fact that from the same fresh tea shoots, two products, instant tea powder ad granules were obtained and there is no waste. Whereas in the existing methods pressed leaf residue after extraction of brew is discarded as waste. From 1 kg of fresh tea shoots 29-30 g of instant tea and 208-210 g of green tea granules are obtained (Figure 10).

3.3 Instant black tea powder and granules

After plucking the leaves, directly it can be crushed and juice was expressed out in the same way as described for instant green tea production. Steaming step prior to crushing is avoided in this case as it will arrest the fermentation process which is very much essential for black tea preparation. The juices as well as the pressed leaf residue samples were separately subjected to fermentation under ambient conditions (30–32 °C and 80–85% RH). During fermentation, oxygen gas was incorporated into the juice (0.5m^3/h). Fermentation was carried out for different fermentation periods starting from 0.5 to 3 h. Under or over fermentation causes poor quality teas (Owuor and Obanda, 2001; Wherkoven, 1974).

Optimum fermentation time can be found out by a method based on the amount of theaflavin formed during fermentation (Muthumani and Kumar, 2007). For this, samples were collected at 10 min interval during fermentation and the optical density values for these samples at 460 nm were determined by a spectrophotometer (Spectronic Genesys 2, Model No. CAT 33600902 SN3NE 9107006). Two samples taken at each time interval were used to calculate the mean optical density value corresponding to that time. The optimum fermentation time corresponds to the maximum theaflavin production (Lakshminarayanan and Ramaswamy, 1978). Level of theaflavin formed during fermentation of juice and pressed leaf residue can be seen in Figure 11 and 12 respectively.

In another experiment, instant tea and tea granules were prepared with different fermentation times, and the prepared samples were analyzed for the quality by the sensory evaluation panel. The overall acceptability was found to be maximum for the sampleswith 1-h fermentation in the case of instant tea and thesamples with 1.5-h fermentation time was having maximum acceptability in the case of tea granules. This confirmed the results obtained for optimum fermentation timeby the method based on the amount of theaflavin formed during fermentation.

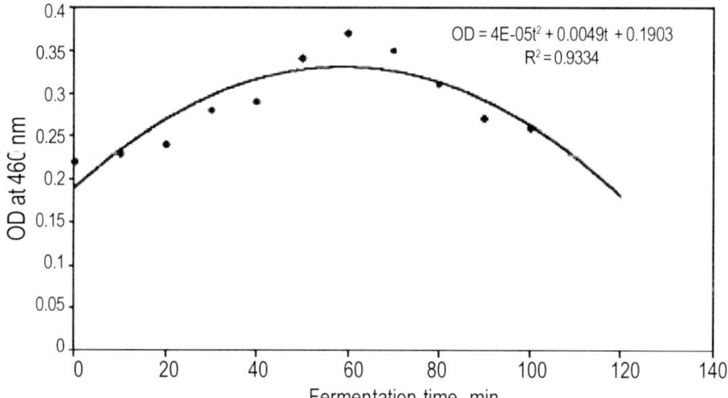

Fig. 11: Level of theaflavin formed during fermentation of juice

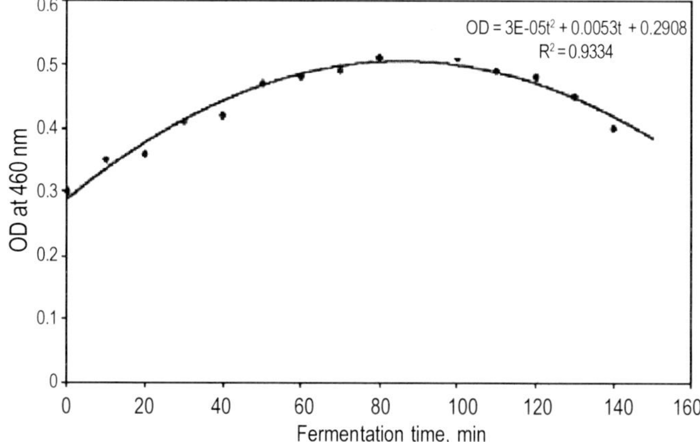

Fig. 12: Level of theaflavin formed during fermentation of pressed leaf residue

In order to arrest the fermentation at the right time steam at a pressure of 1–1.2 kg/cm^2 (temperature 104– 106 °C) was introduced into the fermenting juice for 1 min. This increased the temperature of the juice to approximately 70–80 °C, which caused the inactivation of the enzymes responsible for the oxidation reactions. The juice was then subjected to centrifugation in a laboratory centrifuge (R-24 type) at 10000 rpm for 20 min, to remove the colloidal and other suspended solid particles, which cause turbidity in the tea brew. Finally, the juice was freeze-dried (6–8 h, Labconco FreeZone, Model 79480) to Instant tea powder having a final moisture content of 3–5% (w.b.).The pressed leaf residue was dried to black tea granules in a convective air drier. From 1 kg fresh tea shoots, 20 ± 5 g free flowing instant black tea powder by freeze drying and 225±5 g of black tea granules by convective air drying was obtained.

4. Quality Analysis & Characterization

4.1 Instant green tea powder and granules

The instant green tea powder with a moisture content of 3.4% wb was prepared by freeze drying and green tea granules with 4.1% wb moisture content was produced by recirculatory convective air drying as described above. Water activity of instant green tea powder and green tea granules at 29.5°C was 0.354 and 0.337, respectively. The bulk density values were respectively, 333 and 345 kg/m^3 for instant tea and tea granules. Hygroscopicity and degree of caking values were determined for instant green tea powder and the values were 12% and 30% respectively which indicate that the powder is highly hygroscopic. Total polyphenols and caffeine content in instant green tea power were determined and the values are 245.8 mg/g and 41.6 mg/100 mL respectively. The corresponding values for green tea granules are 261.5 mg/g and 46.3 mg/ 100 mL respectively (Table 1). Green tea sample available in the local market were also analyzed for total polyphenols and caffeine content and the values were 256.32 mg/g and 56.16 mg/100 mL, respectively. Even after extraction of a part of juice from fresh tea leaves the green tea granules produced by this method is having higher total polyphenol content than the market sample and the instant tea produced is also having comparable amount of total polyphenols and caffeine content.

Table 1: Physico-chemical characteristics of prepared green tea samples

Characteristics	Instant green tea powder	Green tea granules
Chemical		
Moisture content (% wb)	3.4	4.1
Total polyphenols (mg/g)	245.8	261.5
Epicatechin (mg/mL)	0.023	0.014
Epicatechin gallate (mg/mL)	0.067	0.048
Epigallocatechin (mg/mL)	0.039	0.062
Epigallocatechin gallate (mg/mL)	0.073	0.084
Caffeine content (mg/100 mL)	41.60	46.30
Protein (%)	8.29	12.05
Crude fibre (%)	0.101	10.80
Ash (%)	3.90	8.20
Physical		
Water activity at 29 °C	0.354	0.337
Bulk density (kg/m^3)	333.0	345.0
Hygroscopicity (%)	12.0	-
Degree of caking (%)	30.0	-

4.1.1 Organoleptic evaluation of brew

Quality of instant green tea and green tea granules prepared from fresh tea leaves were compared with two green tea samples available in the market and the data were analyzed using fuzzy logic. On comparison of highest similarity values for all samples, their ranking was done as green tea granules > instant green tea > market sample 1 > market sample 2. Thus it indicates that even after extracting a part of the juice from fresh tea leaves the quality of green tea granules produced from the residue is of highest quality when compared to the market samples. Also the instant green tea produced from the extracted juice had high scores for all the quality attributes than the market samples. So by this new method of instant tea production, two quality products are obtained, hence it is a value added process technology.

The order of preference of quality attributes for green tea in general was also found out and is taste > flavour > colour > strength. That is taste, flavour and colour are the major quality attributes for a green tea drink which is also reported by (Liang et al., 2002). Strength is the least important quality attribute for green tea as per this evaluation. Quality attribute ranking of individual tea samples were also determined and found that taste is the strongest quality for instant green tea as well as green tea granules samples while strength is the weakest. The numerical scores corresponding to various quality attributes of individual samples were also analyzed and the result is same as the ranking obtained by fuzzy analysis.

4.2 Instant black tea powder and granules

The quality attributes for instant black tea and tea granules were rated from good to very good during sensory evaluation. It was seen from the results that the color and pungency was highest for the instant tea sample, whereas the flavor was relatively less than the CTC tea and tea granules (Table 2). The tea granules exhibited almost same or slightly higher score for all the quality attributes as compared to the CTC tea. The overall acceptability for both the instant tea and tea granules were comparable to that of CTC tea. The removal of a part of juice from the green leaves does not have a marked effect on the quality of black tea. Thus, we can get both instant tea and tea granules from the green shoots simultaneously in this method.

Table 2: Chemical analysis results of instant black tea powder and granules

Constituent	Instant black tea	Black tea granules
Total polyphenols, (%)	19.63	6.61
Catechins, (%)	12.20	3.73
Theaflavins (TF), (%)	0.92	0.57
Thearubigins (TR), (%)	9.88	6.86
Caffeine, (%)	2.02	3.36
High polymerized substances, (%)	9.15	8.58
Total liquor color, AE	3.30	3.70

5. Storage Studies and Evaluation of Shelf Life

Storage studies for the prepared tea samples were conducted under accelerated storage conditions (90% RH and 40 °C) in metalized polyester pouches. The results obtained showed that the quality of both instant green tea powder and green tea granules deteriorated due to the accelerated storage conditions. The change in moisture content, colour, total polyphenols and caffeine contents were recorded during storage and the results showed that all the changes followed zero order reaction kinetics. The moisture content of instant green tea powder after 70 days of storage was 10.5% (db) and the water activity increased from 0.35 to 0.65 (Figure 13). The corresponding values for green tea granules after 90 days of storage were 13.4% (db) for moisture content and water activity value varies from 0.33 to 0.67. The instant tea powder samples exhibited caking tendency at moisture content of 10–10.5% (db) after 70 days of storage. Shelf life was also determined theoretically based on the initial and critical moisture contents and the value obtained was 74 days for instant green tea powder. For green tea granules as the moisture content reached 13.4% db mold growth was seen and hence this is taken as the criteria for deciding the shelf life. Shelf life was estimated as 96 days for green tea granules under accelerated storage conditions.

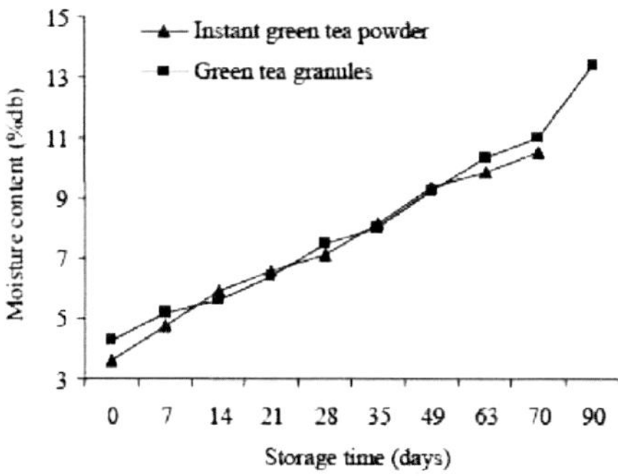

Fig. 13: Moisture uptake by green tea samples accelerated storage

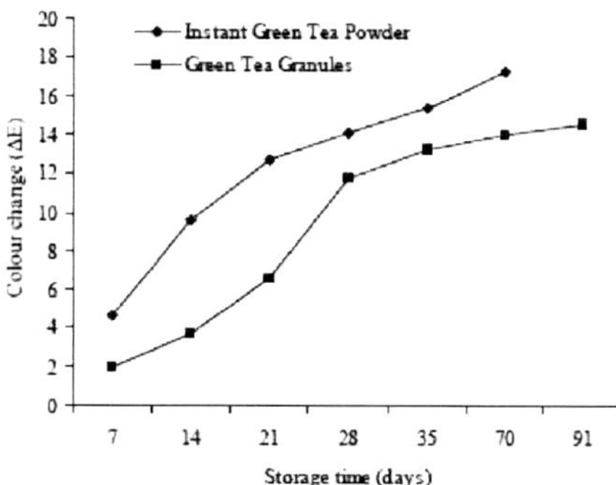

Fig. 14: Colour change during accelerated during storage of green tea samples

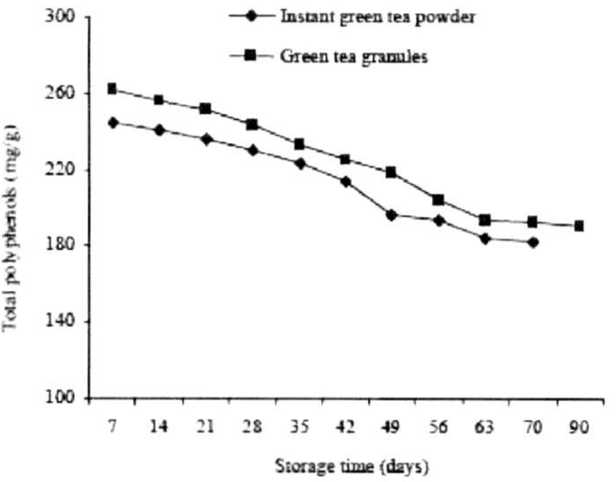

Fig. 15: Change in total polyphenol content of green tea samples during accelerated storage

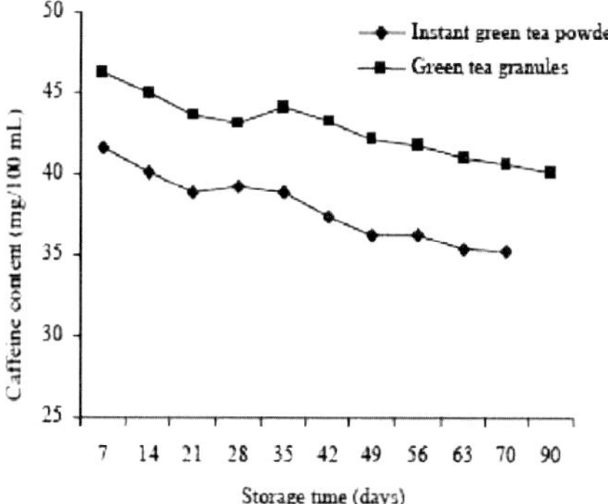

Fig.16: Change in caffeine content of green tea samples during accelerated storage

During storage the color of the brew prepared from stored samples changed from yellowish green to brownish tinge which is clear from decrease in L* value and increase in 'a*' and 'b*' values. More change in colour (ΔE = 17.29) of instant green tea powder was noticed in comparison to that of green tea granules (ΔE = 14.51) which may be attributed to highly hygroscopic nature of instant green tea powder (Figure 14). Total polyphenols and caffeine contents also decreased significantly during the storage. Total polyphenols for instant green tea powder and green tea granules reduced from 245.85 to 182.46 mg/g and from 261 to 191.34 mg/g respectively (Figure 15). The reduction in caffeine is from 41.6 to 35.3 mg/10 mL for instant green tea powder and from 46.3 to 40.2 mg/100 mL for green tea granules (Figure 16).

6. Instant Soluble Tea Tablets

The addition of milk may inhibit some antioxidant components of tea such as the theaflavins and thearubigins that will affect the total antioxidant capacity of black tea. This is because milk protein (casein) binds with these flavonols of tea. Not all scientists are convinced that the effects of milk are strong enough to cancel out potential health benefit of tea. Scotland scientist Alan Crozier explained that teas are loaded with beneficial flavonols, and typically people only add a little milk to a cup of tea, it's about enjoying the distinct flavours of teas on their own. In this regard oat milk can serve as better option due to its reported beneficial effects on human health (Table 3).

Table 3: Nutritional highlights of oat milk (Per serving of 250 mL)

Particulars	Quantity
Energy, (kJ)	605.0
Total carbohydrate, (g)	19.3
Fat (Free of saturated fat), (g)	5.0
Protein, (g)	5.8
Cholesterol, (g)	Nil
Dietary fibre (Soluble β-glucan), (g)	0.8
Sodium, (mg)	90.8
Calcium, (mg)	6.8

6.1 Technology of production of tea tablets

US FDA revised definition suggested that "Effervescent tablet is a tablet intended to be dissolved or dispersed in water before administration". It generally contains in addition to active ingredients, mixture of acids/ acid salt (citric, tartaric or malic acid and any other suitable consumable acid) and carbonates and bicarbonates (sodium, potassium or any other suitable alkali metal carbonates) which releases carbon dioxide when dissolved in water. Sometimes active ingredients itself could act as acid or alkali metal compound necessary for fast disintegration. Through this an alternative material and process to replace instant tea segment i.e. tea bags and such

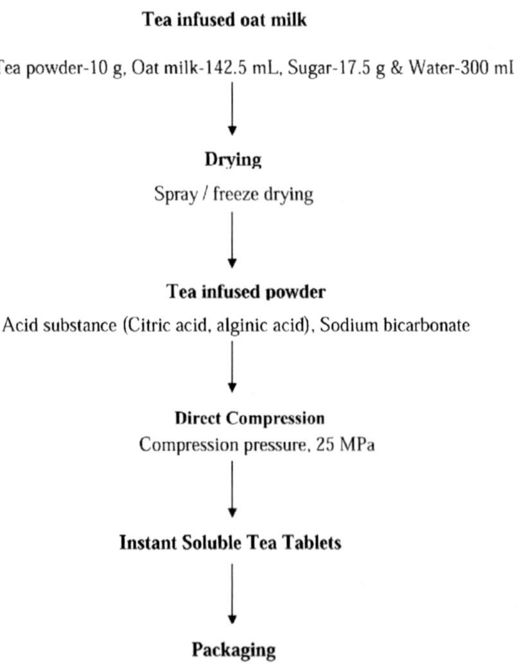

Fig. 17: Flowchart for production of instant soluble tea tablets

other product into instant soluble tea tablets which integrates tea, sugar, oat milk and other flavoring compounds in appropriate proportion was tried.

The science and application of engineering techniques involved in preparation of instant soluble tea, pharmaceutical techniques involved in production of rapidly disintegrable (effervescent) tablets, and characterization of the effervescent tablets are discussed in following paragraphs.The formulation of effervescent tablets contained acidic substance (citric, tartaric. malic acid & alginic acid) and carbonates or bicarbonates (sodium, potassium, etc.) which react rapidly in presence of water by releasing carbon dioxide. The colour value of the blend prepared from oat milk, tea liquor & sugar was measured using Konica Minolta spectrophotometer. The °Brix value of the prepared formulations were in the range of 8-10 and was measured using ERMA Hand Refractometer. Flow chart for production of instant soluble tea tablet has been shown in Figure 17.

6.2 Preparation of oat milk-tea-sugar beverage powder

For optimisation study a three component constraint mixture design was developed using oat milk (X_1), tea powder (X_2) & sugar (X_3).The lower and upper bound constraints were determined based on sensory analysis and consumer acceptability ratings 5.0 (Deshpande et al., 2007). The lower and upper bound constraints for each mixture component (X_1: 10-15 mL; X_2: 0.5-1 g; and X_3: 1-1.75 g) were used to generate the design. These beverage formulations served along with the control in each sensory sub-session resulted in a total of 14 samples for sensory evaluation.

Fig. 18: Oat milk prepared in lab (See colour version on page 334)

The tea brew was prepared by mixing tea powder and sugar, as per the concentration ranges obtained from mixture design, with water (30 mL). The brewing temperature was maintained at 100°C (boiling) for 1-2 min. The brewed liquor was filtered and mixed with oat milk (Figure 18) in proportions suggested by mixture design and heated for 50 sec., as longer heating of oat milk induce stronger flavour. The brew thus prepared were analysed for colour and sensory qualities before converting to powder by suitable drying techniques.

Fig. 19: Oat milk mixed tea beverage (See colour version on page 334)

6.2.1 Quality attributes of oat milk mixed tea

Colour of samples was measured in a Konica Minolta CM-5d spectrophotometer (JISL. Mumbai, India) and soluble solids were determined in a hand refractometer (ERMA, range 0-32%) at room temperature and the results expressed in degrees Brix at 20° C. Trained research scholars and technical staff (n=25) from Food Chemistry and Technology Laboratory (FCTL) were served with 60 mL of prepared tea formulation at 50 ± 1°C in white cups. For each set of samples consumers evaluated five samples per session over three sessions. They evaluated the overall acceptability and the level of suitability of colour & appearance, taste, aroma and mouthfeel of each sample using a 9-point hedonic scale ranging from 1 ("dislike extremely") to 9 (like extremely). This procedure allowed to determine how much the sample varied or approached the intensity of attribute considered to be ideal for the oat milk tea (Villegas et al., 2009).

6.2.2 Drying of oat milk tea brew

The optimized blend of oat milk-tea was dried in Spray Mate spray dryer (JISL, Navi Mumbai, India), which has a centrifugal atomizer (inside diameter 120 mm) and operates in a concurrent manner. The operational conditions for the spray-dryer were: inlet air temperature (110-120 °C) and outlet air temperature (80-85 °C). The feed rate was set as 10 mL/min (Figure 20).

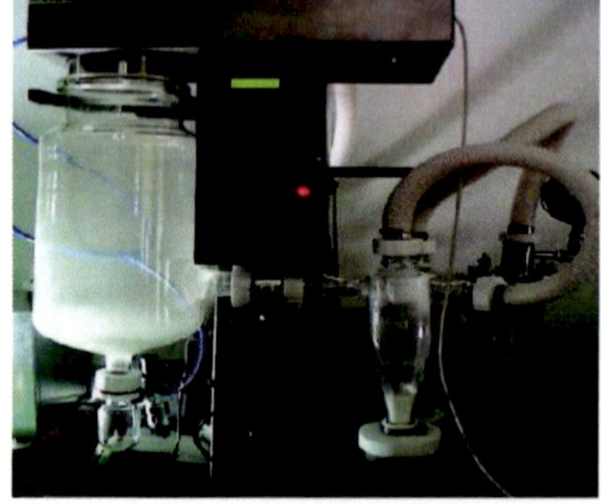

Fig. 20: Spray dryer

6.3 Preparation of effervescent tablets by direct compression

Components prepared using D-optimal mixture design with three components i.e. oat milk mix tea powder (260-320 mg per 500 mg), alginic acid (45-60 mg per 500 mg) and sodium bicarbonate (135-180 mg per 500 mg) were used for the formulation. The constituents were mixed in vortex mixer (Spunk, Germany) and then compressed using a single punch tablet machine using 10 mm flat punches.

Powder tablet ability (dependence of tablet tensile strength on compaction pressure) was assessed using a Universal Testing Machine (Zwick, Germany). Tablets (Figure 22) were prepared using round (10 mm in diameter) flat-faced punches over the pressure range of 25-50 MPa. Prior to each compression run, approximately 500 mg of powder was transferred to a die. The die and punches were lubricated by applying a thin layer of magnesium stearate suspension and air dried. The lower punch remained stationary while the upper punch moved at a speed of 2 mm/min during both loading and unloading phases. No holding was applied at the maximum load. Ejected tablets were relaxed for overnight in a closed glass vial before a standard diametrical breaking test using a texture analyzer (CT3, Brookfield Technologies Corporation, 50 kg load cell, test speed of 0.1 mm/s with a trigger force of 100 g). Tablet thickness and diameter were measured using a digital calliper accurate to 10 µm. Tablet tensile strength was calculated from the tablet breaking force and dimensions using Eq.1.

$$= \frac{2F}{\pi.D.T} \tag{1}$$

Where, F is the breaking force, D and T are the diameter and thickness respectively. Tensile strength of three tablets was determined at each pressure for calculating the mean and standard deviation.

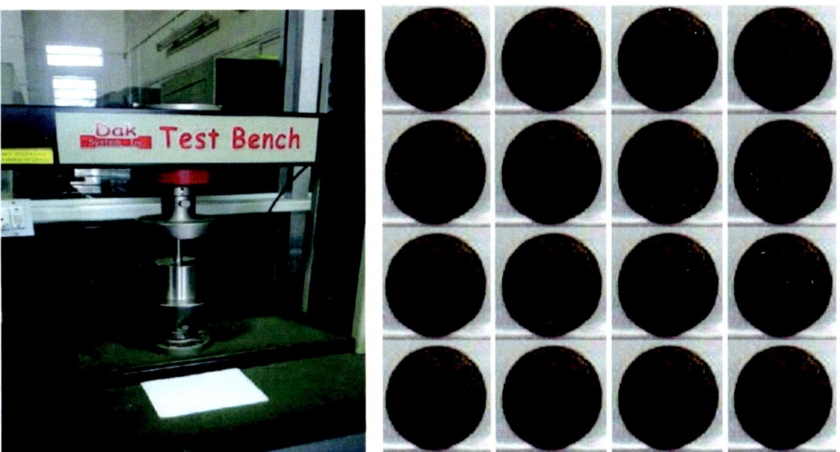

Fig. 21: Set up for tablet production **Fig. 22:** Tea tablets formed during work

(See colour version on page 334)

6.3.1 Evaluation of effervescent tablets

A dial/vernier caliper (Mitutoyo, Mumbai, India) was used to determine thickness of ten randomly selected tablets from each formulation. Ejected tablets were relaxed for overnight in a closed glass vial before a standard diametrical breaking test using a texture analyzer (CT3, Brookfield Technologies Corporation, 50 kg load cell). The hardness was checked at test speed of 0.1 mm/s with a trigger force of 100 g. Two tablets per 20 mL of hot water (were put in (3 beakers) and the effervescence time was measured using a stopwatch. Effervescence time was defined as the moment when a clear solution was obtained. The values for the tablet parameters are given in Table 4

Table 4: Properties of the tea tablet formed

Parameter	Value
Thickness, mm	4.8 -5.2
Weight, mg	480-502
Hardness, kg	4.48 – 5.76
Disintegration time,s	45-60
TSS,Brix	3-4 Brix

Proximate analysis for the tea tablets were performed and the values are shown in Table 5. Percentages of moisture was determined by vacuum oven method, total fat by Soxhlet extraction method, protein by Kjeldahl nitrogen method and ash by direct analysis method according to AOAC methods.

Table 5: Proximate composition per 100 g (50 Tablets)

Proximate composition	Proportion
Energy, kJ	1227.69
Carbohydrate, g	61.45
Protein, g	24.20
Fat, mg	5.70
Mineral, g	1.35
Moisture g (wb)	7.30

7. Quality Issues and Challenges

There are certain difficulties to be solved in producing instant tea, which would give a beverage comparable with an ordinary tea infusion. The problems are connected with the liquoring characteristics, flavour or aroma and with tea quality itself. In the general method of extraction by boiling water, the flavor of this product is generally poor. When hot infusion cools down, it becomes turbid and particles settle down on prolonged cooling. This is called tea cream and decreaming is thus necessary, since it affects the clarity and appearance of cold water soluble instant tea. In addition to the quality problems of instant tea like low aroma, poor taste and insufficient cold water solubility, low productivity is also one of the important problems remaining (Pintaro, 1977). Moreover when instant tea is made from the prepared tea, the process is highly expensive and at the same time energy intensive also.

In most of the existing methods for the production of instant tea, where hot water extraction is used for extracting the constituents from fresh leaves, the residue obtained after the extraction is discarded as waste. Moreover in hot water extraction, the high temperature during processing might cause the degradation of chemical constituents. At high temperature more catechins gets extracted which will produce a tea with greater pungency. Due to the high temperature there is an additional requirement of other processing units to strip off the aromatic volatile components for maintaining the quality of instant tea. Due to the addition of extra processing unit the production cost per unit product is also high. Therefore it is better to have some extraction unit where there is no heat treatment and also low energy consumption.

For black tea manufacturing, tea contains about 300–450 g/kg of extractable solids, but the yield of instant tea is only 200 g/kg prepared tea (black tea) in commercial production scale, leaving the major portion (800 g.) as waste (spent tea leaf) (Chen, 1979), whereas in the present method 20 ± 5 g instant tea powder and 225 ± 5 g. black tea granules were produced from 1 kg of fresh green tea shoots. Even in methods using direct extraction from fresh shoots for production of instant tea, the maximum amount of solids extracted with hot water is around 30% of dry matter present in fresh shoots (Schott, 1988).

With the continuous shifting to a global economy, the international market for value-added products is growing. The main points to be discussed for value addition of tea powder with oat milk are:

Effects of oat milk nutrients- Oat milk is devoid of cholesterol which reduces cardiovascular disease. The most important attribute is soluble fibre (β-glucan) which helps in lowering blood sugar. Moreover, lactose intolerant people can use oat milk as dairy replacer.

Effects of antioxidant- Black tea is rich in numerous antioxidants. Catechin or quercetin concentration in black tea helps in reducing obesity, skin problems. However, caffeine content of tea has great importance in human health. In addition, oat milk contains numerous antioxidants viz. avenanthramides, help in preventing free radicals from damaging LDL cholesterol, thus reducing the risk of cardiovascular disease.

Effects of micronutrient fortification-Multi micronutrient fortified milk and cereal products can be an effective option to reduce anemia of children up to three years of age in developing countries.

Keeping in view of above points, oat milk flavoured instant soluble tea tablets can be potentially used as tea pouch replacer and other form of milk tea.

The difficulties associated with usage of dipping a tea bag in hot water and sachets of creamer and sugar for making tea while travelling in buses, trains or airlines leading to spillage and problem of discarding tea bag has led to the idea of developing instant soluble tea tablets. The idea of combining all the ingredients for making tea in one tablet/cube such that it is convenient for consumers to use is an acceptable solution for this problem.

8. Summary

Tea is the most commonly used, cheap and refreshing beverage used in India. Both black tea and green tea are enjoyed in all groups of people.The technology described above for production of instant tea powder from fresh tea shoots is a novel approach, can be followed by Industries, in which we are getting both quality products simultaneously without any waste. From 1 kg fresh tea shoots (76% mc wb), 477 mL juice (9 °Brix) and 537 g pressed leaf residue (62% mc wb) were obtained. The juice was subsequently converted into 29 g free flowing instant (soluble) tea powder (3.4% mc wb) by freeze drying and the pressed leaf residue was processed into 208 g tea granules (4.1% mc wb) by convective air drying. Both the instant tea powder and the tea granules (black as well as green) were rated as very good in terms of liquoring characteristics and colour. However, the tea granule was found to be superior in sensory attributes (except colour) than instant tea powder. Chemical quality is also comparable or slightly higher than the existing market samples. The developed patented technology (Indian Patent No.195073) was found to be economically feasible as well.

Instant tea in powder form is more convenient to use when compared to tea bags, however, it still presents a problem in measuring and cannot be conveniently dispensed without spillage. Also, powdered tea will absorb moisture on repeated exposure of the opened jar and thereby, lose valuable flavor and aroma. To solve these problems, tea tablets or cubes, which are usually sufficient to make one cup of the beverage, is a probable solution. Tea tablets/cubes can be prepared by combining all the tea ingredients in the appropriate proportions which is then subjected to direct compression. The advantages of tea cubes includes no spillage, convenience to use, no disposable bags, easy solubility, consistent taste and ease of handling. People can avoid the trouble of boiling hot water for brewing tea. Instantly soluble tea tablets would be light, easy to be utilized, and sanitary to handle. The production of instant tea cubes/tablets will provide one step solution to the tea consumers as they would no longer need separate packets of tea bags, sugar and milk/creamer to make that refreshing cup of tea.

References

Anderson, R.A. and Polansky. 2002. Tea enhances insulin activity. *Journal of Agricultural and Food Chemistry,* 50:7182-7186.

Bordoloi, P.K. 2012.Global tea production and export trend with special reference to India. *Two and a Bud.*, 59(2): 152-156.

Chen, Y. 1979. Tea Manufacture. Agricultural Publication House of China: Beijing, pp. 364 – 381.

Cheng, T.O. 2000. Tea is good for the heart. *Arch Int Medicine,* 160: 2397–2401.

Cheng, T.O. 2004. Will green tea be even better than black tea to increase coronary flow velocity reserve? *Am J Cardio*, 194: 1223–1226.

Child, R. 1960. Elementary notes on tea manufacture I. Withering: cited in Kirk- Othmer Encyclopedia of Chemical Technology: 3rd edn. John Wiley & Sons, New York . Vol 22: 628 – 644.

Das, H. 2005. Food processing operations analysis. Asian Books Pvt Ltd., New Delhi pp: 383–402.

Ding, Z., Kuhr, S., Engelhardt, U.H. 1992. Influence of catechins and theaflavins on the astringent taste of black tea brews, *Zeitschrift für Lebensmitteluntersuchung Forschung, A* 195: 108–111.

Deshpande, R.P., Chinnan, M.S. and McWatters, K.H. 2007. Optimization of a chocolate-flavored, peanut–soy beverage using response surface methodology (RSM) as applied to consumer acceptability data. *Food Science and Technology,* 41: 1485–1492.

Dittman, F.W., Cook EM. 1977. Analyzing spray drier. Chemical Engineering, McGraw Hill, New York.

Dryden, G., Song, M. and McClain, C. 2006. Polyphenols and gastrointestinal diseases. *Current Opinion in Gastroenterology,* 22: 165 – 170.

Dufresne, C.J., Farnworth, E.R. 2001. A review of the latest findings on the health promotion properties of tea. *The Journal of Nutritional Biochemistry,* 12: 404 – 421.

Fujiki, H., Suganuma, M., Kurusu, M. 2003. New TNF-alpha releasing inhibitors as cancer preventive agents from traditional herbal medicine and combination cancer prevention study with EGCG and sulindac or tamoxifen. *Mutation Res.,* 523-524: 119–125.

Genin, N., René, F. 1995. Analyse du Rôle de la transition vitreuse dans les procédés deconservation agroalimentaires. *Journal of Food Engineering,* 26: 391–408.

Hara, Y.L.S., Wickremasinghe, R.L. 1995. "Special issue on tea. IX. Uses and benefits of tea." *Food Reviews International,* 11(3): 527-542.

Harler, C.R. 1963. Tea manufacture. Oxford University Press, New York, Toronto.

Hart, A. 2008. Hot Water Soluble Tea. http://tea-beverage.blogspot.com, accessed on 15th December 2008.

Herath, H.M.U.N. and De Silva, S. 2011Strategies for competitive advantage in value added tea marketing.*Tropical Agricultural Research,* 22: 251–262.

Hirasawa, M., Takada, K. 2004. Multiple effects of green tea catechin on the antifungal activity of antimycotics against Candida albicans. *Journal of Antimicrobial Chemotherapy,* 53: 225–229.

http://www.bruha.com, accessed on 14th December 2008

http://www.teacoffeespiceofindia.com/tea/tea-statistics accessed on 5th February 2015.

Irzyniec, Z., Klimczak, J., Michalowski, S. 1995. Freeze-drying of the black currant juice. *Drying Technology,* 13 (1, 2): 417–424.

Katiyar, S.K., Elmets, C.A. 2001. Green tea polyphenolic antioxidants and skin photo protection (review). *International J Oncol,* 18(6): 1307–1313.

Keegal, E.L. 1958. Cited: Jha A, Mann, Balachandram R 1996. Tea: A refreshing beverage. *Indian Food Ind,* 15(2): 22 – 29.

Kris-Etherton, P.M., Keen, C.L. 2002. Evidence that the antioxidant flavonoids in tea and cocoa are beneficial for cardiovascular health. *Current Opinion in Lipidology,*13(1): 41–49.

Lakshminarayanan, K. and Ramaswamy, S.R. 1978. Towards optimizing fermentation during manufacture. UPASI. T. Sc. Dept. Bulletin, 38(35), 34–47.

Lee, M.J., Maliakal, P., Chen, L. 2002. Pharmacokinetics of tea catechins after ingestion of green tea and (-)-epigallocatechin-3-gallate by humans: formation of different metabolites and individual variability. *Cancer Epidemiology Biomarkers and Prevention,* 11: 1025 – 1032

Liang, J.L., Lu, Zhang, L.Y. 2002. Comparative study of cream in infusions of black tea and green tea (*Camellia sinensis* (L.) O. Kuntze) *Journal of Food Science and Technology,* 37: 627–634.

Liapis, A.L., Sadikoglu, H. 1997. Mathematical modeling of the primary and secondary drying stages of bulk solution freeze-drying in trays: parameter estimation and model discrimination by comparison of the theoretical results with experimental data. *Drying Technology,* 15(3-4): 791 – 810.

Luczaj, W., Skrzydlewska, E. 2005. Antioxidative properties of black tea. *Preventive Medicine*, 40: 910 – 918.

McKay, D.L. and Blumberg, J.B. 2002. The role of tea in human health: an update. *Journal of the American College of Nutrition,* 21:1–13.

Muthumani, T. and Kumar, R.S.S. 2007. Influence of fermentation time on the development of compounds responsible for quality in black tea. *Journal of Food Chemistry*, 101(1): 98–102.

Nance, C.L. and Shearer, W.T. 2003. Is green tea good for HIV-1 infection? *Journal of Allergy and Clinical Immunology,* 112: 851 – 853

Owuor, P.O. and Obanda, M. 2001. Comparative responses in plain black tea quality parameters of different tea clones to fermentation temperature and duration. *Food Chemistry*, 72: 319–327.

Pintaro, N. 1977. Tea and soluble tea products manufacture. *Food Technology Review.* 328. Noyes Data Corporation, USA

Sanderson, G.W. 1972. The chemistry of tea and tea manufacturing. In: VC Reneckles and TC Tso, Editors, *Structural and functional aspects of phytochemistry: Recent advance in phytochemistry* vol. 5, Academic Press, New York. pp. 247 – 317.

Schott, G. 1988. Instant beverages and instant teas and their manufacture. *Food Science Technology Abstract,* 6 6V49: 175

Sinija VR, Mishra HN. 2008. Green Tea: Health Benefits. *Journal of Nutritional & Environmental Medicine*, 17(4): 232 – 242

Sinija, V.R., Mishra, H.N. and Bal, S. 2007. Process Technology for Production of Soluble Tea Powder. *Journal of Food Engineering*, 82(3): 276–283.

Song, J.M., Lee, K.H., Seong, B.L. 2005. Antiviral effect of catechins in green tea on influenza virus. *Antiviral Research*, 68: 66 – 74

Takatoshi, M., Satoshi, H., Akira, S., Ichiro, T., Tadashi, H. 2006. Green tea extract improves running endurance in mice by stimulating lipid utilization during exercise. *American Journal of Physiology-Regulatory, Integrative and Comparative Physiology,* 290: R1550 – R1556.

Taylor, J.R., Wilt, V.M. 1999. Probable antagonism of warfarin by green tea. *Annals of Pharmacotherapy* 33: 426 – 428.

Villegas, B., Tárrega, A. , Carbonell, I. and Costell, E. 2009, Optimising acceptability of new prebiotic low-fat milk beverages. *Food Quality and Preference*, 21: 234–242.

Wherkoven, J. 1974. Tea processing. FAO Agricultural Bulletin No. 26. FAO and the United Nations Rome.

Wickremasinghe, R.L. 1977. Process of making cold water soluble tea concentrates and powders (patented work). Tea Research Institute of Ceylon, Sri Lanka.

Yamamoto, T., Juneja, L.R. Chu D-C, Kim M. 1997. Chemistry and Applications of Green Tea. CRC Press LLC: Boca Raton, USA.

Yang, C.S. and Landau, J.M. .2000. Effects of Tea Consumption on Nutrition and Health. *The Journal of Nutrition,* 130: 2409–2412.

3

Functional Fruit Toffees and Candies

S Sehwag, R Upadhyay, H N Mishra

1. Introduction

A confection is sweet tasting delicacies developed using sugar and other carbohydrates with optionally selected chocolate, nuts, fruits, vegetables or gums. Depending on the formulation, the confection may vary from sugar boiled hard lozenges to gum based soft gummy toffees, from chewing gums to mouth melting chocolates. The versatile and broad variation in the product created by differentiating flavours, ingredients and processing has greatly influenced the market demand of these products. Although confection is non-essential commodity but still relished by people of all economic strata and hence creating a great market space and scope of new products.

In confectionery research and development area, current popular sector is functional confections. The sector has been on the trends radar globally, accounting for 8% of the global confectionery market, and most accepted and successful in Japan. Functional confection is a confection which has been modified/fortified with functional additives to exhibit health promotional and disease preventing actions along with sensory acceptance (Figure 1).

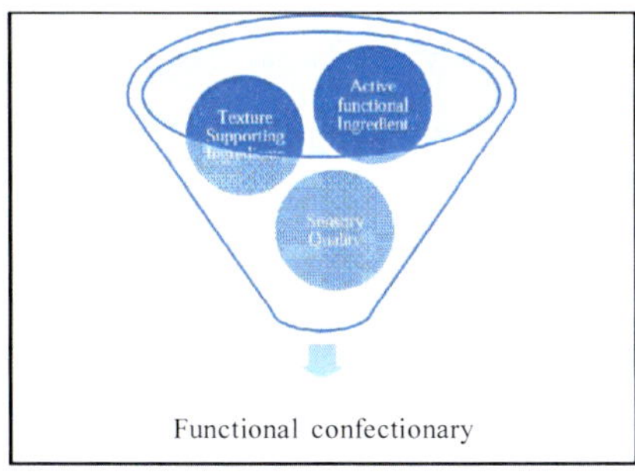

Fig. 1: Functional confection concept

Fig. 2: Example of Functional confection for sports person

There are many well-known functional confectioneries in the market in the form of medicated candy, for instance curing sore throat, digestion, bad breath, etc. These products contain specific curative ingredients. There are also products occupying small niches in the market containing active ingredients aimed at specific consumers, for example for sports people an interesting product came to market in 2008, is Sport Beans (Figure 2) – a range of jelly beans targeted both at athletes and mainstream consumers as an alternative to energy drinks or gels. Functional chewing gums (mainly promoting dental health) currently dominate the functional confectionery market. However, there have also been considerable developments recently in other areas such as low calorie, low fat and sugar free. These products are all designed to provide benefits to distinguish them from standard products, even if specific health claims are not made.

One of the important ingredients in formulation of functional confection is the active functional ingredient. These ingredients can vary from the pure isolated compound to extracts or the parent plant source like fruits. Fruits, especially berries, are rich source of phytochemicals like phenolic compounds, dietary fiber, pigments like carotene. These phytochemicals are proven to possess health benefits like antioxidant, antidiabetic, anticarcenogenic and many more. Thus, the fruits possess therapeutic potential which can be explored as an active functional ingredient in the formulation of functional confection. This idea provides a platform to utilization of fruits as such in a convenient, tasty, healthy and cost effective product.

2. Market for Functional Confectionery

The popularity of confectionery among consumers has led a runway to the industry with a world-wide business of 200 billion USD in 2014, and grew by an annual rate of 2% in real terms over the previous five years.The functional confections are at budding stages in market, consequently few quantified data available to market analysts, and information tends to be fragmentary. A survey (Table 1) largely using data compiled by the Leatherhead Food Research Association (Pettit, 1999) and Agriculture and Agri-Food Canada (2012), can only give a very partial snapshot of developments in some of the most visible markets.

According to Leatherhead Food Research (2011), UK, Spain, Germany and United States are among the strongest markets for functional confectionery, although most innovations have originated from Japan. Among major confectionery industries, Wrigley is the world's leading player in functional confectionery, with its medicated candy and fortified gums.

Table 1: Main functional confectionery varieties by country and value

Country	Chewing gum (sugar free) (Million US$)	Sugar free sweets (Million US$)	Other types* (Million US$)
Australia	57	8	Medicated (64)
China†	69.6	n.a.	Functional chewing gum (1191.4) Medicated (114.2)
France†	528.8	72	Fortified sweets 3.5 Medicated 146.0
Germany	392	298	Cough sweets 104 Fortified sweets 56
Italy	158	150	Herbal 38
Japan†	306	320	Medicated 33.79 Functional chewing gum 27.86
Scandinavia	284	165	n.a.
Spain	138	26	n.a.
United Kingdom	227	67	Low calorie chocolate 43 Fortified sweets 10
United States	856	160	n.a.

Compiled from Pettit (1999) and †Agriculture and Agri-Food Canada (2012); n.a. not available
*May overlap with chewing gum

The size of confectionery industry in India is about Rs 120 billion. According to the leading online newspaper "Food and Beverage" the market is shifting towards value-added confectionery i.e., functional confectionery, a healthier alternative. The major players of functional confection in India enlist of: Cadbury India Ltd., Perfetti Van Melle Ltd., Nestle India Ltd., Lotte India Ltd., Nutrine Confectionery Co. Ltd., Candico India Ltd., Parle Products Pvt. Ltd., Wrigley India Pvt. Ltd, ITC Foods, Hindustan Unilever Ltd and Glaxo SmithKline Pvt. Ltd.. On Indian market shelf, medicated sugar based lozenges and sugar free functional chewing gums for dental care are most popular functional confections.

The whole market scenario, globally and domestically, calls for innovation in the section of functional confectionery addressing the need and economic balance at consumers' end.

3. Fruit Based Functional Confection

The category of fruit based functional confection has fruit as whole or extract or isolated part of fruit as an active functional ingredient imparting health benefits to the consumers. As reported by Kalbag (2012) that the market of functional confectionery is in the infancy stages and clear categories of the sector are vague. However, the confection can be of different types by altering the ingredients and processing of the existing confectionery classes.

In general, functional confection can be prepared by modification of formulation of confection classes: chocolate confectionery, flour confectionery (made from starch and modified starches) and sugar based confections i.e., lozenges, caramel, toffees, fudge, gums, jellies, high and low boiling sugar candy, and chewing gums (Pickford and Jardine 2000). The formulations can be modified in two ways viz.

(i) addition of bioactive compounds having therapeutic values, and

(ii) replacing sugar with other carbohydrates possessing health promising effects.

Therapeutic fruit possess promising future in functional confectionery sector. The type of fruit based functional confection and principle involved is enlisted below.

3.1 Sugar candied fruit

Candied fruits are chewy sugar based confectionery prepared from fruits preferably drupe like cherry, raw papaya, pumpkin, amla, ber, etc. the major principle involved in preparation of these candies is impregnation of sugar into the fruit tissue. The impregnation increases the shelf life of the candied fruit by reducing the water activity, and improves flavor (Nunes et al., 2008).

Bonacina (1977) revealed the technology of manufacturing candied fruits and fruit rinds by submerging cooked fruits in sugar syrup. The key principle of this technology is step wise increase in total solid content of the fruit by slow and empirical increase in sugar syrup strength upto 60 to 75°Brix. The empirical increase in sugar syrup is dependent on the rate of sugar diffusion into the fruit tissue. The technology has been improvised from the nutritional as well as economic point by researchers. Kresic′ et al., (2004) reported the positive influence of partial re-use of sugar syrup on color quality of candied celeriac with reduction in hydroxymethyl furfural formation. Also, Sawate et al., (2005) observed significant influence of method of syruping and drying on the quality of papaya candy. They reported the synergistic effect of cold syruping and cabinet drying on the quality of the candy w.r.t. sensory attributes such as colour, flavour, texture and taste. Scientists have also studied the effect of food additives in candied fruit preparation. In this context, Kaiju et al., (2005) reported that treating ginger with $CaCl_2$, EDTA and phytic acid is an effective alternative to sulphite as antibrowining agent.

Enrichment of sugar syrup with ascorbic acid is capable to enhance the nutritional quality as well as shelf life of blackchoke berry and blackcurrant candies, as reported by Pietrzyk et al., (2010). The pretreatment of the fruit prior to submerging in sugar syrup was also found to affect the quality of the end product (Dar et al., 2011, Priyadarshini, 2013). The most important key of improvising the technology is optimization of the treatment parameters using statistical approach. Many researchers have applied the response surface methodology to optimize the candy process for preparation of honey-ginger candy, beet root candy and pumpkin candy (Gupta et al., 2012; Singh and Hathan, 2013; Singh et al., 2014). The researchers identified the main independent parameters for optimization to process fruit into candy as pretreatment conditions such as blanching time for ginger candy and lime water concentration and temperature for pumpkin candy, osmotic solution temperature, immersion time, and drying temperature.

The functional aspect of the candy processed by this technology is wholly due to the fruit used for candy preparation. One of the fine example is Indian gooseberry (*Phyllanthus emblica*) candy, the berry is rich in Vitamin C content and possess antioxidant potential, antibacterial, astringent, chemopreventive and gastroprotective (Al-Rehaily et al., 2002; Krishnaveni and Mirunalini, 2012).

3.2 Gum based soft confection

Gum based soft confection are hydrocolloid candics which means they gel and thicken but also stabilize. Various hydrocolloids can produce candies with very different eating characteristics. For example, agar agar gives a short breaking jelly with good clarity, whereas gum acacia will produce a long-lasting, chewy sweet (Minifiw 1982). Functional ingredients for inclusion within hydrocolloids should be carefully selected so that they do not disrupt the colloidal characteristics that produce the correct texture, clarity and gelling properties. Recipes require modification so that the addition of functional ingredients results in a solids content that is appropriate for the hydrocolloid in question. Otherwise mould growth, convex sweet backs, crystallisation or drying out/stickiness may occur. Added ingredients, in some cases, may be required to withstand very high temperatures. For example, a starch based confection must be typically at 90°C to ensure depositing without gelation (Burey et al., 2012). If the functional ingredient is less soluble in the formulation then it may produce a cloudy or grainy end product. The pH of some hydrocolloid systems is critical in determining the final set. Acid is the last addition as it has a major effect on gel strength. The acid stability of functional ingredients should be determined. Functional ingredients themselves might also affect pH and cause degradation of a product.

The approach of gum based confection is best suited for fleshy and juicy fruits having therapeutic value such as strawberry, grape, jamun, bael, phalsa, raspberry, blackberry, mulberry etc. Fisher et al., (2014) developed novel strawberry confection using starch as gelling agent. The confection was prepared from freeze dried strawberry powder and reported to possess chemopreventive compounds. Also, a confection having high antioxidant potential was developed by Cappa et al., (2014) using the approach of incorporating grape skin powder into jelly medium. Sehwag and Das (2015) developed functional confection possessing antidiabetic properties with no added sugar using whole jamun fruit including skin, seed and pulp.The judicial selection of hydrocolloids can yield confections without any externally added sugar. Thus, the approach is most suitable to formulate antidiabetic or low calorie confectionery.

3.3 Fruit bars

Fruit bar is a concentrated fruit and dense in nutrition. It is prepared by a single major operation, i.e. drying the fruit pulp after mixing with suitable ingredients. The process is used to prepare bars using pulpy fruits like mango, banana, guava or mixed fruit bars. The major bulking agent used is sugar, however a reduced sugar or no added sugar variant can also be prepared by replacing sugar with other sweetener and functional carbohydrates. Apart from sugar other ingredients like gums, citric acid, preservatives and stabilizers are added to support the processing and obtain sensory acceptable product.

Fortification in fruit bars to enhance the nutritional value and presentation of wholesome product is gaining interests of many researchers. Sarojini et al., (2009) developed protein and mineral fortified fruit bars processed by solar drying. They studied the fortification of mango and guava bars with whey, soy and peas, beta carotene, vitamin C and calcium salts. The study revealed that among protein sources whey fortified bars were highly acceptable for both the fruits where as calcium salts and ascorbic acid was only acceptable for mango bar and beta carotene fortification was sensory acceptable for guava bar. Another successful attempt to develop protein fortified fruit bar was reported by Take et al., (2012) as sapota-papaya fruit bar fortified with skim milk powder. The product was found to have good sensory acceptability with increased protein content.

Fruit bars are nutritionally dense fruit based confection and potential vector for functional active ingredients like fruit extract or freeze-dried fruit powder. Moreover, the product blends in well mixture of fruits to present the synergestic functionality into one product.

4. Raw Materials for Functional Fruit Confection

4.1 Fruits with therapeutic potential

India has wide range of tropical fruits that are underutilized. Most of these fruits can grow even under adverse agroclimatic conditions. A large number of these fruits are known for their characteristic flavor, attractive color and therapeutic/medicinal and nutritive values. As discussed in introduction section, the change in consumer health consciousness and inclination towards healthy and functional food sector. The confectionery market is heading towards functional confectionery. The indigenous fruits of India have an important role to play in satisfying the need of functional confectionery with vast variety in taste and functionality. Some of these fruits are not easy to eat out of hand for instance, Bael fruit that has a hard shell, mucilaginous texture and numerous seeds; Kokum is not acceptable as a fresh fruit due to its high acidity, fresh aonla/amla is also known for its strong astringent taste and being avoided. Most of the underutilized fruits indigenous to India are narrated in Table 2 with their respective health benefits. The table is supportive of the idea to develop functional confectionery using these underutilized fruits

4.2 Gelling and bulking agents

4.2.1 Agar agar (E406)

Agar agar is dried hydrophilic, colloidal polysaccharide from red seaweeds and related marine species. It is available as white to pale yellow agglutinated strips or in flate or powder form. It may have a slightly characteristic odour and mucilaginous flavour. It is soluble in boiling water and insoluble in cold water and most organic solvents. It has a molecular weight of over 20,000. Agar is extracted from a wide range of seaweed varieties which grow in many areas of the world. The main suppliers are Japan, New Zealand, Denmark, Australia, South Africa and Spain. The gel strength varies according to the source and checks should be carried out on each delivery to determine gel strength. Normally agar is dissolved in 30-50 times its weight of water, usually premixed with about 10 times weight of sugar to prevent lumping. Very high viscosities are achieved with concentration upto 10%. Agar provides good gel strength. It forms a firm gel at concentrations as low as 1% and is usually used in confectionery at the level of 1-1.5% of sugar glucose agar recipe. Agar is not absorbed by the body during digestion and can therefore be used in low calorie confections. Agar does not carry flavour well, and as a result of this and its sensitivity to acid and particular types of texture it is being replaced by pectins or modified starches.

Table 2: Compendium of different underutilized fruits possessing therapeutic effects and potential to be employed as raw material for functional confectionery development

Fruits	Therapeutic effects/Functionality	References
Bael/Wood apple (*Aegle marmelos*)	The ripe fruit is capable of curing dyspepsia, diarrhea and dysentery. Ayurveda prescribes the fruit of the herb for heart, chronic constipation, typhoid, cholera, hemorrhoids, hypocondria, melancholia and for heart palpitation.	Sharma et al., 2011
Custard apple (*Annona reticulate*)	Antioxidant, antidiabetics, hepatoprotective, cytotoxic activity, genetoxicity, antitumour activity, antilice agent	Pandey and Barve, 2011
Monkey Jack (*Artocarpus lakoocha*)	Anticariogenic, pancreatic lipase inhibitory and cytotoxic activity	Raghavendra et al., 2012
Star fruit(*Averrhoa carabola*)	Treat throat inflammation, mouth ulcer, toothache, cough, asthma, hiccups, indigestion, food poisoning, colic, diarrhea, jaundice, malarial splenomegaly,hemorrhoids, skin rashes, pruritis, sunstroke and some eye related problems. In women fruit also increases lactation	Dasgupta et al., 2013
Tal fruit (*Borassus flabellifer*)	*Borassus flabellifer* mucilage has gelling property with no syneresis and high stability.	Kumar et al., 2012
Karonda (*Carissa carandas*)	Antioxidant, anti-hypertension, hepatoprotective and antiallergic activity. The fruit possess antimicrobial activity against *B.subtillis*, *Staphylococcus aureus*, *Staphylococcus faecalis*, *E. coli*, *Pseudomonas aeruginosa* and *Salmonella typhi* and *Candida* species	Devmurari et al., 2009
Camito/star apple (*Chrysophyllum cainito*)	The dietary fiber of the fruit sooth inflammation in laryngitis and pneumonia. Antidiabetic and antidiarrhea activity	Morton, 1987
Longan (*Euphoria longan*)	Antioxidant, anti-tyrosinase, anti-glycated and anticancer activities Traditionally in Chinese medicine, the fruit is as a common agent in relief of neural pain and swelling	Yang et al., 2011

Contd.

Table 2 Contd.

Fruits	Therapeutic effects/Functionality	References
Loquat (*Eriobotrya japonica*)	Traditionally, the fruit is used for soothing sore throat. Anti-inflammatory, antidiabetic, antioxidant, antitumor and antiviral Treats chronic bronchitis	Baljinder et al., 2010
Phalse (*Grewia subinaequalis*)	Antioxidant, antipyretic, anti-inflammatory, analgesic, antiviral, antimalarial, antidiabetic and anticancerous activity. The fruit is also reported to have hypoglycemic effect with low glycemic index in healthy humans	Zia-Ul-Haq et al., 2013
Mulberry (*Morus serrata*)	Mulberry fruit have been reported to possess anti-obesity effect with reduced lipogenesisthe fruit is rich in antioxidants studies have reported the neuroprotective effect of the fruit and recommended for treatment and prevention of Parkinson's disease	Peng et al., 2011; Kim et al., 2010
Date Plam (*Phoenix sylvestris*)	Acts as an analgesic Reported to possess hepatoprotective, antioxidant, antimutagenic, antiinflammatory, antiviral and anti diarrhoeal activity	Zafar, 2010; Vyawahare and others, 2009
Amla/Indian gooseberry (*Phyllanthus emblica*)	Possess antioxidant potential, antibacterial, astringent, chemopreventive and gastroprotective	Al-Rehaily et al., 2002; Krishnaveni and Mirunalin, 2012
Cape gooseberry (*Physalis peruviana*)	Used in folk medicine for its properties as anticancer, antimicrobial, antipyretic, diuretic, and anti-inflammatory immunomodulator Also effective in postprandial glucose level control	Ramadan, 2011
Indian Jujube/Ber (*Ziziphus mauritiana*)	Jujubosides (saponin) isolated from Ziziphus possess haemolytic, sedative, anxiolytic and sweetness inhibiting properties. Cyclopeptide alkaloids have sedative, antimicrobial, hypoglycemic, antiplasmodial, anti-infectious, antidiabetic, diuretic, analgesic, anticonvulsant and anti-inflammatory activities.	Goyal et al., 2012

Contd.

Table 2 Contd.

Fruits	Therapeutic effects/Functionality	References
Jamun(*Syzygium cumini*)	Antimicrobial, antioxidant, antidiabetic, chemorpreventive, hepatoprotective, gastroprotective, hypolipidemic, anti-inflamatory, antipyretic, anti-arthritis, hypoglycemic effect	Sehwag and Das, 2015
Safed jamb/ Wax apple	Effective drug on both Gram positive and Gram negative bacteria, Effective in curing diabetes type 2 (*Syzygium samarangense*)	Ratnam and Raju, 2008; Resurreccion Magno et al., 2005
Brazillian cherry (*Syzygium uniflora*)	Anti-diarrheic, diuretic, anti-rheumatic, anti-febrile and anti-diabetic Antimicrobial activity against *S. aureus*, *L. monocytogenes*, *C. lipolytica* and *C. guilliermondii*, anti-Trypanosoma β-adrenergic induced hypotension in rats heart	Costa et al., 2013
Water chestnut (*Trapa natans*)	Antidiabetic effect,Antifungal effect against *Candida tropicalis*.	Das et al., 2011; Mandal et al., 2011
Sea-buckthorn berries (*Hippophae rhamnoides*)	Effective against gastic ulcers cardiovascular diseases, thrombosis and cancer, and Possess antiallergic, antioxidant, hepatoprotective, analgesic effects	Zeb,2004; Xu et al., 2011

4.2.2 Alginate (E401)

Alginate were first is isolated by Stamford by alkaline extraction from brown algae, a process used for iodine production. Commercial extraction is from seaweeds such as *Laminaria digitata*, *Ascophyllum nodosum* and *Fucus serratus*. Each of the seaweeds provides a differing proportion of the main attribute of alginate. It is a white to yellow granular powder, colloidal, insoluble in water, acids and organic solvents. Alginates are comprised of mannuronic and guluroni acids. These can link to form homogeneous segments in which guluronic acid binds to guluronic acid and mannuronic acid binds to mannuronic acid.

4.2.3 Carrageenan (E407)

The name carrageenan is derived from the country of Carraghen on the south coast of Ireland, where Irish moss was used in foods and medicines more than 600 years ago. Red seaweeds were used because of their unique property in gelling milk when they are heated together. The carrageenan coagulates into fibres, leaving impurities in the solution. This product is pressed and washed again with alcohol to complete its dehydration. It is then dried under vacuum, milled and sieved to the exact particle size. The gelation of κ and *i* carrageenans is induced by the association of chains through double helices.

4.2.4 Gelatin (E441)

Gelatin does not exist naturally but is produced by the partial hydrolysis of collagen in the raw material substrate. Collagen is a structural component in animal tissues, present in skin, bone and connective tissue. The raw materials are sourced from slaughter houses, meat-packing plants or tanneries. The products from tanneries have already been salted or limed for preservation. Collagen is made up of films and fibrils. Industrial modification of collagen to produce gelatin is by stepwise destruction of the organized structure to obtain the soluble derivative gelatin. The extraction of gelatin from the raw material is initiated by either liming or acidulation, which disrupts the molecular linkages within the collagen. Gelatin is then extracted by hot-water hydrolysis. This is carried out as a batch operation. Several extracts are produced with a concentration of 5-10% gelatin. A typical analysis of a gelatin would have 14% moisture, 84% protein and 2% ash. Gelatin picks up water in a moist atmosphere and should be stored in a cool dry store. At about 16% moisture mould growth is possible. Hygienic procedures must be implemented when using this product in solution and equipment must be thoroughly cleaned.

4.2.5 Pectin (E440)

Pectic substances are matrix components in the cell walls of higher plants. The compounds are insoluble in aqueous solution and are referred to as protopectins. Protein consists mainly of the partly methylated esters of polygalacturonic acid and their ammonium, sodium, potassium or calcium salts. The molecular weight is between 20,000 and 100,000. The protopectin is hydrolysed using acid in hot aqueous solution. The aqueous extract contains soluble products such as neutral polysaccharides, gums et al., LM pectins are defined as having a degree of methoxylation of less than 50% i.e less than 50% of the functional groups on the molecule are methoxylated. The grade strength of a pectin is defined as the number of grams of sugar with which one gram of pectin will produce a gel of standard firmness, when tested under standard conditions of acidity and soluble solids content. During cooking the product mixture is normally buffered to maintain the pH within controlled limits. Pectins can be purchased pre-buffered or the manufacturers can add citrates, tartrates, etc, to act as the buffering agent.

4.2.6 Xanthan gum (E415)

A plysaccharide gum produced by *Xanthomonas campestris*. It occurs as a cream coloured powder, soluble in water, insoluble in alcohol. Xanthan gum is a secondary metabolite of *Xanthomonas campestris* produced during the commercial aerobic fermentation of carbohydrates. Fermentation is carried out in a batch process. The gum is recovered from the broth by the addition of propan-2-01. The precipitate obtained is washed and pressed to remove residual alcohol. Xanthan gum is a mixed polysaccharide with a molecular weight of approximately 2.5 million. The monomer units are D-glucose, D-mannose and glucoronic acid. Xanthan gum is readily dissolved in hot or cold water to produce an opaque solution of relatively high viscosity. This solution exhibits pseudoplastic flavour characteristics, i.e., the viscosity of the solution decreases rapidly when shear is applied. As the shear rate decreases there is an immediate return to the high original viscosity. This charecteristic makes Xanthan an excellent suspending agent at low concentration. Xanthan forms a thermo reversible cohesive gel system with locust beangum.

4.2.7 Sorbitol (E420)

Sorbitol is primarily used in manufacture of diabetic and sugar free confections. The most important applications includes chewing gum, compressed mints, high boiling, gums, pastilles and chocolates. In chewing gum sorbitol is typically used together with maltitol syrup which provides the liquid phase and saccharin or aspartame which are needed to boost the sweetness. A proportion of mannitol may be included in order to inhibit crystallization if desired chewing gum dragees may be hard coated with sorbitol. Sorbitol cannot be used to manufacture high boiling by conventional means due to its low viscocity. It is however possible to prepare deposited high boiling.

4.2.8 Xylitol (E967)

Xylitol is used in variety sugar free and diabetic confectionary products. Xylitol can be used in the hard panning of chewing dragees or of other confectionery centers. Xylitol can act as the sole sweetner in recrystallized hard candies. Fondant is another application for which xylitol is well suited. Xylitol solutions are lower in both viscocity and water activity then equivalent concentrations of other polyols but do not have particularly good humectants properties.

4.2.9 Maltitol (E965)

Maltitol syrups have been used in a wide variety of applications either alone or in combination with other polyols. Their function is sugarless chewing gums and in high boilings which can be manufactured simply by adding acid, color and flavor to a boiled maltitol syrup. Caramels and chews can be prepared from maltitol syrup. Gums, jellies and pastilles based on gum Arabic or gelatin may be successfully made with malitol syrup. Maltitol syrup is also used to prevent the crystallization of other polyols in hard and soft confections.

4.2.10 Isomalt (E953)

Isomalt can be used in a variety of sugar free confectionery products including high boiling, compressed tablets, marzipan, chews, liquorice and chocolate. Other applications include use of isomaltase in combination or with other polyols to inhibit crystallization as well as to increase the sweetness. Compressed tablets also benefits from the low hygroscopicity of isomalt. A wet granulation stage is required in order to improve the compressibility.

4.2.11 Polydextrose (E1200)

Polydextrose can be used in the manufacture of high boiling since it forms a stable glass structure.It also reduces the viscosity thus improving handling properties. In the context it has been successfully combined with xylitol in reduced calorie chews with sorbitol or xylitol in gelatin jellies and with isomalt in fondant. In addition to its application in sugar-free products, polydextrose is also used in combination with fructose for diabetic lines or with sucrose in standard reduced calorie lines.

5. Formulation and Processing of Functional Confectionery

Formulating a functional confectionery product begins at the same point as standard confectionery development: the generation of a recipe for a good tasting product. Where functional confectionery is concerned, the steps might progress as follows.

5.1 Choosing a suitable confectionery vehicle

This will be defined to some extent by the product concept. Some concepts will fit better with sugar confectionery, others with bars or may be dictated by practical considerations, e.g., an antidiabetic confection has to be prepared without added any external sugar. The cost of added functional ingredients can be quite expensive, so keeping the manufacturing method as simple as possible is desirable. A general scheme of processing a gel based confectionery is presented in Figure 3.

In case of fruit based functional confection, the functional fraction of the fruit with medicinal properties (Table 2) is incorporated in the recipe in most economical choice. For instance, Fisher et al., (2014) developed gel based soft confection using freeze dried strawberry powder as functional ingredient and starch as gelling agent. The confection was reported to possess chemopreventive action. The technology involved was simple including dissolution of starch and corn syrup into a thick mixture with 65-68° Brix followed by addition of fruit powder and depositing to form confectionery. Also, a confection having high antioxidant potential was developed by Cappa et al., (2014) using the approach of incorporating grape skin powder into jelly medium. Another example of fruit based soft confection with health benefits is documented by Sehwag and Das (2015) as antidiabetic and antioxidant confection prepared from whole jamun fruit including skin, seed and pulp without addition of sugar. The steps followed were simple and mainly include heating and mixing as basic unit operations. The process involves dissolution of hydrocolloids by heating, addition of fruit part while heating and setting of the

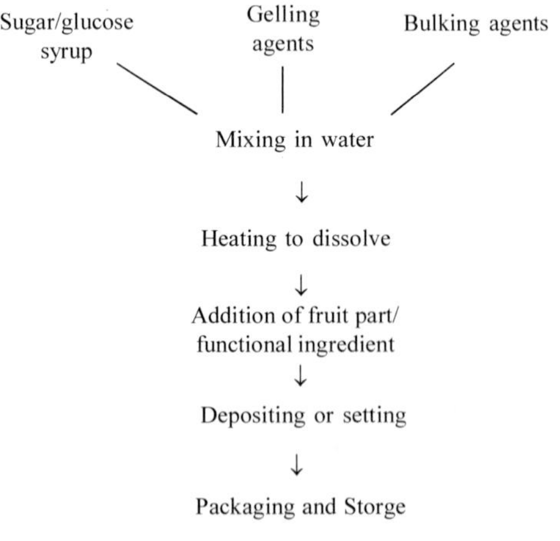

Fig. 3: General scheme of processing a gel based confectionery

mixture using firming agent into confectionery. They optimized the mixture of hydrocolloids, viz-a-viz agar, pectin and polydextrose, with jamun pulp and seed powder to formulate the confectionery with high polyphenolics content and acceptable sensory quality.

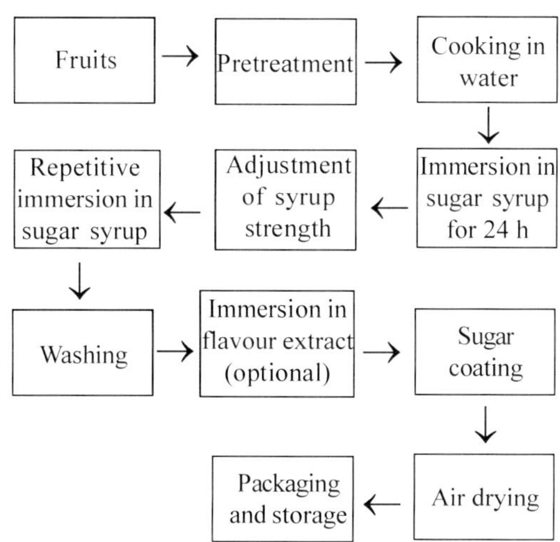

Fig. 4: General scheme for processing a sugar boiled fruit candy

Another way of selecting the processing of fruit into a functional confectionery is sugar boiled fruit. A general scheme of processing a sugar boiled fruit confectionery is presented in Figure 4. One of the fine example is Indian gooseberry (*Phyllanthus emblica*) candy, the berry is rich in Vitamin C content and possess antioxidant potential, antibacterial, astringent, chemopreventive and gastroprotective (Mishra et al., 2011).

5.2 Where to incorporate the functional ingredients in the product

This will depend upon the amount of functional ingredients needed, the nature of the ingredients, such as powder or paste, fine particles or large particles; whether or not they will be affected by certain processing conditions, for example, high temperatures or shearing action. How they behave in a moist or fat-based medium must also be considered and whether the functional ingredients will interact with other ingredients in the recipe.

5.3 Manipulation of ingredients

Reduction or elimination of one or more basic ingredient of fruit based confection with an aim of incorporating bioactive component should not result in deterioration of sensorial and nutritional quality of the end product along with the shelf life.

5.4 Masking agents

The formulation may require the addition of masking agents and taste improvers like salt, citric acid etc. to overcome any undesirable tastes from the functional ingredients.

5.5 Effects of functional ingredients on critical parameters of the end product

The critical parameter of an end product includes moisture content, total solids, water activity, pH, gel strength, viscosity, texture or crystallization. In a formulation functional ingredients often have a dual action, and this must be taken into account. For example, prebiotics, these are oligosaccharides of particular current interest for their potentially beneficial effects on gut health. In addition to health benefits, they also act as bulking sucrose substitutes. Generally, the larger the molecular weight of the oligosaccharide, the weaker its intensity of sweetness. The sweetness of highly purified oligosaccharides is less than half that of sucrose, therefore, the use of an intense sweetener may need to be considered. These bulking agents alike sucrose improves the texture but does not have much effect on water activity of the final product.

5.6 Stability of functional ingredients

Functional ingredients may affect or be affected by certain aspects of processing, for example, high temperatures or shearing action. They may also react differently in a moist or fat-based medium. As discussed previously, other ingredients present in the recipe might affect the action or stability of the functional ingredients. Some functional ingredients could affect critical parameters of the finished product, such as moisture content, total solids, water activity, pH, gel strength, viscosity, texture or crystallization. Conversely, one or more of these factors might affect the action or stability of the functional ingredients themselves.

6. Packaging and Storage

Sugar based fruit confectioneries are intermediate moisture product with water activity ranging between 0.7 and 0.86. When such confectioneries are stored without proper packaging, especially in areas of high humidity, the sucrose may crystallize, making the sweet sticky and grainy. Thus, to provide sufficient protection for a long shelf-life proper barriers to moisture and environmental factors is a requisite. For packaging individual confectioneries waxed paper, aluminium foil, and cellulose film, or a combination of these can be sued. Moreover, metalized biaxially oriented polypropylene (BOPP) and multilayer BOPP with polyethylene terephthalate (PET) and low density polyethylene can also be used. The confectioneries can be wrapped manually for small produce and semi-automatic and form fill and seal machines are available for higher production. For further protection, the individually-wrapped confectioneries may be packed in a heat-sealed polythene bag or PET jars. Confectioneries can also be packaged in glass jars, or tins with close fitting lids.

Muizniece-Brasava et al., (2011) studied the influence of iron based oxygen absorber active packaging on the shelf life of apple-black currant marmalade candies using polymer multibarrier and paper bags as packaging material. In the study an active packaging in combination with modified atmosphere was examined and compared with traditional packaging in air ambiance. They reported that the active packaging in multi barrier pouches (MAP, 100% CO_2) with incorporated oxygen scavenger could be considered as the best variant and paper bags could be used only for candy's short-term packaging on the supermarket shelves.

7. Summary

Confectionery that could be described as functional has a history going back millennia into ancient Egyptian times. Such early functional confectionery had at herapeutic purpose, and this tradition is still apparent today with many confectionery-like products available to cure ailments such as coughs and sore throats. The incorporation of fruits with medicinal effects can provide new horizon to the sector of functional confectioneries with wide range of flavor and health benefits. The increasing demand of functional confectioneries in the market at global as well as Indian market with parallel step-up in life style disorders and concept of functional diet in consumers' are thrust providers in the pertaining area of research.

Confectionery is a group of carbohydrate-rich, sweet taste product with texture varing from hard to soft delicacies. The variety in the texture with alteration in ingredients leads innovative areas of new product development. Fruit based functional confectioneries are broadly classified as: sugar boiled fruit candies, gel based soft candies and fruit bars. All the three categories have texture varing from chewy to gummy and hard to soft. The sugar boiled fruit candies have major constrain of high sucrose content, hence restricting use of this technology for production of functional confection targeting anti-diabetic/reduced or low-calorie/anti-obese/dental protection effects. For such applications, gel based soft candies and fruit bars are most suitable. The process involved is simple including addition of other ingredients, dissolution by heating and setting. Functional confectionery based on tropical medicinal fruits upholds promising future in sector of functional foods. The idea also addresses the problem of post harvest losses of underutilized fruits.

References

Agriculture and Agri-food Canada. 2012. International Market Bureau, Market research survey, Consumer trends: Confectionery Japan, China and France.

Al-Rehaily, A.J., Al-Howiriny, T.S., Al-Sohaibani, M.O., Rafatullah, S. 2002. Gastroprotective effects of 'Amla' *Emblica officinalis* on in vivo test models in rats. *Phytomedicine,* 9(6): 515-522.

Baljinder, S., Seena, G., Dharmendra, K., Vikas, G., Bansal, P. 2010. Pharmacological Potential of *Eriobotrya japonica*-An overview. *International Research Journal of Pharmacy,* 1:95-99.

Bonacia Remigio 1977. Method of candying fruits and fruit rinds, US Patent 4041184.

Burey, P., Bhandari, B.R., Rutgers, R.P.G., Halley, P.J., Torley, P.J. 2012. Confectionery gels: A review on formulation, rheological and structural aspects. *International Journal of Food Properties,* 12(1): 176-210

Cappa, C., Lavelli, V., Mariotti, M. 2014. Fruit candies enriched with grape skin powders: Physicochemical properties. *LWT-Food Science and Technology*, DOI: 10.1016/j.lwt.2014.07.039.

Costa AGV, Garcia-Diaz DF, Jimenez P, Silva PI. 2013. Bioactive compounds and health benefits of exotic tropical red–black berries. *J Func Foods,* 5(2):539-549.

Dar, B.N., Ahsan, H., Wani, S.M., Dalal, M.R. 2011. Effect of $CaCl_2$, citric acid and storage period on physico-chemical characteristics of cherry candy. *Journal of Food Science and Technology,* 1(2):79-85.

Das PK, Bhattacharya S, Pandey JN, Biswas M. 2011. Antidiabetic activity of Trapa natans fruit peel extract against streptozotocin induced diabetic rats. *Global Journal of Pharmacology,* 5(3): 186-190.

Dasgupta, P., Chakraborty, P., Bala, N.N. 2013. Averrhoa carambola: an updated review. *International Journal of Pharma Research & Review,* 2: 54-63.

Devmurari, V., Shivanand, P., Goyani, M.B., Vaghani, S., Jivani, N.P. 2009. A review: Carissa congesta: phytochemical constituents, traditional use and pharmacological properties. *Pharmacognosy Reviews,* 3(6): 375.

Fisher, E.L., Ahn-Jarvis, J., Gu, J., Weghorstc, C.M., Vodovotz, Y. 2014. Assessment of physicochemical properties, dissolution kinetics and storage stability of a novel strawberry confection designed for delivery of chemopreventive agents. *Food Structure,* 1:171-181.

Food Beverage Newz. 2015. Available at http://www.fnbnews.com/article/detnews.asp?articleid=32357§ionid=32 accessed on 21 Feburary, 2015.

Goyal, M., Sasmal, D., Nagori, B.P. 2012. Review on ethnomedicinal uses, pharmacological activity and phytochemical constituents of Ziziphus mauritiana (*Z. jujuba* Lam., non Mill). *Spatula, DD* 2(2):107-116.

Gupta, R., Singh, B., Shivhare, U.S. 2012. Optimization of osmo-convective dehydration process for the development of honey-ginger candy using response surface methodology. *Drying Technology, J* 30:750-759 (DOI: 10.1080/07373937.2012.661818).

Kaiju, M.O., Xingping, W., Cheng, C. 2005. Ginger candy processing technology and antibrowning without sulphite. *Transactions Chinese Soc Agricul Engg*, available at http://en.cnki.com.cn/Article_en/CJFDTotal-NYGU200501035.htm, accessed on 16.8.2014.

Kalbag. 2012. Available at http://www.fnbnews.com/article/detnews.asp? articleid=32110§ionid=7 accessed on 21 Feburary, 2015.

Kim, H.G., Ju, M.S., Shim, J.S., Kim, M.C, Lee SH, Huh Y. Oh MS. 2010. Mulberry fruit protects dopaminergic neurons in toxin-induced Parkinson's disease models. *British Journal of Nutrition,* 104(01):8-16.

Kresic´, G., Lelasb, V., Simundic´, B. 2004. Effects of processing on nutritional composition and quality evaluation of candied celeriac. *S¯adhan¯a*, 29(I):1-12, available at http://www.ias.ac.in/sadhana/Pdf2004Feb/Pe1166.pdf, *accessed on,* 13.08.2014.

Krishnaveni, M., Mirunalini, S. 2012. Chemopreventive efficacy of Phyllanthus emblica L.(amla) fruit extract on 7, 12-dimethylbenz (a) anthracene induced oral carcinogenesis–A dose–response study. *Environmental Toxicology and Pharmacology*, 34(3):801-810.

Kumar, R., Rajarajeshwari, N., Narayana Swamy, V.B. 2012. Exploitation of Borassus flabellifer fruit mucilage as novel natural gelling agent. *Der Pharmacia Lettre*, 4(4): 1202-1213.

Leatherhead Food Research. 2011. In: Innovations in the Global Confectionery Market Published in February 2011.

Mandal, S.M., Migliolo, L., Franco, O.L., Ghosh, A.K. 2011. Identification of an antifungal peptide from *Trapa natans* fruits with inhibitory effects on *Candida tropicalis* biofilm formation. *Peptides*, 32(8):1741-1747.

Minifie, B.W. 1982. Chocolate, Cocoa and Confectionery: Science & Technology, New York, AVI Publishing.

Mishra, S., Verma, A., Prasad, V.M., Sheikh, S. 2011. Development of value added amla candy with rose extract.*The Allahabad Farmer*, 16(2):20-27.

Morton, J. 1987. Star Apple. In: Fruits of warm climates. Julia F. Morton, Miami, FL. p. 408–410.

Muizniece-Brasava, S., Dukalska, L., Kampuse, S., Murniece, I., Sabovics, M., Dabina-Bicka, I., Sarvi S. 2011. Influence of active packaging on the shelf life of apple-black currant marmalade candies. *World Academy of Science, Engineering and Technology*, 56:555-563.

Nunes, C., Coimbra, M.A., Saraiva, J., Rocha, S.M. 2008.Study of the volatile components of a candied plum and estimation of their contribution to the aroma. *Food Chemistry*, 111(44):897-905 (DOI: 10.1016/j.foodchem.2008.05.003).

Pandey, N., Barve, D. 2011. Phytochemical and pharmacological review on Annona squamosa Linn. *International Journal of Research in Pharmaceutical and Biomedical Sciences*, 2(4):1404-1412.

Peng, C.H., Liu, L.K., Chuang, C.M., Chyau, C.C., Huang, C.N., Wang, C.J. 2011. Mulberry water extracts possess an anti-obesity effect and ability to inhibit hepatic lipogenesis and promote lipolysis. *Journal of Agricultural and Food Chemistry*, 59(6):2663-2671.

Pettit B. 1999. The International 'Healthy' Confectionery Market, 2nd edn, Leatherhead Food Research Association.

Pickford, E.F., Jardine, N.J. 2000. Functional confectionery, In: Functional foods, 259-286.

Pietrzyk S, Gakowska D, Fortuna T, Bojdo-Tomasiak I, Wypchol, A. 2010. The influence of storage conditions of candied fruits enriched with vitamin C by different methods on its content. *Potravinarstvo*, 4:65-66 (DOI: 10.5219/55).

Priyadarshini. 2013. Effect of pretreatment on organoleptic attributes of apple candy during storage. *International Journal of Food, Agriculture and Veterinary Sciences*, 3(2):139-148.

Raghavendra, H.L., Mallikarjun, N., Venugopal, T.M. 2012. Elemental composition, anticariogenic, pancreatic lipase inhibitory and cytotoxic activity of Artocarpus lakoocha Roxb pericarp. *International Journal of Drug Development and Research*, 4(1): 330-336.

Ramadan, M.F. 2011. Bioactive phytochemicals, nutritional value, and functional properties of cape gooseberry (*Physalis peruviana*): An overview.*Food Research International*, 44(7): 1830-1836.

Ratnam, K.V., Raju, R.V. 2008. In vitro antimicrobial screening of the fruit extracts of two Syzygium species (Myrtaceae). *Advances in Biological Research*, 2:17-20.

Resurreccion Magno, M., Hanshella, C., Villaseñor, I.M., Harada, N., Monde, K. 2005. Antihyperglycaemic flavonoids from Syzygium samarangense (Blume) merr. And perry. *Phytotherapy Research*, 19(3):246-251.

Sarojini, G., Veena, V., Rao, M.R. 2009. Studies on Fortification of Solar Dried Fruit bars. International Solar Food Processing Conference.

Sawate, A.R., Patil, V.P., Ghatge, P.U., Kshirsagar, R.B., Tapre, A.R. 2005. Studies on effect of syruping and drying methods on quality of papaya-candy. *J Soils Crops*, 15(1):105-110.

Sehwag, S., Das. Inventors. 2015. Whole jamun (*Syzygium cumini*) fruit based functional confection and process of manufacture thereof. India 132/KOL/2015.

Sehwag, S., Das, M. 2015. Nutritive, therapeutic and processing aspects of Jamun, *Syzygium cuminii* (L.) Skeels-An overview. *Indian Journal of Natural Products and Resources*, 5(4):295-307.

Sharma, G.N., Dubey, S.K., Sharma, P., Sati, N. 2011. Medicinal values of bael (*Aegle marmelos*) (L.) corr.: a review. *International Journal of Current Pharmaceutical Review and Research,* 2(1): 12-22.

Singh, B., Hathan, B.S. 2013. Optimization of osmotically dehydrated beetroot candy using response surface methodology. *International Journal of Food and Nutritional Sciences,* 2(1): 15-21.

Singh, S., Kumar, U. and Rai, A. (2014). Process optimization for the manufacture of angoori petha. *Journal of Food Science and Technology,* 51(5): 892-899.

Take, A.M., Bhotmange, M.G., Shastri, P.N. 2012. Studies on Preparation of Fortified Sapota-Papaya Fruit Bar. *Nutrition & Food Sciences,* 2:6.

Vyawahare, N., Pujari, R., Khsirsagar, A., Ingawale, D., Patil, M., Kagathara, V. (2009). Phoenix dactylifera: An update of its indegenous uses, phytochemistry and pharmacology. *The Internet Journal of Pharmacology*, 7(1).

Xu, Y.J., Kaur, M., Dhillon, R.S, Tappia, P.S., Dhalla, N.S. 2011. Health benefits of sea buckthorn for the prevention of cardiovascular diseases. *Journal of Functional Foods,* 3(1): 2-12.

Yang, B., Jiang, Y., Shi, J., Chen, F., Ashraf, M. 2011. Extraction and pharmacological properties of bioactive compounds from longan (*Dimocarpus longan* Lour.) fruit—A review. *Food Research International,* 44(7): 1837-1842.

Zafar, A. 2010. A Review on Analgesic: From Natural Sources. *International Journal of Pharmaceutical & Biological Archive,* 1(2).

Zeb, A. 2004. Important therapeutic uses of sea buckthorn (*Hippophae*): a review. *Journal of Biological Sciences*, 4(5): 687-693.

Zia-Ul-Haq, M., Stankoviæ, M.S., Rizwan, K., Feo, V.D. 2013. *Grewia asiatica* L., A food plant with multiple uses. *Molecules,* 18(3): 2663-2682.

4

Low Cholesterol Dairy Products

J Chitra, M Ghosh, I Dey Paul, H N Mishra

The nutritional richness and the high biological value of dairy milk have made it an important constituent of the balanced human diet. It is associated with better growth, improved status of some micronutrients, cognitive performance, motor development and activity. It is composed of a mixture of nutritive components as well as other bioactive factors such as biogenic amines, conjugated linoleic acid (CLA), with relevant physiological benefits. The consumption of dairy milk and products exert a protective action in human health and have been confirmed in preventing several chronic conditions like cardiovascular diseases (CVDs), some forms of cancer, obesity, immunomodulation and diabetes (Pereira, 2014). The chemical composition of bovine milk consists of 87% water, 4% to 5% lactose, 3% protein, 3% to 4% fat, 0.8% minerals, and 0.1% vitamins (Pereira, 2014).

1. Composition of Milk Fat

Bovine milk contains about 3.5 to 5% total lipid, existing as emulsified globules 2 to 4 μm in diameter and coated with a membrane, casein. The composition of milk fat is somewhat complex. Triglycerides constitute approximately 98% of milk fat, which is found in the milk globule, with the remainder being made up of di- and monoglycerides, phospholipids, cerebrosides, cholesterol, vitamins, tocopherols, carotene, and flavor components. Phospholipids are about 0.5 to 1% of total lipids, and sterols are 0.2 to 0.5%. These are mostly located in the milk fat globule membrane (Jensen et al., 1991). Cholesterol is the major sterol at 10 to 20 mg/dL. On an average, 70% of fat fraction is composed by saturated fatty acids (SFAs) and 30% unsaturated fatty acids (Ohlsson, 2010). Unlike the fats from most animal sources, dairy saturated fats are composed of fatty acids ranging in carbon length from 4 to 20. Bovine milk contains substantial quantities of C4:0 to C10:0, about 2% each of C18:2 and trans-C18:1, almost no other long-chain polyunsaturated fatty acids (Jensen et al., 1991). Total fat content and fatty acid composition of milk from various sources are presented in Table 1.

Table 1: Total fat content and fatty acid composition of milk of various species

Fatty acid	Cattle[d]	Goat[d]	Sheep[c]	Buffalo[c]	Camel Bactrian[b]	Camel Dromader[b]	Donkey[a]	Human[e;f]
C4:0	3.84	1.27	4.06	3.90	0.54	0.34	0.60	0.60[f]
C6:0	2.28	3.28	2.78	2.33	0.46	0.29	1.22	0.07[f]
C8:0	1.69	3.68	3.13	2.41	0.53	0.27	12.80	0.21[f]
C10:0	3.36	11.07	4.97	2.40	0.46	0.27	18.65	1.39[f] - 1.04[e]
C12:0	3.83	4.45	3.35	3.09	1.24	0.80	10.67	4.71[f] - 6.48[e]
C14:0	11.24	9.92	10.16	28.02	15.43	10.10	5.77	3.92[f] - 7.44[e]
C16:0	32.24	25.64	23.10	12.58	32.05	29.74	11.47	18.68[f] - 22.24[e]
C16:1	1.53	0.99	0.68	1.93	7.01	6.60	2.37	1.29[f] - 2.50[e]
C18:0	11.06	9.92	12.88	12.58	14.75	17.82	1.12	5.63[f] - 6.45[e]
C18:1	1.63 + 21.72	0.37 + 23.80	26.01	24.10	18.78	24.66	9.65	31.26[f] - 32.78[e]
C18:2	2.41	2.72	1.61	2.04	1.19	1.61	8.15	17.73[f] - 16.29[e]
C18:3	0.25	0.53	0.92	0.68	0.60	0.51	6.47	1.36[f] - 0.60[e]
C20:4	-	-	0.20	0.35	n/a	n/a	0.07	0.30[f] - 0.51[e]
C20:5	-	-	0.09	0.18	n/a	n/a	0.27	0.10[e]
C22:6	-	-	0.08	0.12	n/a	n/a	0.30	0.19[e]
CLA	0.45	0.68	0.67	0.49	n/a	n/a	n/a	n/a
SFA	71.24	70.42	65.17	65.86	69.90	64.86	67.6	44.30[e]
MUFA	25.56	25.67	24.29	26.43	28.07	33.03	15.8	36.56[e]
PUFA	3.20	4.08	2.45	2.67	1.79	2.12	16.60	19.10[e]
Fat (g/l)[g]	33-54	30-72	50-90	53-90	20-60	3-18	3-42	

2. Cholesterol and its Occurrence in Dairy Products

Cholesterol, (3β)-cholest-5-en-3-ol is a polycyclic steroid compound, with molecular formula $C_{27}H_{43}OH$. It is composed of three regions: a hydrocarbon tail, a ring structure region with four hydrocarbon rings, and a hydroxyl group (Wasowicz and Rudzinska, 2011). Bovine milk contains 0.3-0.4% cholesterol, out of which about 10–15% is available as esters (Mulder and Zuidhof, 1958). Mulder and Zuidhof (1958) also reported that about three quarters of the cholesterol is dissolved in the milk fat, 10% in the fat globule membrane, and the remainder in the skim milk. The chemical structure of cholesterol is given in Figure 1. The cholesterol content in major dairy foods is given in Table 2.

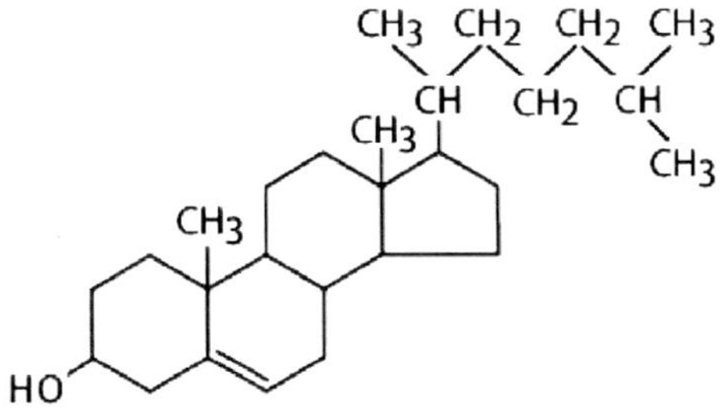

Fig. 1: Chemical structure of cholesterol

Table 2: Cholesterol content in selected dairy products

Dairy Product	Cholesterol (mg/100 g)
Skim milk	1.8
Whole milk	13.6
Curd cheeses	5–37
Processed and hard cheeses	51–99
Cream and sweet cream	35–106
Butter	183–248
Ice cream	30 -46
Yoghurt	7-13

Sources: Wasowicz and Rudzinska (2011); Oh et al., (2001)

2.1 Implications on human health

Cholesterol is an important functional constituent of cellular membrane functions and precursor of important endogenous substances such as corticosteroids, sex hormones, and bile acids (Bragagnolo, 2010). Typically, a human body synthesizes about 600 to 1500 mg cholesterol per day, whereas the diet supplies about 300 to 500 mg cholesterol per day (Sieber, 1993; Bragagnolo, 2010). Inspite of its import physiological functions, numerous studies have established that increase in dietary cholesterol and saturated fat are associated with increased plasma levels of LDL cholesterol and increased risk of cardiovascular diseases (CVDs). High intake of cholesterol results in atherosclerosis susceptibility (plaque formation and constriction of the blood flow), thereby resulting in increased

CVD morbidity and mortality (Sieber, 1993). According to American Heart Association, daily intake of cholesterol should not be more than 300 mg (Huber et al., 1996). Due to these health worries, most consumers have reduced the intake of dairy products rich in cholesterol, affecting the consumption and its use as ingredient in many food products. This has led to the impetus for developing cholesterol free/ reduced dairy products which will have wider consumer appeal and more health benefit.

3. Processes for Cholesterol Removal from Food

In order to satisfy a segment of the consuming public and to increase the general consumption of milk fat, the total cholesterol occurring naturally in milk fat can be reduced by various physical, complexation and biological processes. Some of those processes are explained below in Table 3.

4. Supercritical Fluid Extraction (SCFE)

When a fluid is forced to a pressure and temperature above its critical point, it becomes a supercritical fluid. Supercritical fluid (SCF) exhibits physicochemical properties of both gas and a liquid. The supercritical state of a fluid has been defined as a state in which liquid and gas are indistinguishable from each other, or as a state in which the fluid is compressible (i.e. similar behavior to a gas) even though possessing a density similar to a liquid and, therefore, similar solvating power. Due to their low viscosity and relatively high diffusivity, SCF have better transport properties than liquids, can diffuse easily through solid materials and can therefore give faster extraction yields. The density and the solvent strength of a SCF can be modified easily by changing the temperature and pressure without any phase change and thus used to extract selectively a specific compound or to fractionate mixtures (Herrero et al., 2006). SCFE is advantageous than other conventional solvent extraction, as it possesses higher extraction efficiency with lower extraction times. Also, it mainly uses solvents which are generally recognized as safe (GRAS). The most popular SCF solvent is carbon dioxide (CO_2). It is an inert, cheap, readily available, odourless, tasteless, environment friendly, and a GRAS solvent. CO_2 leaves no residue in the sample as it is gas in ambient temperature. It is

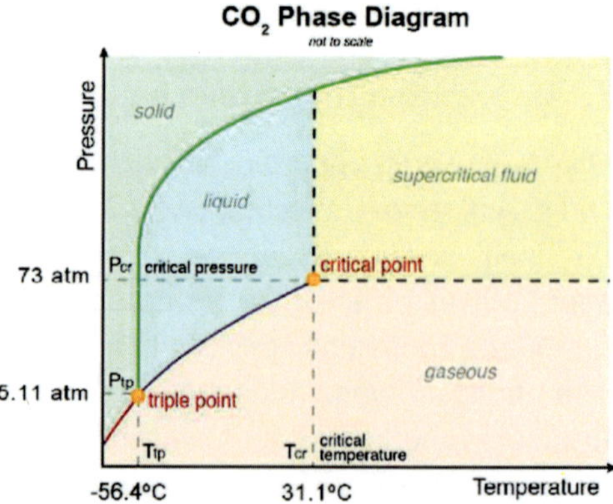

Fig. 2: Pressure vs. temperature graph of supercritical carbon dioxide.

Table 3: Different methods for cholesterol removal from food

Method	Technique	Process	Mechanism	Effectiveness
Biological	Microorganism	*Nocardia,Rhodococcus*	Transformation	Consumer health concern, harmful effects of the cholesterol by-products of this bioconversion.
	Enzyme	Cholesterol reductase, Cholesterol oxidase	Transformation	
Chemical	Solid-liquid extraction	Silica gel, Activated carbon	Absorption	Less effective.
	Complexation	β-cyclodextrin, Bile salt, Digitonin (Polymer-supported saponins)	EmulgationAdsorption Filtration/centrifugation	Extraction can be achieved upto 97% after repeated extraction.
Physical	Distillation	Steam distillation, Short path distillation	Fractions	Quite simple process with more number of processing steps.
	Crystallization	Fractionation	Lipid fractions based on melting points.	
	Supercritical fluid extraction	Extraction with carbon dioxide and co-solvents(optional)		Higher yields and superior quality products because of the use of low temperature.

Source: Sieber (1993)

suitable for thermolabile natural products owing to its near-ambient critical temperature (31.1°C) as depicted in Figure 2 (Palmer and Ting, 1995). By controlling the level of pressure/temperature, supercritical CO_2 (SC-CO_2) can dissolve a broad range of compounds with high extraction rate.

The main disadvantage of SCF $CO_2$is that it behaves as a weak "nonpolar" solvent. It can dissolve nonpolar lipids, hydrocarbons and triglycerides selectively but has a weak affinity to oxygenated or hydroxylated molecules and does not dissolve any polar or hydrophilic compounds like sugars and proteins, and mineral species like salts, metals, etc. However, the CO_2 solvent power and polarity can be significantly increased by adding a polar co-solvent/ entrainer that is generally chosen among short-chain alcohols, esters, or ketones (Perrut, 2003). Ethanol is often preferred as it is abundant and cheap in pure forms (food grade), not environmentally hazardous, and not very toxic. Co-solvent is often added in very small concentration to the SCF solvent in order to change the solvent characteristics, such as polarity and specific interactions without significantly affecting the density and compressibility of the original SCF solvent. The critical properties of different supercritical fluids and co-solvents are given in Table 4.

Table 4: Critical properties of fluids of interest in supercritical processes

Fluid	Critical Temperature (°C)	Critical Pressure (bar)
CO_2	31.1	78.7
Ethane	32.3	48.7
Propane	96.7	42.5
Isopropanol	235.2	47.623
Ethanol	241.7	63.4
Acetone	235.1	47.01
Water	374.2	220.6
Ammonia	132.5	113.5
n-Hexane	234.5	30.2
Methanol	239.6	80.9

Source: Bruno and Svoronos (2003)

The SCFE consists basically of two major steps: (i) extraction of the soluble substances from the solid substrate by the SCF, and (ii) separation of these compounds from the solvent after the expansion.The extraction process is illustrated in Figure 3. In the SCFE process, the solvent is first sub-cooled prior to the pump, assuring a liquid phase to avoid the cavitation issues. Prior to the extraction vessel, the pressurized solvent is heated above its critical temperature to reach the supercritical state. The feed material is placed for extraction with the aid of SCF at specific pressure and temperature keeping the inlet valve open and outlet valve closed, so that the feed material is soaked in the solvent for a certain amount of time. This period is called static extraction period which allows the SCF to penetrate into the matrix providing ample time for reaching equilibrium and extract the targeted compound. After the static time is completed, the outlet valve is opened (adjusted to a required flow rate) for a certain period of time. This is called dynamic extraction period. During this period, both the fluid and compound extracted are passed through the separator and the dissolving capacity of SCF is reduced by reducing/changing pressure and temperature. This results in the separation or fractionation of the compound.

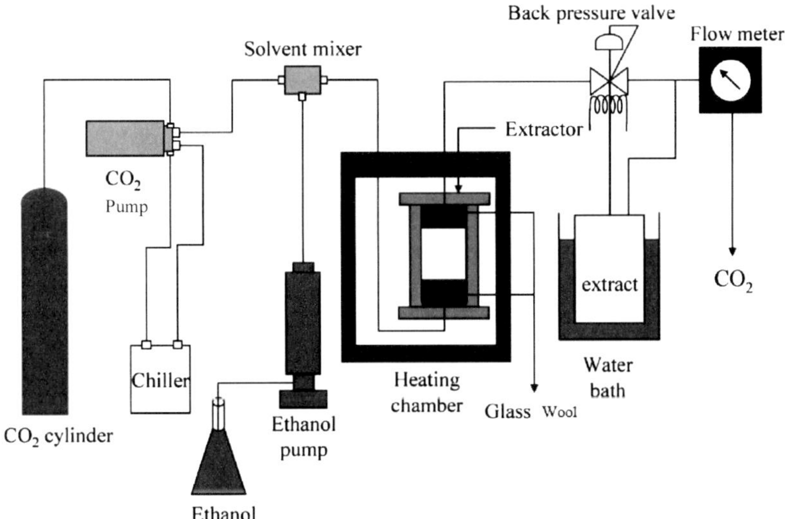

Fig. 3: Schematic diagram of supercritical fluid extraction process using co-solvent

4.1 Decholesterification of dairy products

In the recent years, there has been increasing interest in separating and fractionating cholesterol from food materials to formulate new functional food products. SCFE technology has been used widely for the reduction/ removal of cholesterol from different food products, including dairy products. Shisiiikura et al., (1986) used a single pass SCFE unit with operating pressure between 12.8-24.8 MPa at two different temperatures (40 and 60°C). They reduced cholesterol by 95%, but also removed 75% of the triglycerides from butter oil. They observed that cholesterol selectivity was low and hence, concluded after further studies that cholesterol removal could be increased by passing the solute-rich solvent through a silica gel column or alumina which preferentially adsorbed cholesterol. Using supercritical carbon dioxide at 20 MPa and 80°C, (Kaufmann et al., 1982) fractionated butter oil into a liquid fraction and a solid fraction with different cholesterol contents. Milk fat was fractionated by (Arul et al., (1987) with supercritical CO_2 (SC-CO_2) into 8 fractions at temperatures of 50 and 70 °C, over a pressure range of 100–350 bar, in which the cholesterol removal was highest at 100–250 bar, 50 °C. Hammam et al., (1991) fractionated the butter oil at 40°C, 125 bar and 350 bar along with the characterization of fractionation products, including the redistribution of cholesterol in such products according to their physical properties. Bradley (1989) reported that 90% cholesterol removal efficiency from milk fat at 80°C and pressures from 15.8 to 41.4 MPa is quite feasible for a broad range of dairy foods. Chen et al., (1992) conducted the fractionation of butter oil at 40°C and at 10.3, 13.8, 17.2, 20.7, 24.1, and 27.6 MPa in a continuous flow extraction system. The fractions obtained at 40°C, 17.2 and 20.7 MPa had lower cholesterol (1.07 and 1.76 mg/g), compared with those obtained at 10.3 and 13.8 MPa (2.96 and 2.78 mg/g).

(Bhaskar et al., (1993) designed a continuous pilot-scale supercritical carbon dioxide system for the fractionation of up to 400 g/h of anhydrous milk fat in the pressure range of 24.1-3.4 MPa at 40-75°C. It was seen that the cholesterol had affinity towards low melting and medium melting triglycerides and decreased up to 51% in these fractions. Lim et al., (1991) achieved an overall cholesterol reduction of 92.6% with a process yield of 88.5% using magnesium silicate as an adsorbent after the SCFE extraction at 40°C and 24.1-27.5 MPa. A milk fat fraction enriched with high melting triglyceride was extracted by Shukla et al., (1994) using a continuous, pilot-scale supercritical CO_2 system and recombined into butter. It was observed that the recombined butter fraction high melting had higher contents of unsaturated long-chain fatty acids, high melting triglycerides, and β-carotene. Furthermore, the extracted butter was reported to be richer in unsaturated fatty acids and lower in cholesterol content (117.6 mg/100g) than those found in commercial butter (240.6 mg cholesterol/100 g). Rizvi and Bhaskar (1995) also fractionated and separated milk fat into saturated and unsaturated fatty acids using SC-CO_2, identified the physical properties, and quantified the cholesterol content. They concluded that the fractions obtained with SC-CO_2 were unique exhibiting different characteristic physico-chemical properties. Hierro et al., (1995) studied the extraction of ewe's milk cream and cholesterol by SC-CO_2 in the pressure range 9-30 MPa (90-300 bar) at temperatures of 40 and 50°C. They found that maximum cholesterol reduction occurred at 50°C at moderate pressures of >150 bar. Use of higher pressures >280 bar resulted in higher extraction of long chain triglycerides and thus selectivity of cholesterol removal decreased.

A continuous SC-CO_2 processing system was designed by Lim and Rizvi (1995) for the extraction and fractionation of anhydrous milk fat. The cholesterol and the short chain triglycerides were concentrated in the lower pressure extract fractions. Moreover, the cholesterol removed was higher (>33%) without refluxing as compared to barely 7% with refluxing. Fractionation of butter oil into low, medium and high molecular weight triglycerides was performed by

Mohamed et al., (1998) using SC-CO_2 along with adsorption by alumina. It was observed that the cholesterol content of the fractions reduced from 2.5 to 0.5 mg/g of oil after extraction at 40 °C and 27.6 MPa. Mohamed et al., (2000) used a semi-continuous flow, high-pressure apparatus with supercritical ethane for fractionation of butter oil at 40, 55, and 70 °C and pressure ranging from 8.5 to 24.1 MPa. Supercritical ethane was found to allow higher extraction of cholesterol from butter oil than SC-CO_2. The combined extraction/adsorption process using alumina resulted in better selectivity of cholesterol extraction. The cholesterol content was limited to about 3% of that in the original oil. The increase in cholesterol content in fractions obtained in the later stages was attributed to desorption of cholesterol from the alumina bed to the supercritical fluid stream.

The detailed cost analysis of the industrial scale continuous counter-current processing of milk fat using SCFE was done by Singh and Rizvi (1994). They concluded that the lower cholesterol content and consequently higher quality of milk fat fractions obtained with supercritical CO_2 made this a lucrative economically viable process despite of higher costs as compared to melt crystallization. The operation has also resulted in the generation of butter oil fractions with characteristic properties that are distinctly different from those of the original oil. Detailed economic analysis for continuous counter-current processing of milk fat fractionation was also carried out which concluded that an optimized milk fat fractionation process could be carried out on an industrial scale. The addition of an adsorption step for cholesterol could improve the economics of this process. Semicontinuous systems were not as economically favorable as continuous large-scale systems. Apart from the use of SC-CO_2 in the reduction of cholesterol from dairy products, several other studies can be found in the literature pertaining to the use of SC-CO_2 for the reduction of cholesterol from various meat products, egg products etc. Whole milk powder (WMP) is a commercially available dairy product. It has advantages that it can be preserved longer than liquid milk and does not need to be refrigerated, due to its low moisture content. It contains on average 25-27% protein, 36-38% carbohydrates, 26-40% fat, and 5-7% ash (minerals). Reduction of cholesterol content in WMP can remarkably improve its market value.

Hence, the main objective of the work was to explore the possibility of cholesterol reduction from whole milk powder (WMP) with minimum extraction of milk fat using SCFE under various extraction parameters.

4.2 The supercritical fluid extraction (SCFE) unit

Applied Separations, the *Spe-ed* SFE model 7070 (SFE-2, 120VAC) model, capable of pressures up to 690 bar and oven temperature up to 240 °C, was used for the extraction operation. For each test run, a sample of around 50-60 g of WMP (sourced from Amul Dairy, India), containing about 27.6 % fat with a mean particle size of 250 µm, was fed into the stainless steel extraction vessel for extraction. To ensure proper sealing and avoid leakage of supercritical solvent, suitable sealant (Speed TM SFE Dry release Agent(lube) SFE Miller-Stephens MS-122RB) was sprayed on both the ends of the vessel and screwed tightly. SC-CO_2 was released from CO_2 tank and passed through chiller to lower its temperature to -4 °C before entering the pressure vessel.

4.2.1 Extraction parameters

The extraction pressure was varied in the range of 150-250 bar, and the temperature was in the range of 40-80 °C. The flow rate of SC-CO_2 was kept constant at 6 L/min. Based on preliminary trials, the static and dynamic time of operation were kept as 40 min and 2h for neat extraction (i.e., without addition of entrainer). The static and dynamic time was fixed at 10 min and 80 min, respectively, for co-solvent assisted extraction. Ethanol was used as co-solvent owing to its lesser toxicity and permissibility for food use. The range of ethanol addition was fixed to 10-50 mL; the co-solvent was pumped simultaneously with SC-CO_2 through the feed after the system reached the required temperature and pressure. The cholesterol extracts were collected in glass vials and the ethanol from the residual samples was evaporated under vacuum.

4.3 Extraction of cholesterol from whole milk powder

Extraction of cholesterol from WMP before and after SCFE treatment was carried out using the SB method (Searcy et al., 1960) with slight modifications. Approximately 0.15 g of WMP sample was saponified with 10 mL of 2 (M) methanolic potassium hydroxide (KOH) for 1 h at 75-80 °C followed by hexane extraction (twice) of the unsaponifiable fraction. The hexane extract was washed four times with 10 mL of DW and methanol (4 mL and decreasing concentrations for subsequent extractions), and centrifuged at 2500-3000 rpm for 10 min to remove interferences of impurities from the extract. The clear upper layer was finally evaporated to dryness in vacuum oven at 55 °C.

4.3.1 Determination of cholesterol content by HPLC method

The cholesterol content of the WMP samples was quantified by high performance liquid chromatography (HPLC) according to the modified method of Oh et al.,

(2001). HPLC (Dionex Ultimate 3000, Sunnyvale, USA) was used for separation using C18 column (Dionex Acclaim 120) having dimensions of 4.6x250 mm. The isocratic mobile phase of methanol, acetonitrile and 2-propanol in a ratio of 75: 17.5: 7.5 at flow rate 1.5 mL/min with detection wavelength of 205 nm was used for separation.

4.3.2 Experimental design

Rotatable central composite design (RCCD) was applied (Design Expert Version 7.0.3, Stat-Ease Inc., Minneapolis, USA) for the design of experiments. Efficacy of cholesterol extraction was checked by changing the extraction pressure and temperature at various levels. For studying the effect of co-solvent addition on the efficacy of extraction, ethanol was used at various volumes. Response surface methodology (RSM) was used to investigate the simultaneous effects of the extraction parameters on the cholesterol content and the fat content. The optimization for the best conditions for cholesterol reduction was achieved by RSM.

4.3.3 Neat supercritical reduction of cholesterol

The solvent strength and selectivity of supercritical fluid can be manipulated by changing pressure (P) and/or temperature (T). Hence, the solubility of cholesterol with the changes in operating pressure and temperature was studied and results are presented in Table 5.

Table 5: Effect of SCFE* process variables (temperature & pressure) on the major responses of WMP

Experiment	Independent variables		Responses [a]		
No.	Temperature (°C)	Pressure (bar)	Cholesterol content (mg/100g)	Fat content (%)	Solubility index (%)
1	40	200	63.02±2.36	18.89±0.78	94.97±2.01
2	46	235	92.87±0.84	16.99±1.43	95.85±2.83
3	74	165	21.82±2.12	25.91±1.19	71.81±3.16
4	60	200	47.02±2.54	23.79±1.37	91.98±1.88
5	46	165	52.11±1.38	22.91±0.82	88.21±2.64
6	60	200	44.12±1.22	23.01±1.41	90.82±2.32
7	60	250	89.19±1.98	17.30±1.86	93.98±1.08
8	80	200	11.96±1.57	24.31±0.34	66.29±1.37
9	74	235	37.67±1.35	23.02±1.53	79.82±0.89
10	60	200	47.47±2.79	22.23±1.91	89.07±2.15
11	60	150	55.14±1.03	25.09±1.62	87.79±1.08
12	60	200	43.30±1.46	23.73±0.98	92.53±1.36
13	60	200	47.48±0.67	23.74±1.06	91.02±0.92

*CO_2 flow rate- 6 L/min, static time- 40 min, dynamic time- 2 h; [a] Mean ± S.D., n = 3.

Comparing the responses obtained at different experimental runs, it was found that higher operating pressures resulted in a higher yield of cholesterol than the lower pressure (Figure 4a). Significant increase of the total extracted cholesterol was observed especially in the pressure ranges of 150 to 200 bar. As the operating pressure increased, the density of SC-CO_2 increased. This led to the reduction in the distance between molecules and hence, there was greater interaction possible between solute and CO_2, leading to greater solubility in CO_2. The increase in density also increased the flow Reynolds number leading to a higher mass transfer rate resulting in the improved solubility of cholesterol (Huang et al., 2004; Vedaraman et al., 2005). Mohamed et al., (1998) found that the cholesterol content in butter oil fractions decreased at a faster rate when the pressure was increased from 150-200 bar. The plateau effect in lowering of cholesterol solubility was observed during fractionation above 200 bar. Above 200 bar, the rate of extraction decreased due to the reduction in the diffusion rates of solutes at elevated pressures, contributing to mass transfer resistance (Bimakr et al., 2012; Vedaraman et al., 2005). Chen et al., (1992) also observed that at the range of 20.7 to 24.1 MPa, the rise in solubility of lipids from butter oil was much slower than that at lower pressure zones.

Increased temperatures resulted in marginal increase in the recovery throughout the pressure ranges studied at constant pressure. Higher vapor pressure of solute at elevated temperature increased the penetration ability into the biological matrix and hence improves extractability by improving the diffusion coefficient (Vedaraman et al., 2004; Mohamed et al., 2000). While modelling the solubility of cholesterol and its esters in SC-CO_2 in a binary system, Huang et al., (2004) found that solubility increased with the increase in temperature from 313.15 to 323.15 K in the pressure regions of 150-300 bar. Though the diffusion coefficient increased with temperature, the extraction of cholesterol was limited by the reduction in density of SC-CO_2 especially at lower pressure zones. Below 270 bar of operating pressure, the density effect is observed to be dominant while the sublimation vapor pressure effect is pronounced in the high-pressure domain (Shen et al., 2008; Mohamed et al.,1998). In the present study also, the pressure effects were more dominant in the extractability of cholesterol.

The extraction of milk fat increased as the operating pressure increased which is not desirable (Figure 4b). It was observed that the selectivity of cholesterol over fat decreased with the increase in operating pressure at constant temperature and increased with the increase in temperature at constant pressure. Hence, lower pressures and higher temperature favored higher selective extraction of cholesterol. Around 55.8 % cholesterol could be removed at the optimized extraction conditions of 68 °C, 207 bar, flow rate of 6 mL/min at 2 h dynamic time. The WMP retained good solubility with high milk fat content of 23.68 % (Chitra et al., 2015).

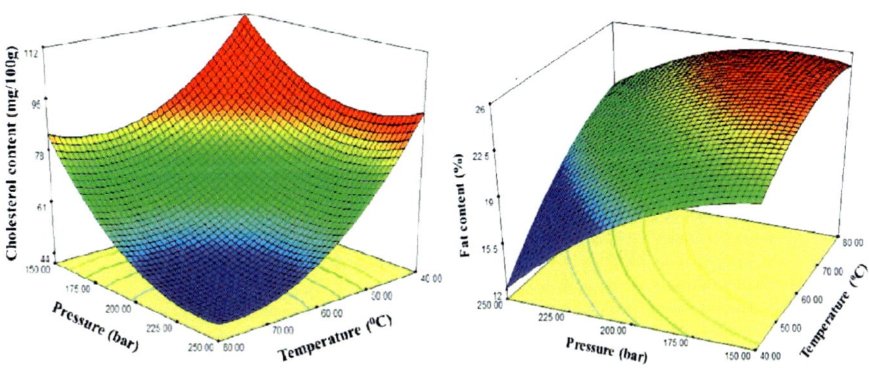

Fig. 4: Response surface plots showing the effects of temperature, °C and pressure, bar on (a) cholesterol content, mg/100g; (b) fat content, % of the SCFE treated WMP (See colour version on page 335)

4.3.4 Effect of co-solvent addition on cholesterol reduction

The effect of co-solvent addition, i.e., addition of food grade ethanol was studied by the addition of various volumes of ethanol at different extraction pressures (150-250 bar) and temperatures (40-80°C). The range of ethanol was fixed from 10 to 50 mL based on the preliminary trials. The results of the experimental runs are presented in Table 6; the unit of extraction pressure is expressed in MPa in the table. From the response surface plot, it was found that with increase of the operating temperature, the cholesterol content of the milk powder increased which means that the cholesterol extraction decreased with increase in operating temperature (Figure 5a). Generally, in a binary extraction system, the solubility of cholesterol increases with extraction temperature which shows that higher vapor pressure of solute at elevated temperature has more pronounced effect than the density of SC-CO_2 (Mohamed et al., 1998; Vedaraman et al., 2004). However, in the present study, an opposite effect was observed which indicated that the solute–co-solvent interaction was affected by temperature and it diminished with the increase in temperature. The reason for the unusual reversal of co-solvent effects is because in ternary solvent system (solute–ethanol– SC-CO_2), the co-solvent affects the crossover of solubility isotherms in SC-CO_2. This leads to opposite effects on the saturation pressure of solute and on the solvent density as compared to binary system, which, in turn decreases the penetration ability into the biological matrix at higher temperatures (Huang et al., 2004).

Table 6: Effect of SCFE* process variables on the major responses

Run	Independent variables			Dependent variables (response)[a]		
	Temp (°C)	Pressure (MPa)	Ethanol (mL)	Cholesterol content (mg/kg)	Fat content (g/kg)	Solubility index (%)
1	72	23	42	611.0	191.0	68.94
2	48	17	18	713.6	218.0	94.20
3	48	23	18	556.6	147.0	97.20
4	60	20	30	669.0	182.3	84.90
5	48	17	42	523.6	130.0	97.00
6	60	15	30	767.9	246.0	88.60
7	60	20	50	532.2	135.0	80.10
8	60	20	30	671.6	200.2	91.80
9	40	20	30	521.7	148.0	93.80
10	60	20	30	687.0	212.6	88.70
11	60	20	30	677.9	215.0	90.63
12	72	17	42	742.9	247.0	55.20
13	80	20	30	822.5	257.7	56.80
14	48	23	42	434.1	85.0	97.50
15	60	20	30	676.0	208.6	90.20
16	72	17	18	954.7	250.0	73.00
17	60	20	30	671.3	211.5	90.05
18	60	20	10	807.4	220.0	87.80
19	60	25	30	492.3	121.3	95.20
20	72	23	18	709.9	186.0	90.00

*CO_2 flow rate 6 L/min, static time 10 min, dynamic time 80 min.

Experimental data suggested that higher pressures promoted the extraction efficiency of cholesterol across all temperatures (Figure 5b). As the operating pressure increases, the density of SC-CO_2 increases resulting in the improved solubility of cholesterol in SC-CO_2 (Mohamed et al., 2000). Although an increase in temperature at constant pressure increases the vapor pressure of the solute, below 270 bar of operating pressure the solubility of cholesterol in SC-CO_2 is predominantly affected by the density of SC-CO_2. Above that pressure, the extraction efficiency of cholesterol is substantially affected by the vapor pressure of cholesterol (Mohamed et al., 1998; Shen et al., 2008). This study also confirmed that high fluid density, as opposed to a high vapor pressure, is most beneficial for the extraction of cholesterol.

Solvent power of SC-CO_2 improved greatly with the incorporation of co-solvent, ethanol (Figure 5c). The addition of co-solvent generally increases the mixture density, which may enhance the overall solubility below 270 bar (Vedaraman et al., 2004). Vedaraman et al., (2008) also observed that although the addition of co-solvents to the SC-CO_2 could not increase the amount of cholesterol extracted from cattle brain, the rate of extraction increased significantly. The higher extraction rates with increasing co-solvent concentrations may be attributed to the hydrogen-bonding interactions between iso-propyl alcohol/ethanol (Vedaraman et al., 2004).

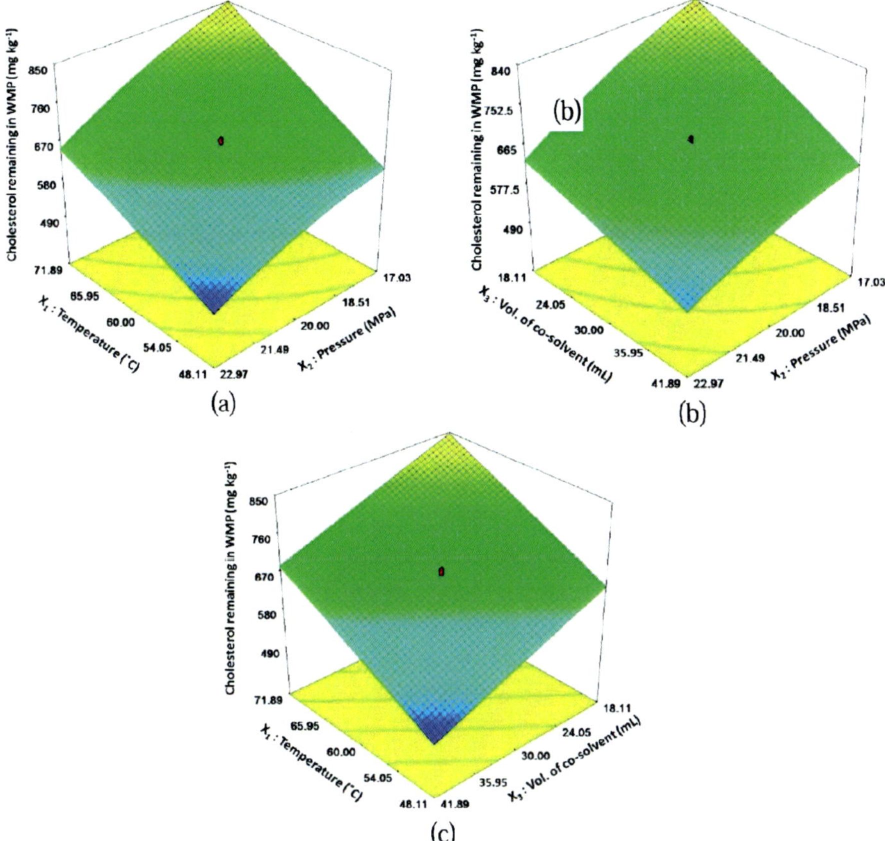

Fig. 5: Response surface plots showing the effects of (a) temperature, °C and pressure, MPa, (b) volume of co-solvent, mL and pressure, MPa, and (c) volume of co-solvent, mL and temperature, °C on the cholesterol content of whole milk powder (See colour version on page 335)

It was observed that though addition of ethanol significantly reduced the extraction time, but the selectivity of cholesterol extraction over milk fat decreased, i.e., extraction of triglycerides also occurred with cholesterol across all pressure and temperature ranges. The fat extraction decreased with the increase in temperature at constant pressure and volume of co-solvent, whereas, it increased with increase in pressure and volume of co-solvent added at constant temperature. Also, ethanol addition resulted in the precipitation of milk proteins, thereby decreasing the solubility index of WMP. The ethanol modified SC-CO_2 was optimized to obtain low cholesterol WMP without considerable changes in its physicochemical properties. The optimized extraction condition of 48 °C, 17

MPa (170 bar), and 31 mL of ethanol as co-solvent led to about 46% removal of cholesterol from WMP (Dey Paul et al., 2016).

5. Beta Cyclodextrin (β-CD) Complexation for Cholesterol Removal

Cyclodextrins (CDs) are a family of cyclic oligosaccharides that are composed of a-1,4-linked glucopyranose subunits, discovered by Villers in 1891 (El-Tahlawy et al., 2006). They are commercially produced by the process of enzymatic degradation of starch using cyclodextrin glycotransferase, which breaks the polysaccharide chain to form cyclic polysaccharide products. Cyclodextrins consisting of six, seven, or eight glucose units are called alpha, beta, and gamma cyclodextrins, respectively (Singh et al., 2002). Out of the three, β–cyclodextrin is the most industrially produced and accessible, the lowest-priced and generally the most useful in food applications.

5.1 Regulatory status of cyclodextrins

β-CDs confer several advantages as they are edible, non-toxic, are not absorbed in the upper GI tract, and are completely metabolized by the colon microflora (Dias et al., 2010). But, the regulatory status of cyclodexrins in foods differs between countries. In the USA cyclodextrins have obtained the GRAS (generally recognized as safe) status and can be commercialized as such. In the Japanese Pharmaceutical Codex, cyclodextrins are recognized as natural products and all three have been approved as food additives. In Australia and New Zealand alpha and beta cyclodextrins are classified as novel foods from 2004 and 2003 respectively. The Joint FAO/WHO Expert Committee on Food Additives (JECFA) has a recommended acceptable daily intake (ADI) of 5 mg/kg/day beta cyclodextrins in food products, but due to their favorable toxicological profile, no ADI was defined for alpha and gamma cyclodextrins. In July 2005 the U.S. Environmental Protection Agency (EPA) did away with the need to establish a maximum permissible level for residues of cyclodextrins (Astray et al., 2009; Kurkov and Loftsson, 2013).

5.2 Structure of β-cyclodextrin

β-cyclodextrin (β-CD) (Schardinger's β-dextrin, cyclomaltoheptaose) is a crystalline, homogeneous and non-hygroscopic oligosaccharide. It comprises of seven glucopyranose units arranged in a doughnut (torus) shaped ring. The glucose units situated in the classical 4C1 conformation of chains are linked through α-1,4 bonds. As a consequence of this conformation, all secondary hydroxyl groups are situated on one of the two edges of the ring, whereas all the primary ones are placed on the other edge. In the structure of β-CD, there

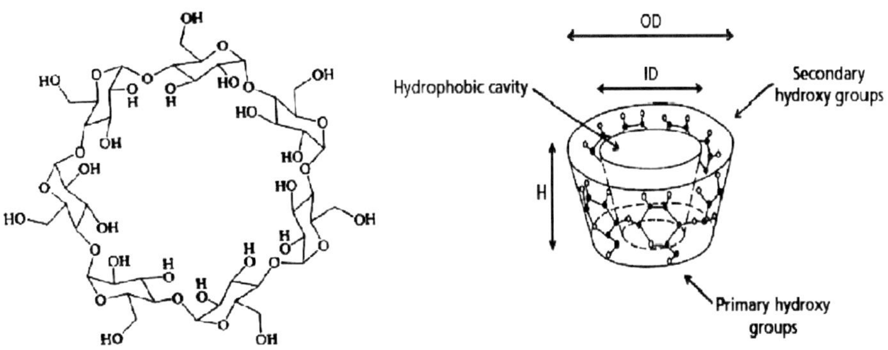

Fig. 6: Schematic representation of the structure of β–cyclodextrin

are seven primary hydroxyl groups (C_6-OH) which rotate freely at the C_6 position and make its diameter narrow; however, there are fourteen secondary hydroxyl groups which are fixed at the C_2 and C_3 positions (C_2-OH and C_3-OH), situated at the wider side of the torus. This arrangement makes the ring, in reality a conical cylinder, which is frequently characterized as a doughnut or wreath-shaped truncated cone. This cyclic ring structure is stabilized by intramolecular hydrogen bonding between adjacent hydroxyl groups at the C_2 and at the C_3 positions. This intramolecular hydrogen bond formation is the explanation for the lowest water solubility of β–CD amongst all cylodextrins. The interior of the cavity, which contains two rings of C-H groups with a ring of glycosidic oxygen in between, is relatively, hydrophobic. Since all of the hydroxyl groups are on the outside of the molecule, the external faces are hydrophilic (Nuñez-Delicado and Gabaldón-Hernández, 2011; Zhao and Yao, 2012). The structure and properties of β-CD in relation to other CDs is explained in Table 7. The structure of β-CD is illustrated in Figure 6.

Table 7: Main Characteristics of α-, β-, and γ-CD

Property	α-CD	β-CD	γ-CD
No. of glucose units	6	7	8
Empirical formula (anhydrous)	$C_{36}H_{60}O_{30}$	$C_{42}H_{70}O_{35}$	$C_{48}H_{80}O_{40}$
Mol. Weight (anhydrous)	972.85	1134.99	1297.14
Cavity diameter, nm	0.47 <" 0.53	0.60 <" 0.65	0.75 <" 0.83
Height of cavity, nm	0.79 ± 0.01	0.79 ± 0.01	0.79 ± 0.01
Diameter of outer periphery, nm	1.46 ± 0.04	1.54 ± 0.04	1.75 ± 0.04
Approx. volume of cavity, nm^3	0.174	0.262	0.427
Approx. cavity volume in 1 mol CD, mL	104	157	256
Approx. cavity volume in 1 g CD, mL	0.10	0.14	0.20
Crystal forms from water	Hexagonal	Monoclinic	Quadratic

Contd.

	plates	parallelograms	prisms
Crystal water, wt%	10.2	13.2 <“ 14.5	8.13 <“ 17.7
[α]D(25°C)	150 ± 0.5	162.5 ± 0.5	177.4 ± 0.5
p*K*a(25°C)	12.33	12.20	12.08
ΔH° (solution), kJ mol^{-1}	32.10	34.78	32.35
ΔS° (solution), J mol^{-1}K^{-1}	57.75	48.96	61.52

Sources: Crini and Morcellet (2002)

5.3 β-cyclodextrin as cholesterol sequestrant

The key structural feature of β-CD is the ability to complex and hold a wide variety of guest/inclusion molecules in its internal hydrophobic cavity. To bind with cyclodextrins, the inclusion molecule must have a size that fits, at least partially, into the cavity, creating the complex. The inclusion compound, however, does not have to be completely contained in the cavity. Complexes can be formed by the insertion of some specific functional groups or part of the molecule to bind in the hydrophobic cavity. The main driving forces behind the formation of these host-guest inclusion complexes are hydrophobic and van der Waals interactions, but also involve weak interactions, including hydrogen bonds, electrostatic interactions, π-π stacking interactions, steric effects, etc., instead of covalent bonding (Zhao and Yao, 2012). In an aqueous solution, the apolar cyclodextrin cavity is occupied by water molecules which are energetically unfavored (polar-apolar interaction). When the hydrophobic guest molecule reaches its vicinity, it has higher affinity for the apolar cyclodextrin cavity and hence, a substitution of the included water molecules by the less polarguest occurs. The guest is included in such a manner as to allow its non-polar portion to get maximum contact with the hydrophobic cavity, while its polar portion interacts with the hydrophilic surface of the β-CD molecule. Thus, the complexation of guest molecules is directly driven with the hydrophobic interaction between the cavity and the less polar part of the guest molecule (Figure 7). β-CD has the capability to form water soluble complexes with lipophilic

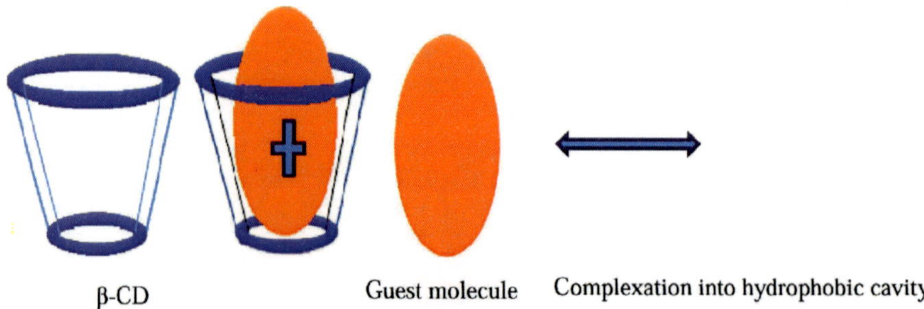

Fig. 7: A schematic representation showing the inclusion complex formation between guest molecule and a CD molecule

water insoluble molecules, such as various steroids, including cholesterol due to the circular hydrophobic space being similar in diameter to a cholesterol molecule (Somogyi et al., 2006). The radius of the cavity is such as to accommodate a cholesterol molecule almost exactly, conferring the high specific nature of β-CD ability to form an inclusion insoluble complex with cholesterol which can be removed by centrifugation. Although there have been innumerable reports of sequestration of cholesterol by use of powdered β-CD, however considerable β-CD is consumed during this process due to the ineffective recovery (Kwak et al., 2004). To overcome this, immobilization of β-CD using chitosan beads and glass beads by silanization have been carried out, but cholesterol recovery was upto 40-50 % only (Dias et al., 2010; Kwak et al., 2004). Hence, crosslinking and polymerization of β-CD can be another way of minimizing losses with efficient regeneration and sequestering capacities.

The general method for synthesis of water-insoluble β-CD polymer (β-CDP) is by crosslinking the hydroxyl groups of β-CDs with bi- or multi-functional molecules to form a stable crosslinked network. The bi- or multi-functional molecules used are normally called crosslinking agents. Effective crosslinking agents that have been reported include: epichlorohydrin. diisocyanates, polycarboxylic acids and anhydrides (Zhao et al., 2009). Crosslinking of β-CD by use of epichlorohydrin and other crosslinking agents have been tried out which converts cyclodextrin from water-soluble monomer into insoluble and regenerable network-shaped polymer, which has high swelling capacity in water as well as multiple binding sites (Liu et al., 2011; Ha et al., 2009).

The advantage of using β-CD is that it is an effective oligosaccharide for cholesterol removal from dairy products without significantly affecting the lipid components and other nutritional components of the milk fat. Results from the study of Alonso et al., (2009a) indicated that the treatment of pasteurized milk with β-CD did not affect the trans C18:1 fatty acid isomers, CLA, PUFA and phospholipids compositions in the milk fat. Also, after the treatment, the β-CD-cholesterol complex can be removed easily by centrifugation to yield cholesterol free dairy products. Several authors have reported the use of β-CD and its polymers to remove cholesterol from various dairy products which is listed in Table 8.

Table 8: Different physical and chemical methods used for the removal of cholesterol from different food products

Product	Reference	Method (s) used	Cholesterol reduction (%)
Homogenised milk (3.6 % fat content)	Lee et al., (1999)	1.5% β-cyclodextrin addition at temperature of 10°C, 10-min mixing time	94.3
Cream	Ahn & Kwak, (1999)	β-cyclodextrin (15%) with stirring speed of 1600 rpm and mixing time of 30 min	97.3
Mozzarella Cheese	Kwak et al., (2001)	1% β-cyclodextrin addition at 70 °C; 70 kg/cm^2 homogenization pressure	63.9
Cheddar Cheese	Kwak et al., (2002a)	10% β-cyclodextrin addition to cream (36 % fat) at800 rpm at 20 °C for 30 min	92.1
Cheddar Cheese slurries	Kwak et al., (2002b)	Treatment A: 1 % β-cyclodextrin, 800 rpm 4 °C for 10 min to homogenized milk Treatment B: 35 % fat cream treated with 15 % β-cyclodextrin with stirring speed of 1600 rpm and mixing time of 30 min	79.3 (Trt A) 91.2 (Trt B)
Cheddar cheese	Kwak et al, (2003)	1 % β-cyclodextrin, 800 rpm 4 °C for 10 min	79.3
Milk	Kwak et al., (2004)	Immobilized β-cyclodextrin in glass beads with 6 h of mixing time in 7 mm diameter tube at 10 °C	40.2
Milk	Kim et al., (2004)	1 % crosslinked β-CD using epichlorohydrin, 400 rpm at 10°C	79.4 - 83.3
Cream cheese	Kim et al., (2005)	Treatment A: cream (35% milk fat) treated with 10% β-CD at 800 rpm at 10°C for 10 min Treatment B: cream (35% milk fat) treated with 10% crosslinked β-CD using epichlorohydrin at 800 rpm at 10°C for 10 min	92.0 (Trt A) 82.6 (Trt B)
Homogenised milk	Han *et al.,* (2005)	1% β-CD crosslinked with adipic acid, and 10 min mixing with 400 rpm speed at 5°C	92.1 – 93.1
Cream	Han et al., (2007)	10% β-CD addition, 30 min mixing at1400 rpm speed at 40°C	91.5
Cream cheese	Han et al., (2008)	10%powdered or crosslinked ^{2}-CD, stirring at 800 rpm at 20°C for 30 min.	91.4-91.8
Camembert cheese	Kim et al., (2008)	cream (36% milk fat) treated with 10% crosslinked β-CD using adipic acid at 800 rpm at 20°C for 30 min	90.6
Chilled Milk	Alonso et al., (2009b)	0.6 % β-cyclodextrin mixing for 20 min and holding for 6 h at 4 °C	95
Processed cheese spread & Butter	Kim et al., (2009)	Treatment of cream with 10% rpm speed at 40°C	Cheese : 91.5 Butter: 90.5

Contd.

Contd.

Product	Reference	Method(s) used	Cholesterol reduction (%)
Ice cream mix	Ha et al., (2009)	Treatment with 10 % β-cyclodextrin	90
Butter	Dias et al., (2010)	Co-precipitation of butter with 10-15% β-cyclodextrin and stirring for 3-6 h	90
Ghee	Kumar et al., (2010)	β-cyclodextrin (15%) with stirring speed of 1600 rpm and mixing time of 30 min	>90 %
Cream cheese	Jeon et al., (2012)	Treatment of blended cream with 3.6% crosslinked β-CD using adipic acid at 800 rpm at 20°C for 30 min	92.19
UHT milk	López-de-Dicastillo et al., (2011)	UHT milk exposed to EVOH films containing 30%β-CD at 4 and 23 °C for 1 week	23
Milk & Cream	Lee et al., (2012)	1% crosslinked β-cyclodextrin using adipic acid addition to milk at 800 rpm at 10 °C for 30 min	90- 92.41
Ghee	Meghwal et al., (2013)	β-cyclodextrin (15%) with stirring speed of 1600 rpm and mixing time of 30 min	83.5
Homogenised milk	Maskooki et al., (2013)	1% β-CD addition and stirring at 8 or 20 °C for 10-15 min	67.2-77.9 for 1- 3% of milk fat

6. Other Processes for Cholesterol Removal or Modification

6.1 Biological process

Microbial treatment of cholesterol provides an alternative way to reduce cholesterol. Some species of *Azotobacter* and S*terolibacterium denitrificans* trans form cholesterol to cholest-4-en-3-one (cholestenone) an intermediate cofactor in the conversion of cholesterol to coprostanol or Δ^7-dehydrocholesterol, andhave the ability to hydrolyse the side-chain of cholesterol generating methyl heptanone (Ooi and Liong, 2010). García et al., (2012) reviewed that bacteria belonging to the genera *Nocardia*, *Azotobacter, Arthrobacter*, *Bacillus*, *Brevibacterium*, *Corynebacterium*, *Streptomyces*, *Mycobacterium*, *Serratia*, *Achromobacter*, *Pseudomonas* or *Protaminobacter, Lactobacillus* among others, have been reported to accomplish partial or complete cholesterol degradation. However not all microorganisms may be suitable for use in foods due to their pathogenicity or spoilage nature. Use of genetic engineering to transfer selective genes for cholesterol degradation from these microorganisms to lactic acid bacteria for use as starter culture will result in more profound acceptance in foods. Also, some strains of these organisms accumulate steroid intermediates, with consequent adverse health effects, however other strains which can degrade cholesterol without accumulating steroid intermediates have also been isolated.

Many probiotics, including *Lactobacillus acidophilus, L. fermentum, L. rhamnosus, L. casei, Bifidobacterium longum* have proven hypocholesterolemic effect both *in-vitro* and *in-vivo* (Ziarno, 2007; Ooi and Liong 2010). (Lye et al., 2010) evaluated the conversion of cholesterolby strains of *lactobacilli* such as *L. acidophilus, L. bulgaricus* and *L. casei* detecting both intracellular and extracellular cholesterol reductasesin these strains, indicating possible intracellular and extra cellular conversions of cholesterol to coprostanol. Noseda et al., (2007) used *Tetrahymena thermophila*, a non-pathogenic ciliate with the ability to desaturate cholesterol into Δ^7-dehydrocholesterol (pro-vitamin D_3), and Δ7,22-bis-dehydrocholesterol (analogue of pro-vitamin D_2) for cholesterol reduction in milk. Optimization of the culture conditions is a necessity before use in dairy products.

Enzymes such as cholesterol oxidase or reductase can be isolated from these microorganisms and use for reduction of cholesterol from dairy products. The enzymes may also be immobilized or attached onto a solid support, or adsorbed to packaging material for efficient recycling of these enzymes. Optimization of critical parameters, such as enzyme stabilization, quality control and instrumentation design is needed before use. The main problem with these enzymes is the accessibility of cholesterol in the milk fat globules to the active site of enzymes as some part of cholesterol is in esterified form. When incubated, only a small proportion of the cholesterol is oxidized. Also, some products of cholesterol oxidase may be toxic (Sieber,1993; Boudreau and Arul 1993). Smith et al., (1991) studied the reactivity of bacterial cholesterol oxidase on the cholesterol reduction in milk. They found that cholesterol oxidation in homogenized milk was dependent on temperature and enzyme concentration. A maximum of 85 % of the initial whole milk cholesterol (initially 100-138 pg/mL) was oxidized in 96 h, giving final cholesterol concentrations of 20-30 pg/mL.

6.2 Physical process

Physical process for cholesterol removal includes distillation and melt crystallization, whereby the dairy fat is deaerated, mixed with steam, heated, flash vaporized, thin-film stripped with counter current steam, cooled (all steps being performed under vacuum), and then stored under oxygen free conditions. The process provides dairy fats from which all or a substantial portion of the free or non-esterified cholesterol has been removed and which, bland in taste. A short path molecular distillation plant, was tested by (Lanzani et al., 1994) to remove cholesterol from anhydrous butter and lard. The experiment showed that cholesterol was almost completely removed during the second hour with minimal loss of low-molecular weight triglycerides at 185 °C and at the maximum operational vacuum. Cholesterol stripping process using vacuum steam distillation process have been developed by General Mills Inc. (Minneapolis, MN); the Omega Source Corporation (Burnsville, MN); Campbell Soup Co.; Land O'Lakes Inc. and Fractionnement Tirtiaux, S. A. (Fleurus, Belgium) (Boudreau and Arul 1993). The process reduced free cholesterol by 93% but gives a milk fat yield of 95%. However, cholesterol removal through the steam stripping process may cause the formation of toxic oxidation products of cholesterol, due to use of elevated temperature, if the process is not done properly (Boudreau and Arul, 1993).

6.3 Other complexation processes

Saponin and digitonin are plant glycosides which can form water-insoluble complexes with cholesterol. Experiments with animals revealed that several saponins such as alfalfa, soybean, quillaja, yucca, karaya and digitonin possess hypocholesterolemic effects (Vinarova et al., 2015). Saponins, however are mildly toxic. As well as irritating the membranes of the respiratory and digestive tract, the aglycones in certain saponins increase the permeability of the membranes of red blood cells. Tomatine, another steroidal saponin, derived from tomatoes and less toxic than digitonin also can from stable complex with cholesterol (Schulz and Sander, 1957). Several authors have reported the use of polymer supported saponin, digitonin and tomatine for the removal of cholesterol. Schwartz et al., (1967) obtained selective extraction of cholesterol from butter oil using digitonin impregnated diatomous earth and a column of digitonin adsorbed on celite. The insoluble cholesterol complexes can be easily removed by filtration or centrifugation. Micich (1990, 1991) successfully removed cholesterol from butteroil using polymer supported digitonin and tomatine. Sundfeld et al., (1993) removed cholesterol from butteroil by treatment with *Quillaja saponins*. Rojas et al., (2006) removed 70% cholesterol using Streamline Phenyl-$^{(R)}$resin in batch suspension at room temperature (25 ^{0}C). (Chang et al., 2001) removed 73 % cholesterol from homogenized milk with 1.5% addition of saponin after 30 min.

7. Summary

The relationship between dietary cholesterol and total serum cholesterol has been extensively investigated along with the suggestion that dietary cholesterol contributes a risk factor in the development of coronary heart disease. Therefore, a lower intake of high-cholesterol foods has been suggested as an effective method for lowering the level of serum cholesterol. The experiments conducted demonstrate that supercritical fluid extraction process using green solvent, SC-CO_2 is a feasible process for the extraction of cholesterol without any considerable change in the physicochemical characteristics of the whole milk powder. Various other physical, chemical, and biological methods can also be suitably employed for reducing cholesterol content in foods.

References

Ahn, J., Kwak, H.S. 1999. Optimizing cholesterol removal in cream using β-cyclodextrin and response surface methodology. *Journal of Food Science,* 64(4): 629-632.

Alonso, L., Cuesta, P., Fontecha, J., Juarez, M., Gilliland, S.E. 2009b. Use of β-cyclodextrin to decrease the level of cholesterol in milk fat. *Journal of Dairy Science,* 92:863–869.

Alonso, L., Cuesta, P., Gilliland, S.E. 2009a. Effect of β-cyclodextrin on trans fats, CLA, PUFA and phospholipids of milk fat. *Journal of the American Oil Chemists' Society,* 86(4): 337-342.

Arsi´c, A., Prekajski, N., Vucic, V., Tepsic, J., Popovic, T., Vrvic, M., Glibetic, M. 2009. Milk in human nutrition: comparison of fatty acid profiles. *Act Vet,* 59:569–78.

Arul, J., Boudreau, A., Makhlouf, J., Tardif, R., Sahasrabudhe, M.R. 1987. Fractionation of anhydrous milk fat by superficial carbon dioxide. *Journal of Food Science,* 52(5): 1231-1236.

Astray, G., Gonzalez-Barreiro, C., Mejuto, J.C., Rial-Otero, R., Simal-Gándara, J. 2009. A review on the use of cyclodextrins in foods. *Food Hydrocoll,* 23(7): 1631-1640.

Bhaskar, A.R., Rizvi, S.S.H., Sherbon, J.W. 1993. Anhydrous milk fat fractionation with continuous countercurrent supercritical carbon dioxide. *Journal of Food Science,* 58(4): 748-752.

Bimakr, M., Rahman, R.A., Ganjloo, A., Taip, F.S., Salleh, L.M., Sarker, M.Z.I. 2012. Optimization of supercritical carbon dioxide extraction of bioactive flavonoid compounds from spearmint (*Menthaspicata* L.) leaves by using response surface methodology. *Food and Bioprocess Technology,* 5(3): 912-920.

Boudreau, A., Arul, J. 1993. Cholesterol reduction and fat fractionation technologies for milk fat: an overview. *Journal of Dairy Science,* 76(6): 1772-1781.

Bradley, R.L. 1989. Removal of cholesterol from milk fat using supercritical carbon dioxide. *Journal of Dairy Science,* 72(10): 2834-2840.

Bragagnolo, N. 2010. Cholesterol and cholesterol oxides in meat and meat products. In: Nollet LM, Toldrá F, editors. Handbook of muscle foods analysis. Florida: CRC Press. p 187-220.

Bruno, T.J., Svoronos, P.D. 2003. Supercritical fluid extraction and chromatography. In: CRC Handbook of basic tables for chemical analysis. CRC Press (Taylor & Francis Group), Florida, USA.

Ceballos, L.S., Morales, E.R., de la Torre Adarve, G., Castro, J.D., Mart'ýnez, L.P., Sampelayo MRS. 2009. Composition of goat and cow milk produced under similar conditions and analyzed by identical methodology. *Journal of Food Composition and Analysis,* 22(4):322–9.

Chang, E.J., Oh, H.I., Kwak, H.S. 2001. Optimization of cholesterol removal conditions from homogenized milk by treatment with saponin. *Asian Australasian Journal of Animal Sciences,* 14(6): 844-849.

Chen, H., Schwartz, S.J., Spanos, G.A. 1992. Fractionation of butter oil by supercritical carbon dioxide 1, 2. *Journal of Dairy Science,* 75(10): 2659-2669.

Chitra, J., Deb, S., Mishra, H.N. 2015. Selective fractionation of cholesterol from whole milk powder: optimisation of supercritical process conditions. *International Journal of Food Science and Technology,* 50: 2467-2474.

Claeys, W.L., Verraes, C., Cardoen, S., De Block, J., Huyghebaert, A., Raes, K., Dewettinck, K. and Herman, L. 2014. Consumption of raw or heated milk from different species: An evaluation of the nutritional and potential health benefits. *Food Control* 42: 188-201.

Crini, G., Morcellet, M. 2002. Synthesis and applications of adsorbents containing cyclodextrins. *J Sep Sci,* 25(13): 789-813.

Dey Paul, I., Chitra, J., Mishra, H.N. 2016. Optimization of process parameters for supercritical fluid extraction of cholesterol from whole milk powder using ethanol as co-solvent. *Journal of the Science of Food and Agriculture,* DOI 10.1002/jsfa.7760.

Dias Hmam, Berbicz, F., Pedrochi, F., Baesso, M.L., Matioli, G. 2010. Butter cholesterol removal using different complexation methods with beta-cyclodextrin, and the contribution of photoacoustic spectroscopy to the evaluation of the complex. *Food Research Internationa,* 43(4): 1104-1110.

El-Tahlawy, K., Gaffar, M.A., El-Rafie, S. 2006. Novel method for preparation of â-cyclodextrin/grafted chitosan and its application. *Carbohydrate Polymers,* 63(3): 385-392.

García, J.L., Uhía, I., Galán, B. 2012. Catabolism and biotechnological applications of cholesterol degrading bacteria. *Microbial biotechnol,* 5(6): 679-699.

Ha, H.J., Ahn, J., Min, S.G., Kwak, H.S. 2009. Properties of cholesterol-reduced ice cream made with cross-linked β-cyclodextrin. *International Journal of Dairy Technology,* 62(3): 452-457.

Hammam, H., Söderberg, I., Sivik, B. 1991. Physical properties of butter oil fractions obtained by supercritical carbon dioxide extraction. Lipid/Fett93(10): 374-378.

Han, E.M., Kim, S.H., Ahn, J., Kwak, H.S. 2005. Cholesterol removal from homogenized milk with crosslinked beta-cyclodextrin by adipic acid. *Asian Australasian Journal of Animal Sciences,* 18(12): 1794-1799.

Han, E.M., Kim, S.H., Ahn, J., Kwak, H.S. 2007. Optimizing cholesterol removal from cream using â cyclodextrin cross linked with adipic acid. *International Journal of Dairy Technology,* 60(1): 31-36.

Han, E.M., Kim, S.H., Ahn, J., Kwak, H.S. 2008. Comparison of cholesterol-reduced cream cheese manufactured using crosslinked beta-Cyclodextrin to regular cream cheese. *Asian Australasian Journal of Animal Sciences,* 21(1): 131-137.

Herrero, M., Cifuentes, A., Ibanez, E. 2006. Sub-and supercritical fluid extraction of functional ingredients from different natural sources: Plants, food-by-products, algae and microalgae: A review. *Food Chemistry,* 98(1): 136-148.

Hierro, M.T.G., Ruiz-Sala, P., Alonso, L., Santa-María, G. 1995. Extraction of ewe's milk cream with supercritical carbon dioxide. *Zeitschrift für Lebensmittel-Untersuchung und Forschung,* 200(4): 297-300.

Huang, Z., Kawi, S., Chiew, Y.C. 2004. Solubility of cholesterol and its esters in supercritical carbon dioxide with and without co-solvents. *The Journal of Supercritical Fluids,* 30(1): 25-39.

Huber, W., Molero, A., Pereyra, C., Martinez de la Ossa. 1996. Dynamic supercritical carbon dioxide extraction for removal of cholesterol from anhydrous milk fat. *International Journal of Food Science and Technology,* 31: 143-151.

Jensen, R.G., Ferris, A.M., Lammi-Keefe, C.J. 1991. The composition of milk fat. *Journal of Dairy Science,* 74(9): 3228-3243.

Jeon, S.S., Lee, S.J., Ganesan, P., Kwak, H.S. 2012. Comparative study of flavor, texture, and sensory in cream cheese and cholesterol-removed cream cheese. *Food Science and Biotechnology,* 21(1): 159-165.

Kaufmann, W., Biernoth, G., Frede, E., Merk, W., Precht, D., Timmen, H. 1982. Fractionation of butterfat by extraction with supercritical CO_2. *Milchwissenschaft,* 37(2): 92-96.

Kim, S.H., Ahn, J., Kwak, H.S. 2004. Crosslinking of β-cyclodextrin on cholesterol removal from milk. *Arch Pharm Res,* 27(11): 1183-1187.

Kim, S.H., Han, E.M., Ahn, J., Kwak, H.S. 2005. Effect of crosslinked β-cyclodextrin on quality of cholesterol-reduced cream cheese. *Asian Australasian Journal of Animal Sciences,* 18(4): 584-589.

Kim, S.Y., Bae, H.Y., Kim, H.Y., Ahn, J., Kwak, H.S. 2008. Properties of cholesterol reduced Camembert cheese made by crosslinked β cyclodextrin. *International Journal of Dairy Technology,* 61(4): 364-371.

Kim, S.Y., Hong, E.K., Ahn, J., Kwak, H.S. 2009. Chemical and sensory properties of cholesterol reduced processed cheese spread. *International Journal of Dairy Technology,* 62(3): 348-353.

Konuspayeva, G., Lemarie, ´E., Faye, B., Loiseau, G., Montet, D. 2008. Fatty acid and cholesterol composition of camel's (*Camelus bactrianus, Camelus dromedarius* and hybrids) milk in Kazakhstan. *Dairy Science and Technology,* 88(3):327–40.

Kumar, M., Sharma, V., Lal, D., Kumar, A., Seth, R. 2010. A comparison of the physico chemical properties of low cholesterol ghee with standard ghee from cow and buffalo creams. *International Journal of Dairy Technology,* 63(2): 252-255.

Kurkov, S.V., Loftsson, T. 2013. Cyclodextrins. *Int J Pharm,* 453(1): 167-180.

Kwak, H.S., Chung, C.S., Ahn, J. 2002b. Flavor compounds of cholesterol-reduced Cheddar cheese slurries. *Asian Australasian Journal of Animal Sciences,* 15(1): 117-123.

Kwak, H.S., Jung, C.S., Seok, J.S., Ahn, J. 2003. Cholesterol removal and flavor development in Cheddar cheese. *Asian Australasian Journal of Animal Sciences,* 16(3): 409-416.

Kwak, H.S., Jung, C.S, Shim, S.Y., Ahn, J. 2002a. Removal of cholesterol from Cheddar cheese by β-cyclodextrin. *Journal of Agricultural and Food Chemistry,* 50(25): 7293-7298.

Kwak, H.S., Kim, S.H., Kim, J.H., Choi, H.J., Kang, J. 2004. Immobilized β-cyclodextrin as a simple and recyclable method for cholesterol removal in milk. *Arch Pharmacal Res,* 27(8): 873-877.

Kwak, H.S., Nam, C.G., Ahn, J. 2001. Low cholesterol mozzarella cheese obtained from homogenized and b—Cyclodextrin—treated milk. *Asian Australasian Journal of Animal Sciences,* 14: 268-275.

Lanzani, A., Bondioli, P., Mariani, C., Folegatti, L., Venturini, S., Fedeli, E., Barreteau, P. 1994. A new short-path distillation system applied to the reduction of cholesterol in butter and lard. *Journal of the American Oil Chemists' Society,* 71(6): 609-614.

Lee, D.K., Ahn, J., Kwak, H.S. 1999. Cholesterol removal from homogenized milk with β-cyclodextrin. *Journal of Dairy Science,* 82(11): 2327-2330.

Lee, Y.K., Ganeshan, P., Kwak, H.S. 2012. Optimisation of cross-linking β -cyclodextrin and its recycling efficiency for cholesterol removal in milk and cream. *International Journal of Food Science and Technology,* 47: 933-939.

Lim, S., Lim, G.B., Rizvi, S.S.H. 1991. Continuous supercritical CO_2 processing of milk fat. MA McHugh, ed. Proceedings of 2nd International Symposium on Supercritical Fluids. Baltimore, May 20–22, Johns Hopkins Univ. Press. p 292–296.

Lim, S., Rizvi, S.S.H. 1995. Continuous supercritical fluid processing of anhydrous milk fat in a packed column. *Journal of Food Science,* 60(5): 889-893.

Liu, H., Cai, X., Wanga, Y, Chen, J. 2011. Adsorption mechanism-based screening of cyclodextrin polymers for adsorption and separation of pesticides from water. *Water Res,* 45:3499-3511.

López-de-Dicastillo, C., Catalá, R., Gavara, R., Hernández-Muñoz, P. 2011.Food applications of active packaging EVOH films containing cyclodextrins for the preferential scavenging of undesirable compounds. *Journal of Food Engineering,* 104(3): 380-386.

Lye, H.S., Rusul, G., Liong, M.T. 2010. Mechanisms of cholesterol removal by Lactobacilli underconditions that mimic the human gastrointestinal tract. *Int Dairy,* J20: 169-175.

Maskooki, A.M., Beheshti, S.H.R., Valibeigi, S., Feizi, J. 2013. Effect of cholesterol removal processing using â-cyclodextrin on components of milk. *International Journal of Food Science,* http://dx.doi.org/10.1155/2013/215305.

Meghwal, K., Sharma, V., Lal, D., Arora, S. 2013. Effect of cholesterol removal on the granulation behaviour of low cholesterol ghee. *International Journal of Dairy Technology,* 66(1): 98-102.

Micich, T.J. 1990. Behavior of polymer-supported digitonin with cholesterol in the absence and presence of butter oil. *Journal of Agricultural and Food Chemistry,* 38(9): 1839-1843.

Micich, T.J. 1991. Behavior of polymer-supported tomatine toward cholesterol in the presence or absence of butter oil. *Journal of Agricultural and Food Chemistry,* 39(9): 1610-1613.

Mohamed, R.S., Neves, G., Kieckbusch, T.G. 1998. Reduction in cholesterol and fractionation of butter oil using supercritical CO_2 with adsorption on alumina. *International Journal of Food Science and Technology,* 33(5): 445-454.

Mohamed, R.S., Saldaña, M.D., Socantaype, F.H., Kieckbusch, T.G. 2000. Reduction in the cholesterol content of butter oil using supercritical ethane extraction and adsorption on alumina. *The Journal of Supercritical Fluids,* 16(3): 225-233.

Mulder, H., Zuidhof, T.A. 1958. The surface layers of milk fat globules. 4. The cholesterol content (free and esterified cholesterol) of the surface layers. *Netherlands Milk and Dairy Journal,* 12(3): 173-179.

Noseda, D.G., Gentili, H.G., Nani, M.L., Nusblat, A., Tiedtke, A., Florin-Christensen, J, Nudel, CB. 2007. A bioreactor model system specifically designed for *Tetrahymena* growth and cholesterol removal from milk. *Applied Microbiology and Biotechnology,* 75(3): 515-520.

Nuñez-Delicado, E., Gabaldón-Hernández, J.A. 2011. Cyclodextrins. Encyclopedia of Biotechnology in Agriculture and Food. New York: Taylor and Francis. p 187-190.

Oh HI, Shin TS, Chang EJ. 2001. Determination of cholesterol in milk and dairy products by high performance liquid chromatography. *Asian Australasian Journal of Animal Sciences,* 14(10):1465- 1469.

Ohlsson, L. 2010. Dairy Products and plasma cholesterol levels. *Food and Nutrition Research,* 54: 5124.

Ooi, L.G., Liong, M.T. 2010. Cholesterol-lowering effects of probiotics and prebiotics: a review of in vivo and *in vitro* findings. *Int J Mol Sci,* 11(6): 2499-2522.

Palmer, M.V., Ting, S.S.T. 1995. Applications for supercritical fluid technology in food processing. *Food Chemistry,* 52(4): 345-352.

Pereira, P.C. 2014. Milk nutritional composition and its role in human health. *Nutrition,* 30(6): 619-627.

Perrut, M. 2003. Applications of supercritical fluid solvents in the pharmaceutical industry. In: Marcus Y. & Sen Gupta A. K. (Eds.). Ion Exchange and Solvent Extraction- A Series of Advances -(Vol.17). New York, USA: Marcel Dekker. p 1-35.

Rizvi, S.S.H., Bhaskar, A.R. 1995. Supercritical fluid processing of milk fat: fractionation, scale-up, and economics. *Food Technology (USA)* 49(2): 90–100.

Rojas EEG, dos Reis Coimbra, J.S., Minim, L.A. 2006. Adsorption of egg yolk plasma cholesterol using a hydrophobic adsorbent. *European Food Research and Technology,* 223(5): 705-709.

Salimei, E., Fantuz, F., Coppola, R., Chiofalo, B., Polidori, P., Varisco, G. 2004. Composition and characteristics of ass's milk. *Animal Research,* 53(1):67–78.

Schulz, G., and Sander, H. (1956). Cholesterol tomatide; a new molecular formation for the analysis and preparative extraction of steroids. *Hoppe-Seyler's Zeitschrift fur physiologische Chemie,* 308(2-4):122-126.

Schwartz, D.P., Brewington, C.R., Burgwald, L.H. 1967. Rapid quantitative procedure for removing cholesterol from butter fat. *Journal of Lipid Research,* 8(1): 54-55.

Searcy, R.L., Bergquist, L.M. 1960. A new color reaction for the quantitation of serum cholesterol. *Clinica Chimica Acta,* 5(2): 192-199.

Shen, C.T., Hsu, S.L., Chang, C.M.J. 2008. Co-solvent-modified supercritical carbon dioxide extractions of cholesterol and free amino acids from soft-shell turtle fish egg. *Separation and Purification Technology,* 60(2): 215-222.

Shisiikura ,A, Fujimoto, K., Kaneda, T., Arai, K., Saito, S. 1986. Modification of butter oil by extraction with supercritical carbon dioxide. *Agricultural and Biological Chemistry,* 50(5): 1209-1215.

Shukla, A., Bhaskar, A.R., Rizvi, S.S.H., Mulvaney, S.J. 1994. Physicochemical and rheological properties of butter made from supercritically fractionated milk fat. *Journal of Dairy Science,* 77(1): 45-54.

Sieber, R. 1993. Cholesterol removal from animal food—Can it be justified? *LWT-Food Science and Technology,* 26(5): 375-387.

Singh, B., Rizvi, S.S.H. 1994. Design and economic analysis for continuous countercurrent processing of milk fat with supercritical carbon dioxide. *Journal of Dairy Science*, 77(6): 1731-1745.

Singh, M., Sharma, R., Banerjee, U.C. 2002. Biotechnological applications of cyclodextrins. *Biotechnol* Adv, 20(5): 341-359.

Smith, M., Sullivan, C., Goodman, N. 1991. Reactivity of milk cholesterol with bacterial cholesterol oxidases. *Journal of Agricultural and Food Chemistry,* 39(12): 2158-2162.

Somogyi, G., Posta, J., Buris, L., Varga, M. 2006.Cyclodextrin (CD) complexes of cholesterol – their potential use in reducing dietary cholesterol intake. *Die Pharmazie-An International Journal of Pharmaceutical Sciences,* 61: 154-156.

Sundfeld, E., Yun, S., Krochta, J.M., Richardson, T. 1993. Separation of cholesterol from butteroil using quillaja saponins. 1. Effects of pH, contact time and adsorbent. *Journal of Food Process Engineering,* 16(3): 191-205.

Talpur, F.N., Bhanger, M.I., Khooharo, A.A., Memon, G.Z. 2008. Seasonal variation in fatty acid composition of milk from ruminants reared under the traditional feeding system of Sindh, Pakistan. *Livestock Science,* 118(1):166–72.

Vedaraman, N., Brunner, G., Muralidharan, C., Rao, P.G., Raghavan, K.V. 2004. Extraction of cholesterol from cattle brain using supercritical carbon dioxide. *The Journal of Supercritical Fluids*, 32(1): 231-242.

Vedaraman, N., Srinivasakannan, C., Brunner, G. and Rao, P.G. 2008. Kinetics of cholesterol extraction using supercritical carbon dioxide with co-solvents. *Industrial & Engineering Chemistry Research*, 47:6727–6733.

Vedaraman, N., Srinivasakannan, C., Brunner, G., Ramabrahmam, B.V., Rao, P.G. 2005. Experimental and modeling studies on extraction of cholesterol from cow brain using supercritical carbon dioxide. *The Journal of Supercritical Fluids*, 34(1): 27-34.

Vinarova, L., Vinarov, Z., Atanasov, V., Pantcheva, I., Tcholakova, S., Denkov, N., Stoyanov, S. 2015. Lowering of cholesterol bioaccessibility and serum concentrations by saponins: in vitro and in vivo studies. *Food & Function*, 6(2): 501-512.

Wan, Z.X., Wang, X.L., Xu, L., Geng, Q., Zhang, Y. 2009. Lipid content and fatty acids composition of mature human milk in rural North China. *British Journal of Nutrition*, 103(06):913–916.

Wasowicz, E., Rudzinka, M. 2011. Cholesterol and phytosterols. In: Chemical, biological, and functional aspects of food lipids. 2nd Ed. Sikorski ZE and Kolakowska A. (Eds.). CRC Press (Taylor & Francis Group), Florida, USA: 113-134.

Zhao, D., Zhao, L., Zhu, C., Tian, Z. Shen, X. 2009. Synthesis and properties of water-insoluble β-cyclodextrin polymer crosslinked by citric acid with PEG-400 as modifier. *Carbohydrate Polymers*, 78:125–130.

Zhao, J., Yao, S.J. 2012. Functional and physicochemical properties of cyclodextrins. In: Ahmed J, Tiwari BK, Imam SH, Rao MA. (Eds.). Starch-based polymeric materials and nanocomposites—Chemistry, processing, and applications. Florida, USA: CRC Press (Taylor & Francis Group). p 183-230.

Ziarno, M. 2007. The influence of cholesterol and biomass concentration on the uptake of cholesterol by *Lactobacillus* from MRS broth. *Acta Scientiarum Polonorum Technologia Alimentaria*, 6(2): 29-40.

5

Baked and Fried Dairy Products

S K Bag, R Kumari, H N Mishra

1. Introduction

Dairy industry size and demand precede all other segments of food industry. Reporting milk production of 146.3 million tonnes in 2014-15, India is currently occupying first position in world scenario (NDDB, 2015). In spite of limited shelf-life, milk offers various opportunities of product diversification, processing optimisation and convenient packaging solutions. All categories of dairy products are unique and vary in market value. The operating margins in value-added dairy products are almost twice of liquid milk business due to reduced need of refrigeration and added shelf-life. The appeal and satiety it offers makes them ambrosia for consumers. Baked and fried dairy products (BFDP) are those class of dairy products in which main processing operation is either baking or frying. In addition to preservation of milk solids for longer time at room temperature, manufacture of baked and fried dairy products add value to milk and also provide considerable employment opportunity (Pal and Raju, 2007; Patil, 2011). The characteristic flavour and colour development of the product rely on this unit operation or cooking method and it cannot be replaced. The unique difference in baked and fried dairy products is that in baking dough is prepared and for frying batter is prepared which vary significantly in characteristics (Table 1). The quality of the finished product greatly depends on how the dough or batter is handled.

Table 1: Distinguished characteristics of dough and batters

Characteristic	Dough	Batter
Basic ingredients in recipe	Flour, sugar, fat	Sugar, eggs, fat, flour
Processing operation	Kneading, mixing	Beating, stirring, mixing, short heating
Raising	Biological, chemical, physical	Chemical, physical
Factors affecting binding of water or consistency	Wheat gluten, pentosans, damaged starch, swelling agents	Fat, sugar, eggs, damaged starch, and in part wheat gluten and swelling agents
Consistency	Elastic to plastic	Foamy, soft-plastic, semi-fluid

Source: Menger and Bretschneider, 1972

1.1 Baking

It is a food cooking method using dry heat by convection normally in an oven. Baking does not contribute any added fat to the finished product, potentially resulting in more healthful products, which explain the recent surge in the popularity of baked goods. Baking step is the most energy intensive stage of the production of baked goods. It converts the dough into the final baked product by firming (stabilization of the structure) and by the formation of the characteristic aroma substances. The increased production of gas and its expansion causes a 40% increase in the volume of the dough piece and a 10% increase of its surface area (Kriems, 1970).

Baking results in various types of chemical reactions and physical changes due to simultaneous heat and mass transfer (Köksel and Gökmen, 2008). Some of these reactions include the following (Potter and Hotchkiss, 1998).

(i) Evolution and expansion of gases,

(ii) Coagulation of proteins and gelatinization of starch,

(iii) Partial dehydration from evaporation of water,

(iv) Development of flavours,

(v) Changes of color due to Maillard browning reactions between protein and reducing sugars, as well as other chemical colour changes,

(vi) Crust formation from surface dehydration, and

(vii) Crust darkening from Maillard reactions and caramelization.

Baked goods are packaged for reasons of hygiene, to prevent loss of moisture (i.e., to preserve freshness), to prevent re-contamination with mold spores, and for promotional reasons.

1.2 Frying

It is one of the oldest methods of food preparation (Stier, 2004). It entails the application of heat from frying oil to achieve cooking and drying as well as flavour, crust, and color development in fried foods. In spite of the high calorific content and related health concerns, fried foods remain among the most consumed foods and their market demand has been increasing steadily (Piper, 2001). During frying, loss of moisture from the product and simultaneous uptake of oil occurs, though not at equal rates.The controllable variables in industrial frying processes are oil variety, frying time, and frying temperature (Hindra and Oon-Doo, 2006). These three parameters are interlinked and finished product quality depends on how these factors are balanced. The oil chosen for deep fat frying should be stable under high temperatures. In general, oil should be kept at a maximum temperature of 160 to 180°C during the frying operation. Frying food at a temperature which is too low results in increased fat uptake. Water, which is contributed by the foods that are fried in oil, enhances the breakdown of fatty acids which occurs during heating. Hydrolysis results in poor-quality oil that has a reduced smoke point, darkened colour and altered flavour. During heating, oils also polymerize, creating viscous oil that is readily absorbed by foods and that produces a greasy product. The more saturated (solid) the oil, the more stable it is to oxidative and hydrolytic breakdown, and the less likely it is to polymerize.

The rheological properties of batters for frying have a decisive influence in determining the final quality of battered food. Raw batter rheology may determine the success of a battered food more than any other characteristic. Batter ingredients bear the main responsibility for the rheological properties of the batter (Sanz and Salvador, 2009). The heat and mass transfer during frying of various foods have been studied and modelled. Frying is of two types: shallow frying (pan frying) and deep fat frying. Shallow frying means partial immersion in oil whereas deep frying consists of complete immersion of food pieces in oil/ ghee. Moisture loss, fat absorption, protein denaturation, starch gelatinization, browning due to caramelization of sugars and Maillard reaction, crust formation, pore development, and shrinkage are among the many changes that occur in foods during frying (Adedeji and Nagadi, 2010).

2. Nutritional Value

Baked and fried dairy products (BFDP) include *chhana podo*, baked *rasgulla*, *gulabjamun*, *pantua*, *khoajalebi*, *chhanajhilli* or *jalebi*, sarvaja and sarpuria. Bakery products are often classified as yeast leavened goods, chemically leavened goods, air leavened goods and partially leavened goods. In either case, the food structure must be such that it can trap the leavened gas and hold its structure (coagulation and fixing of the matrix by the application of heat). Wheat contains gliadin and glutenin which form the principal functional protein, gluten; gluten has the unique property of forming an elastic dough when moistened and worked upon by mechanical action (Potter and Hotchkiss, 1998).*Chhanapodo* is made up of different ingredients; *chhana*, which forms the major fraction, is incapable of holding such a leavened structure. During baking, it was observed that the structure leavened (up to triple the initial volume) but collapsed subsequently on cooling. It is also observed that dairy products in baked goods lower the bread volume (Spicher and Brümmer, 1995).

Baked and fried dairy sweets being derived from milk solids are concentrated source of fat, protein, sugar and minerals. Energy value of 100 g of gulab jamun varies from 250-300 kcal. Not only they are rich in calcium and phosphorus but also it is found that khoa derived fried dairy products have higher iron content (100 ppm) which is minimal in milk (2-4 ppm). Important chemical constituents of some of the traditional Indian fried and baked dairy products are given in Table 2.

Table 2: Major constituents of once important baked and fried dairy products

Constituents (%)	Chhana podo	Gulabjamun	Pantua	Khoa jalebi	Chhanajhilli
Moisture	32.5±0.55	36.10 ± 1.40	40.3 ± 0.62	22.00 ± 0.51	20.23 ± 0.25
Fat	23.0±0.72	12.20 ± 1.05	15.5±0.5	23.20 ±	12.53 ± 0.17
Protein	16.4±1.43	10.50 ± 0.95	8.7±1.07	8.26	5.71 ± 0.2
Lactose	—	12.80 ± 1.20	—	11.35	—
Sucrose	20.0 ± 0.48	—	27.1±0.39	—	40.21 ± 0.30
Other carbohydrates	7.2±0.66	26.80 ± 1.80	7.7±0.49	33.92	67.11 ± 0.19
Ash	0.9±0.05	1.60 ± 0.50	0.67±0.04	1.27 ± 0.04	0.29 ± 0.06
Water activity (a_w)	0.76	0.81	0.80	0.86	0.83

2.1 *Chhana podo*

Chhana podo is the only traditional baked dairy product in India and includes*chhana* (Indian cottage cheese) and sugar as essential ingredients (Figure 1). It was originated in Odisha and closely resembles the north Indian traditional dairy product "milk-cake" in appearance, which is prepared from whole milk by heat desiccation in contrast with heat/acid coagulation employed for *chhana podo* (Karwasra et al., 2001). Local sweet meat makers in Odisha say that the product was invented by Mr. Kelu Behera in Pahel and records are also available stating that it

Fig. 1: *Chhana podo* – the only baked traditional product in India (See colour version on page 335)

was independently produced by the Pratihari family. "*Podo*" in Oriya means burning; this substantiates the term used to define the product, since *chhana podo* is a baked product. Ghosh et al. (2002) reported that traditionally, *chhana podo* is made by smoldering chhana-sugar mix wrapped in sal leaves or other large leaves on slow fire.

Ghosh et al. (1998) stated that *chhana podo* should have a light brown color and a cake-like soft spongy body. It should be sweet, with a rich fat taste and a cooked flavor. Ghosh et al. (2002) attempted characterization of market samples in Odisha with respect to cake height, physical appearance and sensory characteristics as judged by panelists from Dairy Technology Division of National Dairy Research Institute (NDRI), the product varied greatly in all characteristics from district to district. Cake height ranged from 1.5 cm to 10 cm with most places producing *chhana podo* of 4 cm thickness. One-side as well as two side baked *chhana podo* was produced. Product color ranged from creamy white to light brown. Products from some districts had visible semolina particles giving the product a more grainy structure; some samples were moist and had sugar syrup oozing out whereas others were drier; the texture varied from a compact and firm body to a spongy body.

2.2 Gulabjamun

Gulabjamun is a milk based sweet popular in India. The traditional method of preparation involves blending of khoa, refined wheat flour and baking powder to a homogenous mass to obtain smooth dough. The balls of dough are deep-fat fried in ghee or refined vegetable oil to a golden brown colour and subsequently transferred to sugar syrup. The sweet is round or oval in shape and dark brown in colour, and served with sugar syrup (Rangi et al., 1985). Technology for industrial production of gulabjamun has been developed using anassembly line system and is in operation commercially. Canned gulabjamun is mostly exported (Joshi et al., 2008).

2.3 Pantua

Pantua has its origin in the eastern region of India. The dough of pantua is made of a mixture of khoa and chhana, along with other ingredients such as maida and baking powder. The method of manufacture of pantua is almost same as of gulabjamun. Since chhana is added along with khoa, the texture of pantua has more spongy and chewy characteristics than that of gulabjamun. It can also be called fried rasgulla (Figure 2).

Fig. 2: *Gulabjamun and pantua*

2.4 Khoa jalebi

Khoa jalebi is a unique khoa based sweet product of central India. It is relished as snack in the evening time and as breakfast item (Figure 3). Pagote and Rao (2012) observed that khoa jalebi is consumed on religious fasting days such as *Ekadashi* and other festivals. It is generally round or oval, having thicker coils and golden brown to dark brown in colour. The raw materials used for making khoa jalebi are khoa and tikhur though many *halwais* employ arrowroot also. Khoa is a heat desiccated milk product prepared by vigorously boiling and stirring milk continuously and tikhur (*Curcuma angustifolia*), a herb with medicinal properties, is added.

Fig. 3: Khoa-jalebi – fried dairy product

2.5 Chhana jhilli

Chana jhilli or Chana'r jilipi in Bengali is a traditional sweet popular in West Bengal and Odisha. It was first prepared in 1938 at Nimpara in Orissa. It is prepared by frying of batter made from *chhana*, *maida* and water and finally soaking in sugar syrup. It has a coiled structure with syrupy interiors and chewy body. It has close resemblance to *maida* jalebi and khoa jalebi, but has closed and firmer coils (Figure 4).

Fig. 4: Chhana jhilli

2.6 Sarvaja and Sarpuria

It is purely made of cream of the milk. Sometimes khoa and chhana are also mixed with it. In Sarvaja, cake made of milk cream, are deep fried whereas in sarpuria it is baked. Krishna Nagar (Nadia, West Bengal, India) is considered the birth place of Sarvaja since 1902. The manufacturing of this sweet is a tedious process and requires expert skills.

2.7 Fried ice cream

The dessert is commonly made by taking a scoop of ice cream frozen well below the temperature at which ice cream is generally kept, possibly coating it in raw egg, rolling it in cornflakes or cookie crumbs, and briefly deep frying it. The extremely low temperature of the ice cream prevents it from melting while being fried. It may be sprinkled with cinnamon and sugar and a touch of peppermint, though whipped cream or honey may be used as well.

Preparation of fried ice-cream comprises mixing ice-cream to a mixture of hydropholic colloid, then jellifying and freezing to harden to solid form. Flour and egg white are added with excipient to form protective film, which is used as a coating to wrap the ice cream. It is then placed on a tray pressed into enclosed semi-product using mould, and fried in heated edible oil (Karakaya, 2015; Patent No: WO 2015119585 A1).

3. Manufacturing Methodologies

Baked and fried dairy products are made from two base materials – khoa and chhana. Some 9,00,000 tonnes of khoa valued at Rs 45,000 million is produced in India. Approximately 1,20,000 tonnes of chhana (coagulated milk product), valued at Rs 6,000 million is produced (Rao and Pagote, 2013). Manufacturing methodologies of the above products depend mainly on the quality of initial ingredients viz. milk, ingredient ratios, how the main cooking operation was followed and finally sugar syrup strength and soaking. Function of major ingredients of some important backed and fried dain products (BFDP) are summarised in Table 3.

Table 3: Function of ingredients in different BFDP

BFDP	Function of ingredients
Chhana podo	Water makes the product soft and spongy, fat present in chhana mellows the product, sugar imparts sweetness and also aids in colour and flavour development and semolina particles give the product a grainy structure.
Gulabjamun	Khoa acts as base material and refined wheat flour or maida is generally used as a binder for preparation of gulabjamun, baking powder acts as a leavening agent.
Khoa-jalebi	Khoa forms the base material for the jalebi manufacture, and tikhur acts as a binding agent and helps in holding the sugar syrup inside the jalebi coils.
Chhana jhilli	Higher fat content of milk gives soft chhana which is desirable for jalebi, water content of batter affects the integrity of coil and maida acts as binder
Pantua	Khoa forms the base, chhana provides spongy and chewy characteristics and maida helps in obtaining the typical shape and texture

3.1 Process technology for preparation of *chhana podo*

The important steps involved in manufacture of chhana podo are preparation of chhana, mixing and kneading chhana with sugar, semolina and water, and finally baking. Preparation of chhana involves heating of milk to near-boiling temperatures followed by acid coagulation of the milk. The yield of dry-type chhana podo (MC 29-35% wb) is 17 kg and that of the wet type (MC- 40-50% wb) is 18.5 kg from 100 kg of milk with 3.5-4.0% fat. Addition of sugar and semolina increases the profit margin for manufacturers and hence, there is a tendency to add more of these than the optimum quantities. According to Kumar et al. (2002), the desirable product of *chhana podo*, had chhana from milk (Fat: 4.5%), 35% sugar, 5% semolina and 30% added water (all ingredients were added by weight of chhana).The minimum time of baking is that which permits full gelatinization of the starch, denaturation of the protein, and the formation of aroma substances. The upper limit of the bakingtime is determined by a loss of vitamins, excess evaporation of water, and also by economic considerations (Schneeweiss, 1965). The optimum baking condition reported was 200 °C for 60 min. The process flow chart for *chhana podo* preparation is shown in Fig. 5.

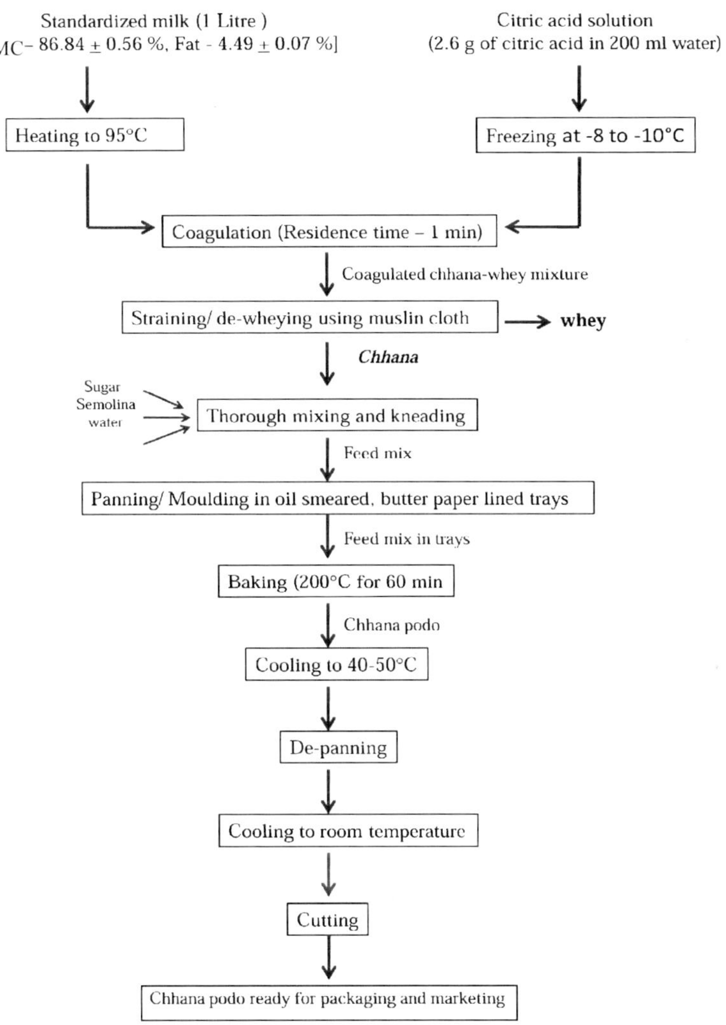

Fig. 5: Process flowchart for preparation of *chhana podo* preparation

3.2 Process technology for preparation of gulabjamun

For gulabjamun, khoa (100 g) is properly blended with 30 g of refined wheat flour and 0.5 g baking powder to obtain homogeneous and smooth dough. Dhapkhoa, having about 40–45% moisture, is used for making gulabjamun (Dodeja and Deep, 2012). Refined wheat flour or maida is generally used as a binder for preparation of gulabjamun. The type of binder and content play an important role in deciding the composition, rheology and sensory attributes of gulabjamun (Joshi et al., 2008).Too much baking soda will cause the gulab jamun to get too soft or they will break apart when frying. The portioned balls of dough are fried and soaked in sugar syrup. Frying is carried out slowly so that the core of gulabjamun gets cooked properly. If the gulabjamun are fried on high heat, they will become hard inside and not fully cooked. Process flow chart for the preparation of gulabjamun is shown in Figure 6.

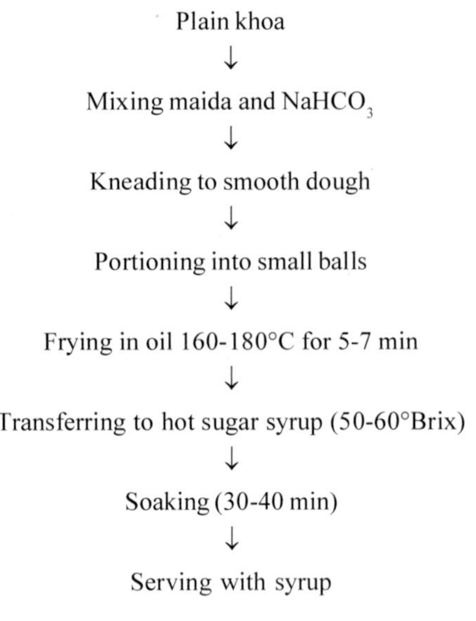

Fig. 6: Process flowchart for preparation of gulabjamun

3.3 Process technology for preparation of pantua

Pantua dough consists of chhana and khoa (in 5: 4 ratio), maida (3 %), arrowroot (3 %), suji (3%), ground sugar (0.7 %) and baking powder (0.3 %). The ingredients are properly kneaded to form dough of smooth consistency with about 40 % moisture (Fugyre 7). Spherical balls of about 12 grams are prepared and fried in vanaspati ghee at 120°C and after obtaining deep brown colour, transferred to hot sugar syrup (60°C) having concentration of 55° Brix (Nath, 1992).

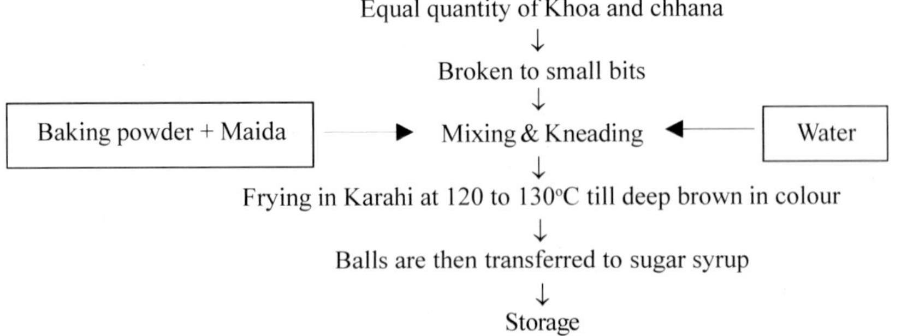

Fig. 7: Process flowchart for preparation of *pantua*

3.4 Process technology for preparation of *khoa-jalebi*

The first and important step in khoa jalebi manufacture is the preparation of batter of right consistency. Cow milk khoa is more suitable for jalebi preparation. This may be scientifically convincing observation because the jalebi needs to have soft texture that can only be imparted by cow milk proteins. Cow milk proteins are less springy and are low in calcium (Ganguli, 1974; Pagote and Rao, 2012). The preparation technique of khoa-jalebi sweet including the proportion of ingredients, batter making, frying and soaking in sugar syrup etc.

varied significantly from sweet makers shop to shop. Khoa and soaked tikhur were properly mixed to form a smooth batter. Small quantity of water was added if the batter was too tight otherwise the mixture was taken in a special woven jalebi making cloth having an opening of 1 cm. The batter was given a round shape and one knot in ghee and fried for 5 to 4 min at medium flame (160-170°C); 4 to 5 jalebi samples were fried at a time. The fried jalebi was soaked in sugar syrup of 65°Brix for 5 min (Kumari et al., 2012). Process technology for the preparation of khoa-jalebi is given in Figure 8.

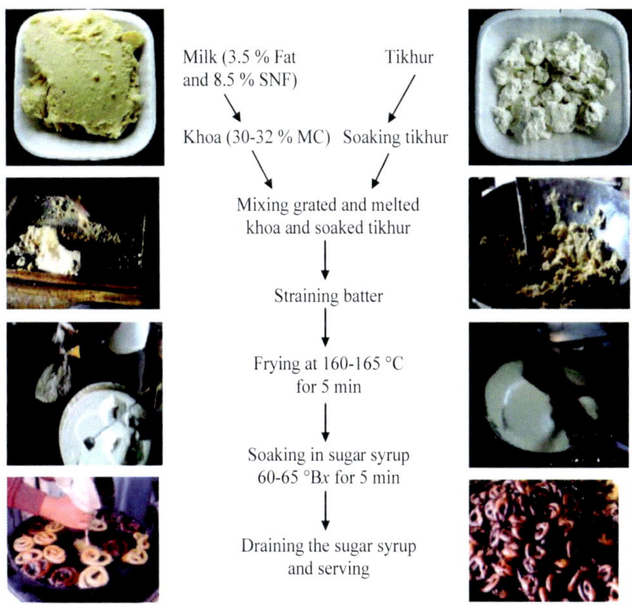

Fig. 8: Process flowchart for preparation of khoa-jalebi (See colour version on page 335)

3.5 Process technology for the preparation of chhana-jhilli

Chhana jhilli production process is successfully optimized by Geetha et al (2013). Traditionally, maida and corn flour were thoroughly mixed with required quantity of water into a thick consistency and left for 3 hours for hydration of the contents. Corn flour was used for providing crispy texture to the product. After hydration, chhana was mixed with vigorous agitation in a mixer to a flowable, but thick consistency. The batter (about 100 g) was filled into a 100 mL capacity flexible plastic bottle and extruded through its narrow opening (4-5 mm diameter) into hot refined sunflower oil in a shallow stainless steel frying pan, with circular movements of hand resulting in formation of coils. However this process requires practice and skill to achieve coils of attractive and regular shape and size. The coils were fried to brown colour, taken out of the hot oil and placed in hot sugar syrup of for 1-2 min. Optimized process recommends 3 % fat level in milk, 1:1

ratio of maida – chhana combination, 45 % of water level in batter, 160-170 °C frying temperature for 2 min, 68 °B sugar syrup concentration for 1 min soaking time (Figure 9).

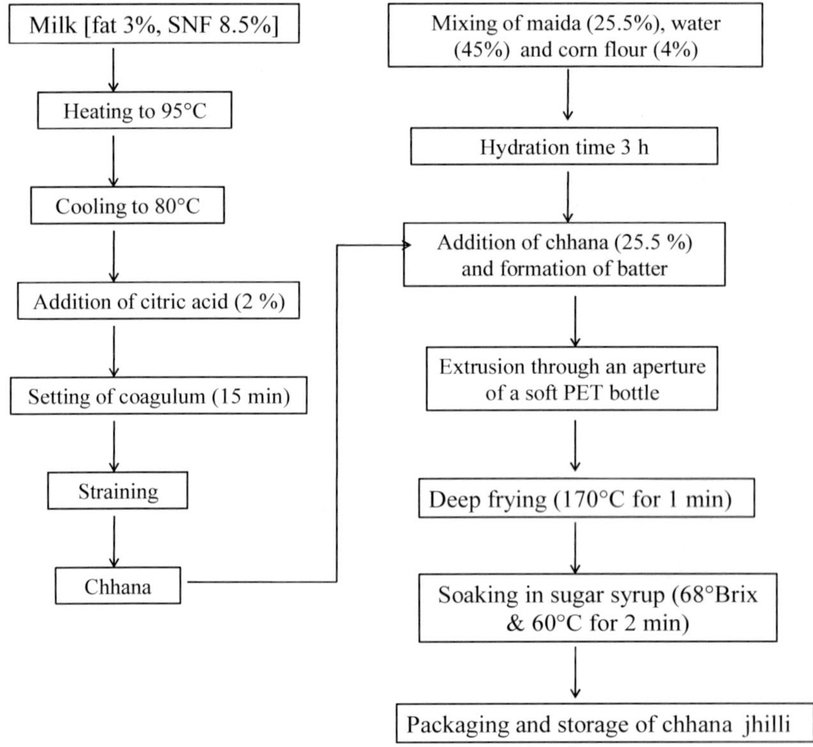

Fig. 9: Process flow chart for preparation of chhana-jhilli

4. Quality Issues and Challenges

Food quality is a consumer-based perceptual/evaluative construct that is relative to person, place and time and that is subject to the same influences of context and expectations as are other perceptual/evaluative phenomena (Cardello, 1995). This definition is apt for baked and fried dairy products as there is huge variation in quality aspects of aforementioned sweets prevalent in India. Few BFDP are soft in texture, some are crispy, porous, varying in degree of caramelisation, and colour development. Understanding the relationship between food texture perception and food structure is of great importance for enterprising manufacturers intending to produce texturally attractive food products (Wilkinson et al,, 2000). Bedekar (2006) mentioned that although we inherit the art of making good BFDP for centuries, but it is unfortunate that our food technology, as of date, is most undeveloped for processing in a big way. Some of the technical challenges with respect to quality of BFDP are summarised in Table 4.

Table 4: Technical challenges vis-a-vis quality of baked and fried dairy product

BFDP	Technical challenges vis-a-vis quality reported	References
Chhana podo	Batch to batch variation in product quality in terms of physical (cake height) as well as sensory characteristics	Mukhopadhyay et al. (2015)
Gulabjamun	Non-availability of khoa all year round, variation in its quality, and poor shelf life.	Chetana et al. (2004)
Khoa-jalebi	Ingredient standardization, variation in colour index, limited shelf-life	Kumari et al. (2012)
Chhanajhilli	Product optimisation is not standard and poor shelf stability (5 days at 28°C)	Geetha et al. (2015)

Technical challenges of baked and fried dairy product are inherent and comprise high fat and sugar content of the product which pose problems in continuous mechanization and shelf-life. The traditional method of gulabjamun preparation has several limitations such as non-availability of *khoa* all year round, variation in its quality, and resultant gulabjamun with poor shelf life. Consequently, many gulabjamun instant mixes have been developed both from roller- and spray-dried skim milk. Gulabjamun of uniformand acceptable quality can be prepared by both housewives and confectioners from these mixes (Ghosh et al., 1986; Chetana et al., 2004). The instant *gulabjamun* mix is a dry product formulated by mixing skim milk powder, *ghee*, refined wheat flour, semolina, citric acid and baking powder (Ghosh et al., 1984). The growth of baked and fried dairy products had been hampered or not commercialized mainly due to the following reasons

(i) Sweet industry is governed like family business and much depends on artisans or *halwais*.

(ii) Lack of investment or subsidy from government; even quality standards are also not established by government institutions like BIS for all sweets.

(iii) Limitation in the scope of increasing the shelf-life of dairy sweets as some techniques like drying and freezing could quash the quality of sweets.

(iv) Finally lack of proper storage, transportation and distribution network.

4.1 Mechanization of baked and fried dairy product production

The current methods for the manufacture of baked and fried dairy products are based on the techniques that remained unchanged over ages. These small scale dairy shops utilize energy inefficiently, produce variable product quality in hygienically poor conditions, and are based on skills of one particular *halwai*. As a result, this product sector is more dominant in unorganized sector and fetches variable profits and losses seasonally. Mechanization will pave the way for organized sector to diversify these product ranges and will favour more investment and reduce labour. It includes process machinery development; conversion from batch mode to continuous as batch methods for product preparation results in higher processing cost and reduced energy efficiency and lastly effective and convenient packaging solutions.

The role of National Dairy Development Board (NDDB) is highly commendable in mechanization of various baked and fried dairy products. The Odissa State Cooperative Milk Producers Federation Limited, Bhubaneswar (OMFED) has adopted the process technology of *chhana podo* from NDDB and presently manufactures 50 – 60 kg of wet type *chhana podo* everyday. A technology was developed for industrial production of *gulabjamun* using an assembly-line system and is in use at Sugam Dairy, Vadodara (Gujarat) since 1981-82.

5. Potential for Value Addition

Value addition will not only multiply the choices of dairy sweets for consumers but also can unravel nutritional issues by increasing the functionality of dairy sweets. The market basket will also expand by value addition resulting in more profit. This valorization includes shelf-life extension, innovative packaging, and functionality enhancement in terms of low fat, low calorie or fortification by dietary fibre addition. Indian government has now permitted use of sweeteners and preservatives in 25 food items including traditional sweets like halwa, gulabjamun, and rosogolla (George et al., 2006). Dietetic *chhana podo* (Kumar 2008) is prepared using sucralose, fructooligosaccharides based low calorie gulabjamun (Renuka et al., 2010), defatted soy flour supplemented gulab jamun (Singh et al., 2009) and khoa-jalebi dry mix (Chaudhary and Pagote, 2015) is prepared for the value-added ease of health-conscious consumers.

5.1 Shelf-life extension

The most serious constraint for shelf-life enhancement is the activity of microorganisms. Troublesome spoilage microorganisms include aerobic psychrotrophic gram-negative bacteria, yeasts, molds, heterofermentative lactobacilli, and spore-forming bacteria. Psychrotrophic bacteria can produce large amounts of extracellular hydrolytic enzymes, and the extent of recontamination of pasteurized fluid milk products with these bacteria is a major determinant of their shelf-life. Fungal spoilage of dairy foods is manifested by the presence of a wide variety of metabolic by-products, causing off-odors and flavors, in addition to visible changes in color or texture. The rate of spoilage of many dairy foods is slowed by the application of one or more of the following treatments: reducing the pH by fermenting the lactose to lactic acid; adding acids or other approved preservatives; introducing a desirable microflora that restricts the growth of undesirable microorganisms; adding sugar or salt to reduce the water activity (a_w); removing water; packaging to limit available oxygen; and freezing (Ledenbach and Marshall, 2009). Bikash et al. (2007) reported that shelf life of Chhana podo has been extended up to 8 days at 37°C by the incorporation of potassium metabisulphite and potassium sorbate. Vijayalakshmi et al. (2005) extended the shelf-life of burfi by packaging under different atmospheres- normal packing, vacuum packing and use of a free-oxygen absorber (FOA) at 65% RH/ 27°C. They recommended packing with FOA in foil laminate or MPET for longer storage of burfi (more than 45 days).

References

Adedeji A. A. and Nagadi M. O. 2010. Physicochemical Changes of Foods during Frying: Novel Evaluation Techniques and Effects of Process Parameters. Chapter 2 In: Physicochemical Aspects of Food Engineering and Processing, Devahastin S. CRC Press, Boca Raton, Page 42.

Bedekar, B.R. 2006. Heritage or traditional processed foods – where is the technology? *Indian Food Industry*, 25(6): 46-47.

Bikash, C., Ghosh, K., Rao, J., Balasubramanyam, B. N.and Kulkarni S. 2007. Process standardization and shelf life evaluation of chhana podo. *Indian Journal of Dairy Science*, 60: 11-19.

Cardello, A. V. (1995) Food quality: relativity, context and consumer expectations. *Food Quality and Preference*, 6(3): 163-170.

Chaudhary, M. B. and Pagote, C. N. 2015. Development of technology for Khoa jalebi dry mix – studies on formulation, packaging and shelf life. *Indian Journal of Dairy Science*, 68(3): 206-217.

Chetana, R., Manohar B. and Reddy S. R.Y. 2004. Process optimisation of Gulab jamun, an Indian traditional sweet using sugar substitutes. *European Food Research and Technology*, 219: 386-392.

Dodeja, A. K. and Deep, A. 2012. Mechanized Manufacture of Danedar Khoa using Three stage SSHE. *Indian Journal of Dairy Science*, 65(4): 274-284.

Ganguli N. C. 1974. Milk proteins. Indian Council of Agricultural Research, New Delhi.

Geetha P., Arivazhagan R., and Palaniswamy P.T. 2013. Optimization of production process and preservation of jalebi by using heat – acid coagulum of milk. *International Journal of Advanced Research in Engineering and Technology*. 4(4):230-241.

Geetha P., Arivazhagan R., Periyar Selvam S. and Ida I M. 2015. Process standardization, characterization and shelf-life studies of Chhana jalebi – A traditional Indian milk sweet. *International Food Research Journal* 22(1):155-162.

George, V., Arora, S., Sharma, V., Wadhwa, B., Sharma, G. and Singh, A. 2006. Sweetener blends and their applications: a review. *Indian Journal of Dairy Science*, 59:131–138.

Ghosh B. C.,Rao, J. K.and Kulkarni S. 1998. Chhana podo – baked indigenous delicacy. *Indian Dairyman.*, 50(1): 13-14.

Ghosh, B.C., Rao, J. K., Balasubramanyam, B. V. and Kulkarni S. 2002. Market survey of chhana podo sold in Orissa, its characterisation and utilisation. *Indian Dairyman*, 54(6): 37-41.

Ghosh B.C., Rajorhia, G.S. and Pal, D. 1984. Formulation and storage studies of *gulabjamun* mix powder. *Indian Journal of Dairy Science*, 37: 362–368.

Ghosh, B.C., Rajorhia, G.S. and Pal, D. 1986. Complete gulabjamun mix. *Indian Dairyman*. 38 (6): 279–282.

Hindra, F. and Oon-Doo, B. D. 2006. Kinetics of Quality Changes during Food Frying. *Critical Reviews in Food Science and Nutrition,* 46 (3): 239-258.

Joshi, M. U., Sarkar A., Rekha S. Singhal and Aniruddha B. Pandit 2008. Optimizing the formulation and processing conditions of GulabJamun: a statistical design. *International Journal of Food Properties*, 12:1, 162-175, DOI: 10.1080/10942910802312249.

Karakaya, Z. 2015. Fried Ice-cream. Patent No: WO 2015119558 A1. http://www.google.com/patents/WO2015119585A1?cl=en Accessed on 21-6-2016.

Karwasra, R. K., Srivastava, D. K. and Hooda, S. 2001. Standardization of the process for manufacture of milk-cake. *Indian Journal of Dairy Science*, 54(5): 280-282.

Köksel and Gökmen 2008. Chapter 3: Chemical reactions in the processing of soft wheat products In Food Engineering Aspects of Baking Sweet Goods edited by Servet Gülüm Sumnu and Serpil Sahin.© Taylor and Francis Group LLC Page 50.

Kriems, P. 1970. Ber. 5. Welt-Getreide- und Brotkongress 5, 125-141In: Chapter 7 Baked Goods by Gottfried Spicher and Jürgen-Michael Brümmer Page 302.

Kumar, R. 2008. Technology of dietetic chhana podo production. M. Sc. Thesis. National Dairy Research Institute, Bangalore, India.

Kumar, S. Khamrui K. and Bandyopadhyay P. 2002. Process optimization for commercial production of chhana podo. *Indian Dairyman,* 54(10): 64-65.

Kumari, R., Shrivastava S. L., and Mishra H. N. 2012. Optimization of khoa and tikhur mix for preparation of khoa-jalebi sweet.1:469-472. doi:10.4172/scientificreports.

Ledenbach, L.H. and Marshall, R.T. 2009. Microbiological spoilage of dairy products. In Compendium of the microbiological spoilage of foods and beverages, Food Microbiology and Food Safety. W. H. Sperber and M. P. Doyle (eds.) pp. 41.

Menger, A. and Bretschneider, F. 1972. Getreide Mehl Brot 26, 120 In Chapter 7 Baked Goods by Gottfried Spicher and Jürgen-Michael Brümmer Page 266.

Mukhopadhyay, S., Mishra H. N., Goswami T. K., and Majumdar G. C. 2015. Neural network modelling and optimization of process parameters for production of chhana cake using genetic algorithm. *International Food Research Journal,* 22(2): 465-475.

NDDB 2015. http://www.nddb.org/information/stats. Accessed on 06-06-2016.

Nath, J. 1992. Standardization method for production of*pantoa*. M. Sc. Thesis, National Dairy Research Institute (Deemed University), Karnal, India.

Pagote, C. N. and Rao K. J. 2012. Khoa-Jalebi-a unique traditional product of Central India. *Indian Journal of Traditional Knowledge,* 11(1): 96-102.

Pal, D., and Raju P.N. 2007 Nov 14-17. Indian traditional dairy products – An overview. In souvenir of the international conference on traditional dairy foods (pp. 1-27). Karnal (Haryana): National Dairy Research Institute.

Patil, G.R. 2011. Traditional dairy products: present status and strategies to promote exports. In Proceedings of the Brainstorming session on "Promotion of Indigenous Dairy Products in International Market" held at National Dairy Research Institute, Karnal on 22nd January.

Piper, R. 2001. Regulation in the European Union. In J. B. Rossell (Ed.), Frying: Improving Quality. (pp. 19-43). Woodhead Publishing Ltd, Cambridge, U. K.

Potter, N. N. and Hotchkiss, J. H. 1998. Food Science. Chapman and Hill,New York, USA.

Rangi, A. S., Minhas K. S. and Sidhu J. S. 1985 Indigenous milk products. I. Standardization of recipe for Gulabjamun. *Journal of Food Science and Technology*, 22(3): 191-193.

Rao, K. J. and Pagote C. N. 2013. Indian dairy sweets - Use of innovative technologies. F&B Specials Accessed on 11 May 2016.

Renuka, B., Prakash, M. and Prapulla, S. G. 2010. Fructooligosaccharides based low calorie Gulab Jamun: Studies on the Texture, Microstructure and Sensory Attributes. *Journal of Texture Studies*, 41: 594–610.

Sanz, T. and Salvador, A. 2009. Rheology of batters used in frying. In: Sahin S, Sumnu S. G., editors. Advances in deep fat frying of foods. Boca Raton, FL: CRC Press.

Schneeweiss, R. 1965. Ber. 2. Tagung Int. *Probleme der modernen Getreideverarbeitung und Getreidechemie* pp. 260-278.

Singh, A. K., Kadam D. M., Saxena M. and Singh R. P. 2009. Efficacy of defatted soy flour supplement in gulabjamun. *African Journal of Biochemistry Research*, 3(4): 130-135.

Spicher, G. and Brümmer, J. M. 1995. Baked goods. In: Biotechnology, Vol 9 (Rehm H. J. and Reed G. eds.) Weinheim: VCH, pp. 241–319.

Stier R. F. 2004. Frying as a science – An introduction. *European Journal of Lipid Science and Technology*, 106(11): 715-721.

Vijayalakshmi, N. S., Indiramma A.R., Viswanath P., Dattatreya A. and Kumar K. R. 2005. Extension of the shelf-life of burfi by packaging. *Journal of Food Quality*, 28: 121-136.

Wilkinson C., Dijksterhuis G. B. and Minekus M. 2000. From food structure to texture. *Trends in Food Science & Technology*, 11(12): 442-450.

6

Safe Storage of Food Grains

R Pande, G Mishra, S Srivastava, H N Mishra

1. Introduction

Storage is an important part of any food grain processing chain. Storage of raw and processed food grain has been used by humans since time immemorial as a prerequisite for insuring food security due to off time availability, and for holding seed grain for long periods (Pande and Mishra, 2012). After harvesting grains may be required to be stored for long periods before these are marketed or used as food or seed. The grain storage is mainly carried out for three purposes viz (i) to retain a supply of food (ii) to service a trading system and (iii) to retain seed for planting the following season (Hall, 1980). Storage is, therefore, carried out by the producer, trader, processor, and the exporter and at all these levels the methods adopted affect the quality of food stuffs. The length of time grains can be safely stored will depend on the condition it was harvested and the type of storage facility being utilized. Grain storage and handling is a major concern for food grain growers and processers worldwide (Mohan et al., 2011; Pande and Mishra, 2012). Depending upon the condition and uses food grains may be stored in different storage structures using traditional or modern storage methods. Different storage structures are used by farmers with the traditional names like Gumma, Oliya and Jute bag. Caswell (1973) recorded 50 to 60% of insect damage on food grain after six months of traditional storage. A typical range of storage methods encountered in tropics and subtropics region is given in Table 1.

Table 1: Traditional storage methods for food grains (*Source*: D.W Hall, 1980)

Storage method	Special measures	Product
Without Cover		
No structure	Heaped on ground	Paddy, Groundnuts
Vertical pole	Tied to poles	Maize
Horizontal cords or concepts	Hung on these strands which are tied between poles or trees	Maize
Vertical racks	Hung on horizontal poles fixed to vertical poles	Paddy
Platform (timber and grass)	Heaped on platform	Maize, Pulses, Groundnuts
Open baskets (grass)	Raised 1 metre or above ground	Paddy, Maize, Groundnuts
Sacks (woven plant material)	Placed on platform 1 metre high	Paddy
With Cover		
Horizontal grid	Hung on horizontal poles; covered with loose thatch roof	Paddy
Platform	Heaped on platform, covered with 'straw hat' which rests on platform	Paddy, Maize
Granary		
(a) Simple type (usually cylindrical)	Constructed of plant material raised above ground, with thatch roof	All types
(b) Structure incorporating clay	As (a) but with mud or clay worked into floor and walls	All types
(c) Wall of clay mixed with plant material supported by timber frame	Cylindrical or elliptical, raised above ground	Cereals, Paddy, Millet, Sorghum
(d) As (c) but not supported by timber frame	Jar-shaped, raised on "foot" of clay or log. Sometimes divided into compartments	Maize, Sorghum, Millet
(e) Wall of clay only	Various shapes, "straw hat"	Cereals, groundnuts
Clay jar (usually kept in living hut)	Sealed with damp earth, sealed withflat stone and clay, partially baked before storage, products mixed with ash, jar sealed with clay, produce mixed with ash	Maize seed, All types of grain, Maize meal, Maize, Sorghum
Gourds	Sealed with clay plugged with stems of plant	All types Maize grain
Baskets	Usually placed in kitchen	Groundnuts, Paddy (seed), Beans
Commodity wrapped	Kept in living hut in matting	All types
Stored under roof of living hut	Small bundles hung from roof above fire	Cereals
Stored on floor of living hut	Temporary storage	Paddy

Contd.

Table 1: Contd.

Storage method	Special measures	Product
Underground storage	Sometimes lined with cow dung and fired. Opening sealed with clay or grass thatch and thorns	Paddy
	Large crib of millet stalks or bamboo (12 ton capacity)	Paddy
Improved traditional	Square sided crib with timber frame and walls of wire netting	Maize

2. Storage Structures and System

Improved storage structures and practices are now popularized for effective and safe storage of food grains. Subramanya *et al.* (1999) summarized storage practices for safe storage of pulses like coaltar coating and polythene sandwiching proved to be effective in providing protective layer on mud structures making them moisture proof and also, hermatic storage of cowpea in plastic bags with cotton lining proved to be very effective against *Callosobruchus maculatus*. Some of the improved storage structures reported in literature are as follows:

Improved bins : These are structures for scientific storage of food grains designed moisture resistant and rodent proof (Figure 1a).

Brick-build godowns : These are made by brick-walls with cemented flooring for storing paddy/rice in bulk and bags (Figure 1b).

Cement plastered bamboo bin : In this bin bamboo strips are used to form the skeleton of the bin and cement-sand mortar (1:2.5 ratio) is plastered on outer and inner surface of the bin (Figure 1c).

CAP (Cover and plinth) storage : It is an economical way of storage on a large scale. The plinth is made by cement concrete and bags are staked on open and covered by polythene cover (Figure 1d).

Silos : These are made from concrete, bricks and metallic materials with loading and unloading arrangements (Figure 1e).

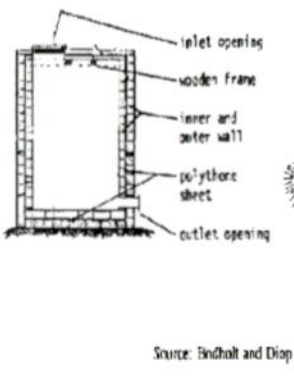
inlet opening
wooden frame
inner and
outer wall
polythene
sheet
outlet opening
Source: Bodholt and Diop (1987)

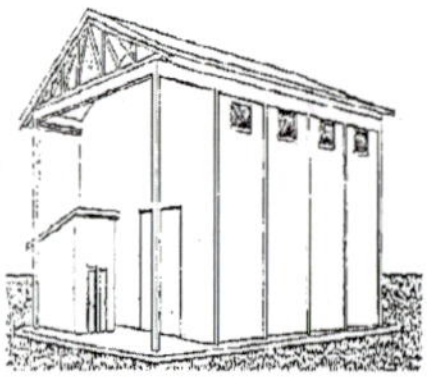

a. Improved bins

b. Brick-build godowns

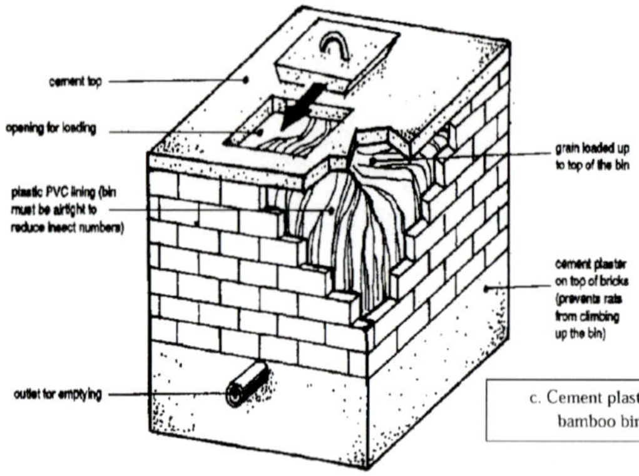
cement top
opening for loading
plastic PVC lining (bin
must be airtight to
reduce insect numbers)
outlet for emptying
grain loaded up
to top of the bin
cement plaster
on top of bricks
(prevents rats
from climbing
up the bin)

c. Cement plastered bamboo bin

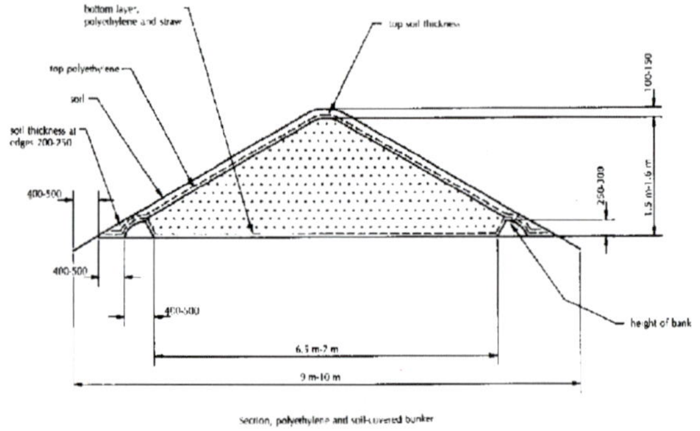
bottom layer,
polyethylene and straw
top soil thickness
top polyethylene
soil
soil thickness at
edges 200-250
100-150
1.5 m-1.6 m
250-300
400-500
400-500
400-500
height of bank
6.5 m-7 m
9 m-10 m
Section, polyethylene and soil-covered bunker

d. CAP (Cover and Plinth)

All dimensions in mm unless stated

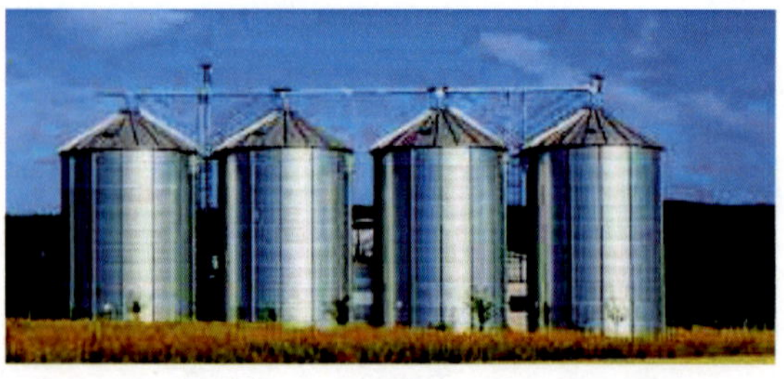

e. silo

Fig. 1: Different types of grain storage structures

A food grain storage system is an artificial ecological system in which deterioration of the stored product results from interactions among physical (temperature, moisture, and storage structure), chemical (carbon dioxide and oxygen) and biological (grain characteristics, microorganisms, insects, mites, rodents, birds) factors (Jayas and White, 2003). Quality and quantity of food grain are affected by intrinsic and extrinsic factors (Figure 2). Among all the factors temperature and moisture content greatly influences the shelf life of grain. Grain viability and reproduction of biological agents in grain are dependent to great extent on the temperature and moisture level (White, 1995).

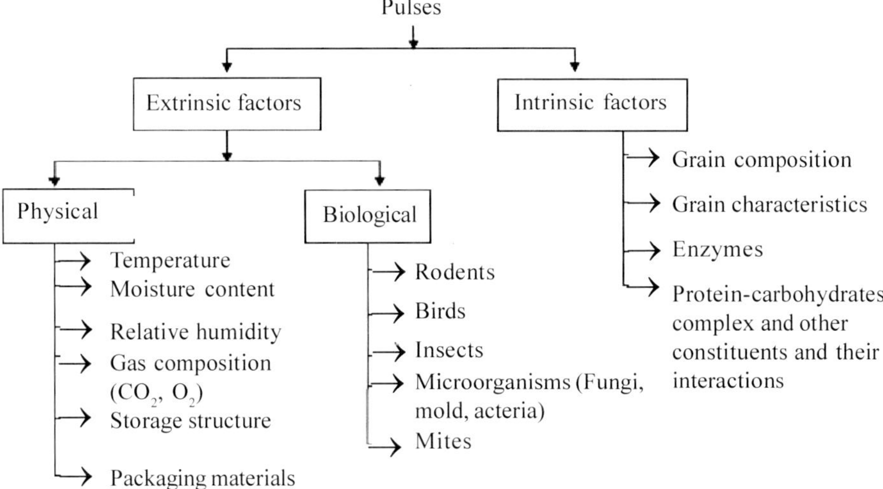

Fig. 2: Factors influencing grain quality and quantity
(*Source:* Mohan *et al.*, 2011; Pande and Mishra, 2012)

2.1 Structure designs and building materials

The engineering and construction details of the modern storage structures and there working principles are briefly narrated as follows.

2.1.1 Traditional structures

There are two main types, viz., storage in bags Sisal, Kenaf, Jute in warehouses and bulk storage in various types of silo bins. Such methods are most commonly used by traders, cooperatives, etc and at large-scale central storage depots, but they are also being adopted by framers in number of countries.

2.1.2 Improved storage structures

There are two improved types storage structures viz (i) bag, and (ii) bulk. Their usage depends upon on the type of grain stored, length of storage period required and kinds of grains to be stored.

Bag storage: In Indian godowns (Figure 3) the food grains (rice, pulses, and other milled products) are stored in bags. Due to lack of sufficient bulk storage facilities bag storage occupied a dominant place in country because of small capital investment for godowns.

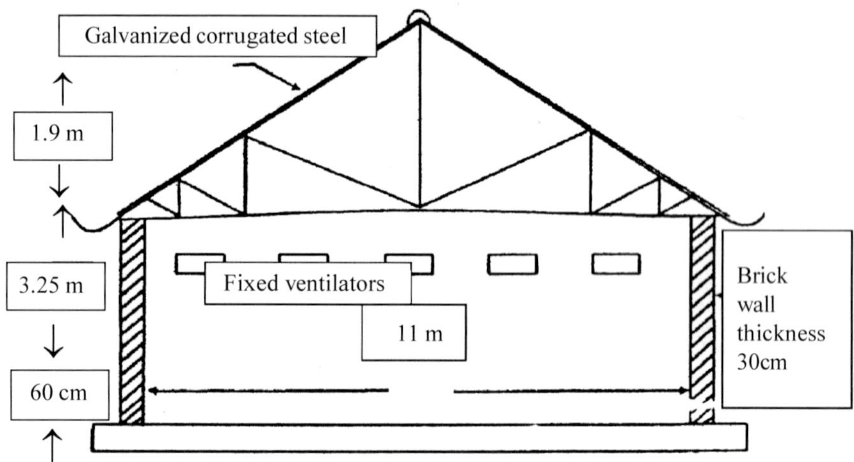

Fig. 3: A typical godown structure (L = 28.5 m, H = 3.25 m)
(*Source:* Chakraverty and Singh, 2014)

Location of godowns : The structure should be located on a land where there is no chance of any flood, it must be at least 0.5 km away from kilns, bone crushing mill, garbage dumping ground and tanneries and at least 30 m away from factories and other possible sources of fire. There should be no tree near the structures so that its root can affect the foundation. The structures must always be kept clean.

Stack plan : The floor space is marked into equal rectangle to ensure uniformity in stacking and in counting. A minimum space of 1 m is left around each stack. Stack dimensions are on the basis of bag dimension (0.7 m × 0.6 m). Bags are placed in blocks containing lengthwise and widthwise both layers. The common dimensions stacks are 9.14 m × 6.09 m, 7.43 m × 12.2 m, etc. Various stack plans used for different capacities are shown in Figure 4

Dunnage: It comprises wooden planks or polythene sheets. Wooden plank of 1.5 m × 0.9 m size are convenient to place below and are easy to carry. The height from ground is 0.2 m. These should be strong to bear a load of about 0.07 tonne/m^2. Polythene sheet of 0.03 mm thickness is used. It should be black in color.

Foundation: Depth and width of foundation depends on soil condition of each locality. In all cases, it is necessary to dig down to a point where the soil-bearing pressure is 150 kN/m^2 or better. Column and walls depth is 1.22 m and platform retaining walls depth is 0.9 m below the ground level.

Walls: The walls of the warehouse are built between the supporting pillars. The walls may be made of breezeblocks, or stabilized earth bricks 15 to 20 cm thick, and should be rendered smooth on both sides. They should be painted white, on the inside to facilitate the detection of insect pests, and on the outside to help keep the warehouse as cool as possible. A vapour proof barrier should be incorporated into the base of the walls, to prevent damp causing damage to the warehouse structure and its contents. A height of wall 5.64 m from plinth level allows stacking of food grains to a height of 20 bags. The height of walls will be decided by the number of layers of bag to be stored. Thickness of walls 0.39 m will be needed.

Roofing: Internal pillars supporting roof frames should be avoided because, as they can interfere with pest control. Roof cladding may be of galvanized steel or aluminum sheeting, or asbestos cement; the latter being more fragile but having better insulating properties. Tiles are not recommended, especially for large warehouses.

Flooring: Floor should be damp proof, rigid, and durable.

Doors: Double sliding doors are recommended; preferably made of steel, or at least reinforced along their lower edges with metal plate as protection against rodents, they should be sufficiently large (at least 2.5 x 2.5 m) and close fitting. If swing doors are fitted they should open outwards in order not to reduce the storage capacity of the warehouse.

Ventilators: Ventilation openings are necessary for allowing the renewal of air and reducing the temperature in the warehouse, they also allow some light to enter it. If such openings are located too low down they can be the source of numerous problems, viz., entry of water, rodents, thieves, etc. They should be fitted on the outside with anti-bird grills (20 mm mesh) and on the inside (10 cm behind the grills) with 1 mm mesh screens (removable for cleaning) which will deter most insects.

Illumination: Adequate light in a warehouse is an important factor as far as the safety of workers inside it is concerned. Many warehouses are fitted with translucent sheets in the roof. However this is considered inadvisable, because it may involve the risk of spot heating of produce in the top layers of stacks underneath. Artificial lighting is justified only in warehouses which are regularly worked in during hours of darkness.

Figure 4: Bag storage floor plan
(*Source*: General Mills India Private Limited Nashik, Maharashtra)

Silos A silo is a tall storage structure which is nothing but a deep bin (Figure 5 and 6). Concrete is a durable and economical material for construction of high capacity silos. But it should be used in pre-fabricated or precast form, Steel is perhaps the most common building material used for the construction of modern grain storage facilities. However, the use of steel silo is limited in India and in many countries. Advantages of concrete and steel silos are given in Table 2.

Table 2: Comparison between concrete and steel silos

Concrete Silo	Steel Silo
Moisture, vemin and insect proof	High strength and used as prefabricated members
High strength, durability and workability	Gas and water tight and long service life
For stored material with a reasonable cost of storage there is maximum space utilization per tone	Readily disassembled or replaced
While filling and emptying there is no food grain loss	
Fire proof and long life	

Source: Chakraverty and Singh, 2014

Fig. 5: Galvanized silo storage system
Source: http://www.fowlerwestrup.com/galvanized-silo-storage-system.php

Fig. 6: Steel silos storage system
Source: http://dir.indiamart.com/dewas/storage-tank.html

3. Novel Methods For Safe Storage of Grains

Safe storage is important to maintain physical, nutritional and processing quality of food grain. Storage limits are influenced by moisture content, storage temperature and length of storage period (Mills and Woods, 1994). These help in stabilization of prices by regularizing demand and supply (Pande and Mishra, 2012). Various chemical, physical and biological methods of grain storage are discussed below.

3.1 Disinfestation and pest control

3.1.1 Chemical method

Synthetic chemical insecticides are the primary pest control strategy, with annual worldwide sales reaching 7.7 billion US $ (Hunter- Fujita et al., 1998). Fumigants such as methyl bromide, aluminium phosphide or sulfuryl fluoride rapidly kill all life stages of stored grain insect. The use of pesticides is one means of preventing some losses during storage. Despite their usefulness, the choice of pesticides for storage pest control is fairly limited because of the strict requirements imposed for the safe use of synthetic insecticides on or near food. The continuous use of chemical pesticides for control of stored-grain pests has resulted in serious problems such as insecticide resistance (Suchita et al., 1989). Chemicals used for control of stored product pests or as protectants need also to be compared with the suitability and effectiveness of alternative methods of control. Non-chemical methods are attractive since they neither leave chemical residues in the commodity nor do they cause resistance in insects. The public awareness and concern for environmental quality, has led to more focused attention on research and development of alternative stored product protectants (Pande and Mishra, 2012).

3.1.2 Physical method

Disinfestation on stored grain and cereals by elevated temperature are carried out by using radiant heating processes such as infrared, microwave, and dielectric heating (Boulanger et al., 1971; Nelson, 1972; Kirkpatric and Tilton, 1972). Essentially, these studies have evaluated the effectiveness of rapidly heating infested grain to temperature ranges of 48–85 °C for a few seconds to 2 min, and then allowing it to cool passively to ambient temperature. Microwave energy has been considered for controlling insects since world war-II. The use of ionizing radiations has been recommended as a possible alternative or a supplement to conventional control methods (Waters, 1976). Combination of gamma radiation treatments with microwave, infrared radiation and insecticides has also been evaluated by Cogburn et al. (1972). Dermott and Evans (1977) reported the disinfestations of wheat in a continuous flow fluidized bed. Mahroof et al. (2003) determined time-mortality relationships for *Tribolium castaneum* (*Coleoptera: Tenebrionidae*) life stages exposed to elevated temperatures. Beckett and Morton (2003) studied the mortality of three species of Psocoptera (*Liposcelis bostrychophila),* Bandonnel (*Liposcelis decolour),* Pearman (*Liposcelis paeta)* at moderately elevated temperatures. Vadivambal *et al.* (2008) reported mortality of different life stages of *Tribolium castaneum* (*Coleoptera: Tenebrionidae*) in stored barley using microwave radiation. The microwave drying results in a high thermal efficiency, shorter drying time, improved product quality and also potential means of replacing chemical fumigation in pest control. The major advantage of disinfestations by physical means is that it does not leave any toxic residue on the grain (Pande et al., 2012; and Pande and Mishra, 2012).

Physical control also includes storage methods such as controlled and modified atmosphere storage. Controlled atmosphere (CA) storage involves maintaining an atmospheric composition that is different from air composition (about 78% N_2, 21% O_2, and 0.03% CO_2); generally, O_2 below 8% and CO_2 above 1% are used. In CA storage, the gas atmosphere is continuously controlled throughout the storage period. In most of the laboratory experiments, the effects of different atmosphere compositions on the storage life and quality of food products are evaluated. In such cases, the atmosphere composition is controlled continuously and hence the CA term is more frequently used (Jayas and Jeyamkondan, 2002).Two major classes of commodity can be stored in controlled atmosphere:

(i) Dry commodities are grains, legumes and oilseed. In these commodities the primary aim of the atmosphere is usually to control insect pests. Most insects cannot exist indefinitely without oxygen or in conditions of raised (greater than approximately 30%) carbon dioxide. Controlled atmosphere treatments of grains can be a fairly slow process taking up to several weeks at lower temperatures (less than 15 °C). A typical schedule for complete disinfestations of dry grain (<13% moisture content) at about 25 °C, with carbon dioxide concentration above 35 % (v/v) in air for at least 15 days have been provided by researchers (Annis and Morton, 1997; Hashem et al., 2012) These atmospheres can be created either by:

- adding pure gases carbon dioxide or nitrogen or the low oxygen exhaust of hydrocarbon combustion, or
- using the natural effects of respiration (grain, moulds or insects) to reduce oxygen and increase carbon dioxide, hermetic storage (Annis and Banks, 1993; Mujumdar and Beke, 2003).

(ii) Fresh fruits, most commonly apples and pears, where the combination of altered atmospheric conditions and reduced temperature allow prolonged storage with only a slow loss of quality (http://www.bestapples.com/facts/facts_controlled.asp, assessed on 19th March, 2015).

Modified atmosphere storage is one of the food preservation methods extended by modifying the atmosphere surrounding the food i.e alteration of natural storage gases CO_2, O_2 and N_2 which reduces the respiration rate of food products and activity of microorganisms in food. The MA may be achieved in several ways, i.e., by adding gaseous or solid CO_2, by adding a gas of low O_2 content (e.g. pure N_2 or output from a hydrocarbon burner) or by allowing metabolic processes within an airtight storage to remove O_2, usually with associated release of CO_2. Such atmospheres are referred to as 'high-CO_2', 'low-O_2', and 'hermetic storage' atmospheres, respectively. They are collectively known as 'modified atmospheres' (Banks and Fields, 1995; Jayas and Jeyamkondan, 2002; Caleb et al., 2012).

3.1.3 Biological method

Biological control agents (BCA) offer more environmentally safe alternatives to chemical pesticides. Microbial agents (bacteria and fungi) are used for controlling insect pest and storage fungi (Lacey *et al.*, 2001). Microbial pesticides are probably the most widely used and less costly source for pest bioregulation. Insect can be infected with many species of bacteria but the species belonging to the genus *Bacillus* are the most widely used pesticides. Shah and Pell, (2003) reviewed entomopathogenic fungi as biological control agents. Minimal effects of entomopathogenic fungi on non-target species, and they offer a safer alternative for use in Integrated pest management (IPM) than chemical insecticides (Goettel and Hajek, 2000; Pell et al., 2001). Approximately 750 species, belonging to 100 genera of entomopathogenic fungi have been reported, but only about 10 of them have been or are now being developed for insect control (Hajek and Lager 1994). Sabboour and Shadia E-Abd-El-Aziz (2007) studied the efficiency of some bioinsecticides against broad bean beetle, *Bruchus rufimanus* and they used different concentration of *Beauveriabassiana* (Biover) and *Metarhizium anisopliae* (Bionaz) (Pande and Mishra, 2012).

3.2 Grain disinfection methods

3.2.1 Microwave heating

Microwaves are electromagnetic waves with frequencies ranging from about 300MHz to 300GHz and corresponding wavelengths from 1 to 0.001m (Decareau, 1985). In electromagnetic spectrum, microwaves lie between radio frequencies and infrared radiation. Microwave heating is based on the transformation of alternating electromagnetic field energy into thermal energy by affecting polar molecules of a material. Many molecules in food (such as water and fat) are electric dipoles, meaning that they have a positive charge at one end and a negative charge at the other and, therefore they rotate as they try to align themselves with the alternating electric field induced by the microwave beam. The rapid movement of the bipolar molecules creates friction and results in heat dissipation in the material exposed to the microwave radiation. Microwave heating is most efficient on liquid water and much less on fats and sugars (which have less molecular dipole moment) (Sutar and Prasad 2008). The most important characteristic of microwave heating is volumetric heating which is different from conventional heating. Conventional heating occurs by convection or conduction where heat must diffuse from the surface of the material. Volumetric heating means that materials can absorb microwave energy directly and internally convert it into heat.

3.2.2 Infrared heating

Thermal energy obtained from either infrared or from heat by convection can be used to control insects infesting stored products. The time required to obtain temperatures that are lethal to insects depends not only on the insect species but also on the moisture content and heat retention characteristics of the grain or other commodities. Heat can be transmitted most efficiently by infrared that occurs below 200°C; many stored grain commodities such as wheat, corn, rice are heated during the drying or manufacturing process. Infrared heating is often used in these processes since energy can be readily absorbed by the commodities. Rice and wheat can be successfully treated with infrared energy for insect control because the temperature of these two cereal grains can be increased to 80°C without damage to either baking or milling qualities.

Frost and others (1944) observed that commodities such as corn (maize), peanuts, are damaged if their temperatures are increased to 65°C by using infrared radiation. In earlier studies, (Yeomans, 1952) both grain and insects were heated by radiant energy from infrared lamps, but in recent years the energy from gas fired infrared heaters has been used to control stored-product insects.

3.2.3 Ozonation

Ozone is a powerful oxidant that has numerous beneficial applications. Ozone has been used as a water treatment to disinfect, eliminate odors, taste, and color, and to remove pesticides, inorganic and organic compounds (Legeron, 1984). Ozone is attractive because its degradation product is oxygen, thus leaving no undesirable residue, and ozone can be generated on-site, eliminating the need to store or dispose of chemical containers. About 92–100% mortality of adult maize weevils has been reported. Ironically, the gas is being touted as a fumigant alternative in response to an international treaty banning the use of ozone layer harming chemicals currently used to rid food storage facilities of insects. When ozone is used for killing grain insects, it lasts for a very short period of time without damaging the environment or the grain. The ozone treatment of grain included two applications of ozone. In the first, the ozone moves through the grain slowly because the gas reacts, or bonds, with matter on the grain surface. This first treatment allows ozone to react with most of the grain surface and degrades the ozone. With the second ozone application, the gas moves through the grain more quickly because it isn't slowed by reactions with the grain. This allows the ozone to kill the insects by reacting with them rather than the grain. The highly reactive nature of ozone poses some unique challenges to achieve effective movement of ozone through a grain mass. Two distinct phases of ozone movement in maize have been described (Strait 1998; Kellsand et al., 2001).

When grain is treated with ozone for the first time, the concentration of ozone is reduced as it moves through the grain because interaction with matter on and around the grain surface rapidly degrades the ozone. Once these reactive sites are eliminated, the ozone moves freely through the maize with little degradation caused by the grain mass. Ozone as a fumigant to disinfest stored maize, had two distinct phases. Phase 1 was characterized by rapid degradation of the ozone and slow movement through the grain. In phase 2, the ozone flowed freely through the grain with little degradation and occurred once the molecular sites responsible for ozone degradation became saturated. The rate of saturation depended on the velocity of the ozone/air stream. Ozone (O_3) can be generated by electrical discharges in air and is currently used in the medical industry to disinfect against microorganisms and viruses, as a means of reducing odor, and for removing taste, color, and environmental pollutants in industrial applications (Kim et al., 1999). The attractive aspect of ozone is that it decomposes rapidly (half-life 20–50 min) to molecular oxygen without leaving a residue. In 1982, the US Food and Drug Administration (FDA) classified ozone for treating bottled water as "generally recognized as safe" (GRAS) (FDA, 1982). An extension of this GRAS declaration to other agricultural uses was recently filed with the FDA and is awaiting affirmation (Graham, 1997). Ozone is a strong oxidizing agent that has been effectively used to control fungal growth and reduce mycotoxin contamination (Kim et al., 1999). At low concentrations ozone protected clean surfaces from subsequent fungal contamination and growth, although higher doses were required to kill fungi on contaminated surfaces (Rice et al., 1982).

Aflatoxin is a toxin created by a specific mold that can grow in the field prior to harvesting crops, most commonly we see this occurring in corn during wet years. High levels of alfatoxin in grain are not allowed for human consumption or use in some animal feeds. This can dramatically lower the value of crop, especially if that crop is corn, as most corn is used for consumption by humans or animals. Ozone has the ability to oxidize and break down alfatoxin in grains. Due to the fact that aflatoxin is a result of mold that is most likely not present anymore in the grain, when the aflatoxin levels are reduced to acceptable levels they will stay at those levels and not rebound. While it is challenging to eliminate 100% of the aflatoxin in grains, reducing the level to normal and safe levels is very reasonable. Insect control in grains using ozone as a fumigant has been researched and studied widely. Practical application has been limited mostly due to the unknowns that still exist about this application. Ozone can definitely keep insects from infesting grain if used all the time. Low levels of ozone gas (0.1 ppm or below) used at all times will keep insects out of grains, and other produce as well. The use of ozone for food storage and insect control in this manner is proven and successful. In grains ozone gas can be added to already existing airflow through the grain to keep insects from damaging the grain. Once grain has been infested with insects it is more difficult to eliminate them. High levels of ozone are required to kill these insects. This is shown in research done by Purdue University and others. The levels of ozone required should be scrutinized to ensure this is a cost effective application, and that all safety precautions are taken for mold removal in grain. Ozone is proven as a great method to keep mold from growing and killing mold spore. If used over long periods of time, ozone will also kill mold. As there may be damage from mold that cannot be reversed, ozone use for this application should be evaluated carefully to ensure the benefits will still outweigh the costs.

3.3 Novel methods for evaluation of stored grain quality

Quality can be defined as the possession by a product of the conditions that make it suitable to meet the expressed or potential needs of its users (Giusti et al., 2008). In this definition both the consumers and the producers needs are considered: the first is interested in health, safety, organoleptic characteristic and utilization modalities, the second in parameters more related to the market. Consumers have become accustomed over the years to demanding grain with particular qualities.

To tackle the post-harvest processing and storage issues, quality control analyses are carried out using reference methods (wet chemistry) that have limitations in terms of adequation for the optimum implementation at the different steps of the food chain and for the control of the end-products and are also time consuming. Performed in the laboratory, although the management control needs to be at the production level (on-line measurement) or at the field level (in-field measurement) and inflexible, whereas rapid methods that can perform online measurement and allow the simultaneous analyses of different compounds are needed. The limitations of the reference methods for food grain safety and quality control have prompted research teams from public centers, universities and private companies to develop new analytical solutions based on spectroscopic technologies, such as integral near-infrared spectroscopy (NIR) multispectral or hyperspectral imaging. Besides computer vision, E-noses, E-tongues are intended for grading and controlling quality of the food grains by sensing the flavour and aroma rapidly and accurately than human trained sensory panelists. Using this technology allows the overcoming of the limitations of human sensing, including individual variability, inability to conduct online monitoring, subjectivity, adaptation, infections, harmful exposure to hazardous compounds and effect on the mental state.

3.3.1 Near infrared (NIR) spectrophotometery

In recent years, near-infrared (NIR) spectroscopy has gained wide acceptance in agricultural field by virtue of its advantages over other analytical techniques. It can rapidly determine the physico-chemical properties as well as detection of foreign materials, insect and fungal infestation in the cereal grains by its ability to record spectra with no prior manipulation. NIR covers the wavelength range adjacent to the mid infrared and extends up to the visible region. Although Herschel discovered light in the near-infrared (NIR) region as early as 1800, even spectroscopists of the first half of the last century ignored it in the belief that it lacked analytical interest. The earliest applications of NIR spectroscopy were reported in the 1950s, but it was not until the 1970s that the group headed by Norris used it to analyze agricultural food samples (Blanco and Villarroya, 2002).

A NIR spectrometer is generally composed of a light source, a monochromator, a sample holder or a sample presentation interface, and a detector, allowing for transmittance or reflectance measurements. The light source is usually a tungsten halogen lamp, since it is small and rugged. There are two types of wavelength selection, namely discrete wavelength and whole spectrum. In discrete-wavelength spectrophotometers, wavelengths can be selected by using as light sources filters that allow the passage of variably broad wavelength bands or light-emitting diodes (LEDs) that emit narrow bands. Whole-spectrum NIR instruments usually include a diffraction grating, although they may be of the Fourier transform (FT)-NIR type. Other wavelength-selection devices incorporated into NIR spectrophotometers in recent years include Acousto-Optic Tunable Filters (AOTFs). The sample presentation generally done by quartz-fiber-optic-based devices that allow the spectra of samples with different characteristics to be recorded simply by selecting the most suitable mode for each (usually reflectance for solids, transmittance for liquids, and transflectance for emulsions and turbid liquids). Detection in NIR spectroscopy uses devices comprising semiconductors (PbS or InGaAs). In multi-channel detectors, several detection elements are arranged in rows (diode arrays) or planes [charged coupled devices (CCDs)] in order to record many wavelengths at once, so as to increase the speed at which spectral information can be acquired. To perform qualitative or quantitative NIR analysis, i.e. to relate spectral variables to properties of the analyte, mathematical and statistical methods (i.e. chemometrics) are required that extract relevant information and reduce irrelevant information, i.e. interfering parameters. Data pretreatments, prior to multivariate modeling are required to reduce, eliminate or standardize their impact on the spectra. Since multivariate NIR spectral data contain a huge number of correlated variables, there is a need for reduction of variables, i.e. to describe data variability by a few uncorrelated variables containing the relevant information for calibration modeling. The best known and most widely used variable-reduction method is principal component analysis (PCA).

Before a NIR spectrometer can do any quantitative analysis, it has to be trained, i.e. calibrated using multivariate methods, by first selecting representative calibration sample set, secondly, spectra acquisition and determination of reference values, then multivariate modeling to relate the spectral variations to the reference values of the analytical target property, lastly validation of the model by cross validation, set validation or external validation.

3.3.2 Electronic nose

The characterization of olfactory quality as a method for quality assurance is of great interest to the food industry. On-line detection of drift or a defect in a process or in a discrete product can be possible by using gas sensors. Current 'electronic nose' technologies could perhaps be most accurately defined as odour sensors, but should not be confused with gas chromatography or sensory analysis. These electronic odour-detection systems are incorporated with sensors, which are conceptually analogous to human olfactory receptors, and a data-processing system, that conceptually simulates the brain. Gardner and Bartlett'(1993) defined an 'electronic nose' as: an instrument which comprises an array of electronic chemical sensors with partial specificity and an appropriate pattern recognition system capable of recognizing simple or complex odors.

Electronic nose systems comprise sensors, electronics, pumps, air conditioner, flow controller, and software for monitoring, data pre-processing, statistical analysis, etc. The ideal sensors to be integrated in an electronic nose should fulfill criteria such as high sensitivity towards chemical compounds, that is, similar to that of the human nose (down to 10^{-12}g/mL); low sensitivity towards humidity and temperature; they must respond to different compounds present in the headspace of the sample; should have high reproducibility and reliability. Four type of sensors are currently used in electronic noses, *viz.*, metal oxide semiconductors (MOS); metal oxide semiconductor field effect transistors (MOSFET), conducting organic polymers (CP), piezoelectric crystals (bulk acoustic wave – BAW). Volatile organic compounds (VOCs) presented to the sensor array produces a signature or pattern which is characteristic of the vapor. By presenting many different chemicals to the sensor array, a database of signatures can be build up. Data analysis and pattern recognition (PARC) particular, are also fundamental parts of any sensor array system.

3.3.3 Machine vision and image processing

Color immediately stands out as quality indicator for foods, as color provides much information on the condition of both the external surface and the whole food product. Color is also a very convenient quality indicator that is very easy to observe with a computer vision system, and it can also be reliably verified independently with a colorimeter. A computer vision system is a simulation of the human assessment process (Du and Sun, 2006), with the camera replacing the eye and a learning algorithm replacing the brain. The camera records objective and consistent quality indication data free of confounding noise. Then a trained learning algorithm matches the quality indication data to the correct quality class or level. The case, where the object surface does not contain enough quality information and need to find more quality information, penetration imaging can be used for finding internal object outlines for shape and size evaluation.

All computer vision devices involve a food image being recorded by an array of light-sensitive devices, with a CCD camera as the most common one. The received light is passed through a special prism to separate out the red, green, and blue elements. These are then directed onto photosensitive capacitors. Other types of imaging systems can be used thermal imaging, x ray imaging, hyper spectral and multispectral imaging are used to control the quality parameters of cereal grains. Thermal imaging is a technique which converts the radiation emitted by an object into temperature data without establishing contact with the object. X ray, hyperspectral and multispectral imaging are other popular nonvisible imaging technologies that record images under a range of light frequencies one by one. They follow the principle of seeing how a pixel changes under light of different frequencies (Chen, 1999). The difference is that hyperspectral involves much higher band density. A food product can thus be scanned at various wavelengths, and a three-dimensional hyperspectral "image cube" is formed, which is denser than the multispectral one. Plain X-ray imaging uses a single kilovoltage (kv) beam generated by a normal x-ray generator. The beam passes through the target and is attenuated by the surface therein. It then strikes radiographic film where it causes a chemical reaction to occur, leading to image production. The radiographic film is covered by a filter which reduces the amount of scattered x-rays which can cause a chemical reaction (coherent scattering is more common at very low kv energies and can reduce image contrast). The film itself is a polyester sheet covered with a silver bromide emulsion. When ionising radiation interacts with the emulsion, free electrons may be captured by the silver ion, forming a stable silver atom. This atom comes out of the solution and attaches to the polyester film. When treated, areas exposed to ionising radiation will appear darker on the film due to the silver atoms.

Regardless of what imaging equipment is used, there is an important question of image precision. This is the number of rows and columns of pixels. The finer the precision, the more details can be observed. However, the finer the image precision the greater the equipment cost and image processing computational load.

3.4 Disinfected and deinfested grain storage

3.4.1 Controlled / modified atmosphere (CA/MA) storage

Controlled atmosphere, a disinfestation technology wherein the normal composition of atmospheric air, that is, 21% O_2, 0.03% CO_2, and 78% nitrogen, is altered appropriately for disinfestation process (Banks and Fields, 1955). An atmosphere containing more than 35% CO_2 (known as carbon dioxide atmosphere) and the atmosphere containing less than 1% oxygen, that is, low-oxygen atmosphere are lethal to insects. Controlled atmospheres are mainly based on the establishment of a low-oxygen environment which kills pests. The oxygen levels vary between 0% and 2%. It can be applied in airtight environments ranging from $1M^3$ to $1000M^3$ depending upon food commodity. Insects in all stages are eliminated because of the lack of oxygen which causes the insect to dry out and suffocate. This controlled atmosphere (CA) storage has been shown to be promising in creating lethal conditions for insects and fungi in stored food commodities. Annis and Morton (1997) studied the effect of 15% to 100% CO_2 on developmental stages of insects in wheat at 25°C and 60% RH. They found that pupae were the most tolerant stages for all CO_2 concentrations and eggs were the only stages with 100% mortality at 20% CO_2 for less than 30 days. Gunasekaran and Rajendran (2005) also found that the pupae stages were the most tolerant stages when exposed to different concentrations of CO_2. Controlled atmospheres are always being seen as average alternative such as longer treatment time, usability and availability, and being not suitable for dealing with high level of infestation.

High concentrations of carbon dioxide (Table 3) in air can kill insect pests and rodents in storages and control the growth of moulds in wheat with moisture contents between 12-16% (Banks and Sharpe, 1979). The toxic action of CO_2 rich atmospheres is independent of CO_2 concentration above 60% in air, while it declines between 60-35%, and some species are capable of surviving or are tolerant to CO_2 levels immediately below 35%. Reproduction rates are, however much reduced even at concentrations approaching 10% CO_2 in air. One advantage with CO_2 atmospheres as compared to low oxygen in nitrogen atmospheres is that CO_2 may be applied in "One shot" operation in welded steel bins after gas tightness specifications have been adequately fulfilled. No further gas needs to be maintained while the target regime for insecticidal activity can still be achieved (Banks and Sharpe, 1979; Wilson et al., 1980).Monro (1967) has summarized some of the original observations of Dendy and Elkington (1920) concerning the effects of airtight (hermetic) storage on grain insects.

Table 3: Carbon dioxide concentrations as an index of infestation

CO_2 Concentration (% per 24 h)	Indication of Infestation
0.25	Uninfested grain with less than 15% M.C.
0.30	Clean grain
0.35- 0.50	A slight insect infestation or a rater high infection of microorganisms
0.50-0.90	Close observation required
1.0	Limit of dangerous storage conditions
1.0	Highly unsuitable storage conditions

Grain insects in hermetic closures succumb as soon as the oxygen has been used up with a corresponding increase in the amount of CO_2 being produced. The gases present under normal sealed conditions are oxygen, nitrogen and carbon dioxide. The amount of CO_2 given off by live cereal grains in hermetic enclosures is directly related to moisture content and temperature; above a critical point of moisture, the production of CO_2 increases very rapidly. Above a certain critical level of moisture content, grain stored under hermetic conditions creates an environment lethal to insects (and other aerobic organisms) quite rapidly, but at lower moistures, the rate of "self-sterilization" takes a comparatively longer time to achieve. The amount of oxygen consumed by live cereal grains of low moisture is greater than the amount of CO_2 given off (i.e., a respiration quotient less than unity). At approximately 30°C, 100 *Sitophilus oryzae* give off about 30 milligrams (i.e., nearly a fifth of their own bodyweight) of CO_2 in 24 hours, while at lower temperatures (20 to 21°C), the amount of CO_2 evolved is reduced to about 9.4 mg in 24 hours. *Sitophilus granarius* is a less active weevil as compared to *S. oryzae*, and on a weight for weight basis, the amount of CO_2 evolved is less. The respiratory quotient (RQ or ratio of CO_2 given off to the amount of O_2 consumed) for *S. oryzae* is 0.773, while for *S. granarius*, it is around 0.815.

It is generally accepted that the absence or lack of oxygen is sufficiently lethal to weevils, without taking into account the presence of increased concentrations of CO_2, although they may remain alive for extended periods under low oxygen tensions. However, CO_2 does exert a lethal effect on weevils irrespective of depleted oxygen tensions such that at approximately 30°C, *S. oryzae* was killed in less than 12 days in an atmosphere containing 14-23 % CO_2, although around 14% O_2 still remained. Pure (moist) CO_2 is less fatal than CO_2 with a small admixture of O_2; where pure CO_2 acts almost instantaneously as a narcotic under the influence that *Sitophilus* spp. remain in a state of "suspended animation" without losing their power of recovery. The objective of MA (modified atmosphere) treatment is to attain a composition of atmospheric gases rich in CO_2 and low in O_2, within the storage enclosure or treatment chamber for long enough to control the storage pests. At present, the most widely used source for production of such atmospheric gas compositions is tanker-delivered liquefied CO_2 or N_2.

4. Methods for Safe Storage of Selected Grains

Once a cereal crop is harvested, it may have to be stored for a period of time before it can be marketed or used as food or seed. The length of time grain can be safely stored will depend on the condition it was harvested and the type of storage facility being utilized. Grain binned at lower temperatures and moisture contents can be kept in storage for longer periods of time before its quality will deteriorate. The presence and build up of insects, mites, molds and fungi, which are all affected by grain temperature and grain moisture content, will affect the grain quality and duration of grain storage.

4.1 Conditioning of grain before storage

Conditioning of grain has the single purpose of preserving it quality. Low moisture content and low temperature have been shown to be essential for successful storage of grain for long period of time. A number of processes are available for conditioning of grain thereby ensuring safe storage.

Proper conditions to store grain effectively are those which prevent or discourage the growth of microorganisms and insects. Such conditions involve control and maintenance of moisture content, temperature, and soundness of the grain. Oxygen supply of storing environment and conditioning processes of cereal grain also effect storage of grains. Conditioning of grain before storage offer several advantages as follows.

- Allows for harvesting tough grain and thereby reduces losses from weather and wildlife.
- Extends available harvest period.
- Earlier harvest is possible.
- Drying tough or damp grain can reduce or eliminate spoilage in storage.
- May improve market grade and acceptability of grain.
- May afford alternative market outlets for grain.
- May eliminate necessity of swathing to obtain "dry" grain.
- May improve malting quality by reducing kernel peeling and cracking during combining.

4.2 Safe storage of paddy and rice

Rice is the staple food for 65% of the population in India. It is the largest consumed calorie source among the food grains. With a per capita availability of 73.8 kg it meets 31% of the total calorie requirement of the population. India is the second largest producer of rice in the world next to China. In India paddy occupies the first place both in area and production. Indian Basmati Rice has been a favorite among international rice buyers. Following liberalization of international trade after World Trade Agreement, Indian rice has become highly competitive and has been identified as one of the major commodities for export. This provides us with ample opportunity for development of rice based value-added products for earning more foreign exchange. Apart from rice milling, processing of rice bran for oil extraction is also an important agro processing activity for value addition, income and employment generation. It has been reported that about 9 % of paddy is lost due to use of old and outdated methods of drying and milling, improper and unscientific methods of storage, transport and handling (Basavaraj et al., 2015).

To minimize losses, precautions should be taken to follow proper post-harvest practices. They include timely harvest at optimum moisture percentage (20 to 22%), use of proper method of harvesting; avoid excessive drying, fast drying and rewetting of grains; ensure drying of wet grain after harvest, preferably within 24 hours to avoid heat accumulation; uniform drying to avoid hot and wet spots and mechanical damage due to handling. The losses in threshing and winnowing can be avoided using better mechanical methods. Proper sanitation during drying, milling and after milling to avoid contamination of grains and protect from insects, rodents and birds and use of proper technique of processing i.e. cleaning, parboiling and milling helps in reduction in post-harvest losses. To avoid storage losses maintaining optimum moisture content i.e. 12 % for longer period and 14 % for shorter storage period is essential.

Grains must be stored safely to meet the quality and quantity requirements meant for the various end uses. The factors affecting safe storage are broadly classified as abiotic factors and biotic factors (Table 4).

Table 4: List of abiotic and biotic factors

Abiotic factors	Biotic factors
Moisture content and temperature	Physical seed characteristics
Storage period	Microorganisms
Intergranular atmospheric composition	Insects and mites
Types of storage structure	Birds and rodents

Source: Mills, 1996

Strong correlation exists between the biotic variables and their abiotic environment. The abiotic variables influence the presence and development of biotic variables (Sinha and others 1973; Wallace et al., 1983) and the improper interaction among biotic and abiotic variables causes grain deterioration. Insect pests also play a pivotal role in transportation of storage fungi (Sinha and Sinha 1990).The survival and reproduction of biological agents in grain is dependent largely on the temperature and moisture levels (White 1995; Wilson 2013). Prakash and others (1985) earlier reported that kernel hardness in rice and ratio of length and breadth of grain were positively correlated with the number of adult insects emerged, and, rice grain damaged. This shows the importance of some seed agronomic traits such as length, breadth and texture for tolerance to storage pests. In moist and warm grain, inscets, mites and fungi can increase rapidly and produce moisture, heat and carbon dioxide by respiration, which further leads to deterioration of the grain bulk. The stored grains maintained at a sufficiently low moisture level can be stored for many years without any significant loss in quality. Storage conditions had a significant effect on important nutritional values of stored rice, wheat and maize; and the change was severe at elevated storage temperatures (Zia-Ur-Rehman, 2006).

Paddy seed is sun dried for 2-3 days continuously and then stored in gunny bags on indigenously made 4-5 feet stand to prevent pest infestation as reported by Lakshmana (2000). Most of paddy parboiled in the traditional mills is sun dried on a drying floor. The drying floor is normally cemented floor consisting of layer of brick bats with cement plaster on it. The parboiled paddy is spread on the floor 10 to 30 mm thick layer, then it is continuously stirred by a spiked plank or by feet. After 4 hours of drying in sun, paddy is heaped and covered by mats for tempering for a period of 2 to 3 hours. Then the paddy is spread again for 2 hours in the sun to dry it to desirable moisture content. Therefore, drying of parboiled paddy from 30% to 13% moisture content takes about 6 hours during summer and relatively long period about 8 to 10 hours during winter for proper storage.

4.3 Safe storage of wheat

Wheat under storage is likely to undergo infestation by insects and fungi. There are large numbers of insects which cause damage to the grain during storage. Out of total grain losses above 5% are caused by insects. Insect and fungal infestation cause both quantitative and qualitative damage in grain. Quantitative damage is physical damage due to loss in grain weight by direct feeding on grain by adults and larvae. Qualitative damage is due to product alteration such as loss of nutritional value, aesthetic value and loss of industrial characteristics for the preparation of good qualities bread and other products). The more common insects viz. granary weevil (*Sitophilus granarius*), lesser grain borer (*Rhyzopertha dominica*), rust-red flour beetle (*Tribolium castaneum*), sawtoothed grain beetle (*Oryzaephilus surinamensis*) and fungus viz. *Aspergillus* spp., *Penicillium* spp. and *Fusarium* sp. and mycotoxins such as aflatoxins (AFB_1, AFB_2 and AFG_1, AFG_2), ochratoxin A (OTA), zearalenone (ZON), deoxinivalenol (DON) and fumonisins (FB_1 and FB_2) resulting in a reduction of the grain quality; natural toxins can cause human and animal intoxication (Birck *et al*. 2006). The main concern is, therefore, disinfection of wheat grain and determination of quality parameters before storage. Rapid detection and disinfection technologies play important role for quick quality control of grains before it goes for storage in improved storage structures resulting very less chance of disinfection; have better storage value and consistent quality.

4.3.1 Rapid method for disinfection of wheat grain

Since losses of grain due to insect infestation are high, disinfestations of grain is very important for the safe storage. Acceptable insect disinfestations, usually has done by chemical fumigation, heat and cold treatment, or combination of these treatments. Chemical fumigants are widely used such as methyl bromine and phosphine, but the former is considered to pose a health risk and is being phased out in many countries. Also, the continuous and indiscriminate use of chemicals has resulted in the development of resistance strains of pests that are not killed. Ionizing irradiation has, therefore, been proposed as a good alternative to methyl bromide and other fumigants for insect and pest control (Ignatowicz, 2004). Ionizing radiation is a physical technique of food sterilization that has the potential to protect such grains from insect's infestation and microbial contamination. Ionizing irradiation include the use of ultraviolet rays, x-rays, and gamma rays. Ionizing radiation is a promising new tool for insect control and can supplement the existing chemical control measures. Not only does radiation offer a method for complete control of infestation, it also ensures partial protection against re-infestation through reproductive sterilization (Enu and Enu, 2014).

Microwaves and radiofrequency energy can be used to control insects in stored wheat grains. Microwaves are electromagnetic waves with frequencies ranging from about 300 MHz to 300 GHz and corresponding wavelengths from 1 to 0.001m (Decareau 1985). Microwave heating is based on the transformation of alternating electromagnetic field energy into thermal energy by affecting polar molecules of a material. The use of microwave energy to control insects was initiated by Webber et al. (1946) and the literature prior to 1980 has been reviewed by Tilton and Vardell (1982). Mortality of insects increased with either power or exposure time or both was found when studied by Vadivambal et al. (2007) studied mortality of three common species of stored-grain insects, namely *Tribolium castaneum*, *Cryptolestes ferrugineus*, and *Sitophilus granaries* in a pilot-scale industrial microwave system operating at 2.45 GHz. Mortality of the stored-product insects *Tribolium confusum* and *Ploasa interpunctella* exposed, either intermittently or continuously, to microwave radiation (2450 MHz) was evaluated as a function of exposure time and power, as well as at different growth stages and showed that Insects were killed by overheating and the lethal medium temperature for both species was estimated at 80% for the most resilient growth stage (Shayesteh and Barthakur, 1996). The insects in the mobile state were observed to move towards the surface from inside the nutrient medium during irradiation. They also curled up and in some cases aggregation was observed by the researchers. Power level and exposure time of microwave have been identified as two important parameters to provide 100% insect mortality (Bedi and Singh, 1992). Hamid and Boulanger (1969) had found that the bread making quality of wheat was affected with increase in treatment temperature when exposed to power of 1.2 kW. They reported that 70% and 100% mortality were obtained when the wheat temperature was 55°C and 65°C, respectively. Locatelli and Traversa (1989) reported that temperature of grain has to reach 80°C for achieving complete mortality of insect's infected using microwave. In fact, most infesting biological agents do not survive over a certain temperature called lethal temperature, generally between 55°C and 60°C, which can be rapidly reached through microwave irradiation.

Irradiation process of exposing crops to carefully controlled amounts of energy (Opadokun 1996) reduction of post-harvest losses through suppressing sprouting and contamination, eradication or control of insect pests, reduction of food-borne diseases and in extension of shelf life (Mokobia & Anomohanran, 2005). *Rhyzopertha dominica* (F.), is one of the most resistant of the stored product insect species to gamma radiation and it is also relatively resistant to infrared treatment (Tilton and Schroeder, 1963). Therefore, Kirkpatrick et al. (1973) studied the effect of combination of gamma, infra red and microwave radiations on mortality and found that in emergence averaged 99% for the gamma and infrared combination and 96% for the gamma and microwave combination. The susceptibility of infra red radiation was least for the eggs followed by adults within kernels (28 days old), pupae (24 days old), young larvae (7 days old), larvae that were (14-21 days old), and adults (42 days old) indicated infrared radiation from the flameless catalytic emitter to be a viable option for disinfesting wheat containing various life stages of *Sitophilus oryzae* (Khamis et al., 2010).

Ozone (O_3) in low concentrations can potentially replace insecticides to control insects in stored wheat grain. The advantage is that ozone is an unstable gas and quickly converts to the oxygen molecule (O_2), which is harmless. The ozone is produced in situ by an electric high-voltage process and can be used in, for example, a grain silo. There are variations in the amount of ozone needed to control the distinct stages and types of pests; adult insects are generally more sensitive to ozone. Sousa et al. (2008) investigated the ozone toxicity to 16 populations of *Tribolium castaneum* (Herbst), 11 populations of *Rhyzopertha dominica* (F.) and nine populations of *Oryzaephilus surinamensis* (L.) over phosphine using using time response bioassays at the dosage rate of 150 ppm ozone in a continuous flow of 2 L min^{-1} and found that none of the populations showed resistance to ozone, regardless of their susceptibility to phosphine; ozone is a potential alternative for phosphine resistance management in the insect species. The susceptibility of two stored-product insects, *Ephestia kuehniella* and *Tribolium confusum*, to gaseous ozone was investigated by Isikber and Oztekin (2009). Two ozone fumigation methods were used in that study, empty space fumigation with only one flush of ozone treatment held for 2 h, and a reflush ozone treatment at 30-min intervals for 5 h in the presence of 2 kg wheat, with an initial ozone concentration of 13.9 mg/L. There was a remarkable difference indicated in susceptibility between the life stages of *E. kuehniella* and *T. confusum*. They also observed empty space ozone treatment resulted in complete mortality of adults, pupae and larvae, while only 62.5% of the eggs of *E. kuehniella* were killed. Ozone treatment resulted in very low mortality of adults, pupae and eggs of *T. confusum*, ranging from 4.2 to 14.1% while only larvae had a high mortality (74%). Generally *T. confusum* was more tolerant to ozone treatment than *E. kuehniella*.

4.3.2 Rapid quality control methods for stored wheat grains

There are opportunities to improve the quality and market value of wheat grain at all stages of production, storage and transport. For better storage, the preliminary quality control of wheat grain is necessary such as the grain prior to storage should be free from insect infestation or foreign material such as chaff, straw, weeds or other contaminants. Grain safety is another major quality commitment; it should be free from insecticide residues, mycotoxins and trace elements. To ensure better quality of grain, rapid quality control and disinfection techniques prior to storage are required. For disinfection of wheat microwave treatment is under elusive research for the 30 last years. Except microwave, other surface disinfection sources such as ultraviolet, infra red or ozone are still under research. For analysis of physico chemical properties and detection of insect and fungal infection, rapid methods such as NIR, FT NIR, E-Nose, etc are of great importance.

4.3.3 Rapid quality control systems for stored wheat grain

Rapid methods for the determination of the mycotoxins deoxynivalenol (DON) and ochratoxin A in wheat are highly needed to screen for their occurrence in batches delivered to elevators or mills. Near Infrared (NIR) spectroscopy has been used routinely during many years at elevators and mills for determination of protein and moisture in wheat and other cereals. The technique is rapid taking just a few minutes and has the potential for determination of other compounds. The concentrations of mycotoxins normally found are, however, considered low for this technique and the reason for the limited number of publications concerning mycotoxins and NIR. Near infrared reflectance measurements have been assessed for the detection of scab and estimation of DON and ergosterol in single kernels of highly infected wheat (Dowell et al., 1998).

Pettersson and Aberg (2002) investigated near infrared transmittance instrumentation for determination of the mycotoxin deoxynivalenol (DON) in wheat kernel samples inoculated with *F. culmorum* and *F. graminearum* and determined the detection limit of DON by constructing partial least square regression model. Performances of prediction for DON contamination of wheat makes FT-NIR analysis a useful tool to discriminate between high and low DON-contaminated samples. FT-NIR analysis is suitable for the determination of DON in a specific wheat species (Girolamo et al., 2009). Siuda et al. (2006) studied the usefulness of the changes in different spectral ranges within the near ultraviolet, visible and near infrared ranges (UV-VIS-NIR) region which also can be used for determination of the content of scab-damaged wheat. They found that the content of protein and lipids increased with an increase of the scab-damaged constituent, whereas the content of moisture and starch decreased.

In the past electronic noses have been developed for the classification and recognition of a large variety of foods and can also be useful in detecting the infestation level, degree of fungal growth and presence of odorous compounds, which ultimately affects the quality of wheat. Water plays a crucial importance in the configuration of a chemical profile. Since the fungal species, which are able to grow on a specific substrate, are strongly influenced by water content, the fungal species grown on the seed samples with the same water activity have a significant overlapping. Stetter (1992) studied the classification of 12 samples of wheat into the five odor categories viz. normal, insect, musty, foreign, or sour, by four inspectors. Unanimous agreement was obtained for only four of the samples. However, when all off-odors (insect, musty, and foreign) were lumped together into one category, unanimous agreement as to whether the samples were normal or off-odorous was obtained for eight samples. Perkowski et al. (2008) analyzed four groups of wheat kernels in terms of their volatile metabolite contents using GC/MS and the electronic nose inoculated with *Fusarium culmorum* and simultaneous non-inoculated samples as controls. All sample groups contained significantly varied levels of trichodiene (TRICH), a precursor for the formation of *fusarium* metabolites, with approximately two times higher concentration recorded in triticale; however in inoculated samples TRICH concentration for wheat was found to be on average six times higher and for triticale eight times higher than in non-inoculated samples. Wheat of five storage ages and with 15 degrees of insect damage were evaluated and classified by the static-headspace sampling method using an electronic nose and found that the E-nose could discriminate successfully among wheat of different age and with different degrees of insect damage (Zhang and Wang, 2007).

Visual reference images and interpretive line prints are used as the inspection system for insect infestation and grain grading (USDA, Federal Grain Inspection Service, 1997). In this system individual grain kernels are compared with the photographic print of the representative sample. Computerized image analysis has been shown to have great potential for detecting and identifying various non-grain particles and insects in wheat. Zayas and Flinn (1998) detected *R. dominica* adults in wheat bulks with higher than 90% accuracy using structural and colour information. Karunakaran et al. (2003) correctly identified wheat kernels infested by *Sitophilus oryzae* (L) larvae and pupae-adults with more than 97% accuracy from the soft X-ray images. They also identified sound kernels with 99% accuracy and also indicated that in the future an automated line-scan X-ray system could inspect 1 kg grain in about 15 min compared to 5–6 h using a Berlese funnel. For detection of adult beetles, rodent droppings and ergot in harvested wheat, a machine vision method was developed by Ridgway et al. (2002) which was efficient. Detection rates of artificially prepared samples ranged from 87% for ergot to 100% for adult *Oryzaephilus surinamensis* (saw-toothed grain beetle). They also found that the rate of false positives arising from samples of wheat containing permitted admixture-only was low and the speed of the integrated algorithm package for detection of contaminant types approaches that required to scan 3 kg grain in just 3 min.

4.4 Safe storage of maize

Insect pests are one of the major organisms that are responsible for decline in quantity, quality and germination potential of maize seeds in storage. Adequate and effective storage of maize grains is therefore a major research thrust for enhanced maize productivity in order to reduce the huge economic loss. Boxall (1998) reported that common strategy in many African countries is to sell maize grains immediately after harvest, to avoid losses due to insect pests.

Although, in traditional storage system, losses are usually well contained at about 5% (Tyler and Boxall, 1984), National Stock Product Research Institute (1989) reported effective maize storage using fumigated maize, stored in gourds for a period of ten months; non-fumigated maize stored in similar containers were badly damaged in three months. The more damage suffered by maize grains the higher was the % weight loss. Thapa and Dhakal (1997) estimated 11-25% grain loss due to storage pests using this system. This probably suggests the need for cheaper, more effective and affordable storage methods that may give higher value to stored maize grains. Drying and storing corn on-farm can help producers and farm managers control elevator discounts and improve economic returns to their operation. The use of such facilities requires operators to maintain grain quality from thc ficld to thc point of salc to capturc markct premiums. Treatment of grain soon after harvest often determines the storability of a crop and can strongly influence its quality when delivered to the inducer which may be several weeks, months, or even years after harvest. Thus, it behooves grain farmers to implement sound grain harvest, drying, and storage practices to maintain the reputation of the United States as a reliable supplier of good quality corn to the global market. Successful post-harvest grain processing with on farm facilities requires a thorough understanding of the factors that influence grain quality. The best way to protect dry stored corn from spoilage by mold and insect activity is to apply integrated pest management practices, which are based on an understanding of the ecology of grain pests. The application of a broad range of preventive practices has a cumulative effect on pest control. Examples include cleaning grain bins and the area surrounding them prior to harvest, controlling grain moisture throughout drying, cleaning dried corn prior to storage to remove broken kernels and trash, controlling temperature throughout storage, managing the depth of grain in the bin to permit uniform airflow, and monitoring grain during storage for temperature, moisture, and mold and insect populations. By applying all these practices, a post-harvest integrated pest management (IPM) strategy can be substituted for some or all of the chemicals that have traditionally been used to control pests in stored grain. Corn with a high level of trash and fine material that has been not dried uniformly can develop problems during storage quickly even though the average moisture readings throughout the bin may be 15 %. Thus, it is wise to check the top layer of corn in all storage bins about a week after drying and cooling to be sure that no moisture buildup has occurred. If elevated temperatures or moisture conditions are left unchecked, mold and insect growth can flourish even in cool weather because their activity produces heat, which accelerates grain deterioration further. Controlling the moisture content and temperature of corn throughout the storage period is the most cost effective way to prevent spoilage problems and potential dockage from musty odors, insects, low test weight, and poor condition. These are based on the equilibrium moisture content properties of corn and the fact that mold and insect activity is held in check when grain temperatures are below 13°C and the relative humidity in the air space between corn kernels is below 65% the recommended moisture level but not cooled thoroughly. Uneven grain temperatures can lead to moisture migration (which usually occurs in the top center of the bin), which can promote mold growth and insect activity.

4.5 Safe storage of barley

Barley is one of the most important cereals in the world, stands fourth among the cereals with an estimated global production above 132 million tons over 50 million hectares in 2011 (USDA, 2011). The statistic shows the quantity of barley produced worldwide in crop years 2008 & 2009 to 2014 & 2015. In crop year 2008 to 2009, barley production amounted to approximately 155.3 million metric tons (FAO, 2015). The industrial use of barley grain has experienced continuous growth mainly due to its economic importance for malt production. During storage, the common insects found in barley grain are granary weevil (*Sitophilus granarius*), lesser grain borer (Rhyzopertha dominica), Indian meal moth (*Plodia interpunctella*), *Tribolium*, sawtoothed grain beetle (*Oryzaephilus surinamensis*). Filamentous fungi are a major safety concern in the malting and brewing industry. From a consumer safety viewpoint, they can produce diverse mycotoxins with carcinogenic and mutagenic potential transferable from grains to malt and other processed foods, such as beer (Lancova et al., 2008). Storage moulds can be avoided by good storage practices, although field fungi, such as *Fusarium* species, can still cause problems during grain processing. To ensure better quality barley grain, there is rapid quality control and disinfection techniques prior to storage and during storage are required.

4.5.1 Rapid disinfection of barley grain

Fusarium head blight (FHB) is common disease of cereals all over the world and currently the most important barley disease in Croatia. *Fusarium* species reduces yield and quality and may also contaminate the grain with mycotoxins. Cosic et al., (2012) used a physico chemical method for disinfection of barley grain and studied the influence on fungal population of barley grains. A pilot-scale industrial microwave system operating at 2.45 MHz was used by Vadivambal et al. (2008) to determine the mortality of life stages (egg, larva, pupa, and adult) of *Tribolium castaneum* (Herbst). Barley samples of 50 g each at 14, 16, and 18% moisture content (wet basis) were infested with various life stages of *T. castaneum* and exposed to microwave energy at different power levels and exposure times, and the mortality of the insects was determined. They found that complete mortality of all life stages of *T. castaneum* can be achieved at a power level of 400 W and an exposure time of 56 s or at 500 W for 28 s. Among the life stages of *T. castaneum*, eggs were the most susceptible to microwave energy and adults were the least susceptible. Allen et al. (2007) studied effect of ozonation on fungal infection and germination of barley, Results showed that ozone was very effective in inactivation of fungi associated with the barley regardless of whether the fungi were in the forms of spores or mycelia, however, the mycelia were less resistant to ozone. With 5 minutes of ozonation, 96% of inactivation were achieved for spores as well as for mixtures of spores and small amount of mycelia by applying 0.16 and 0.10 mg of ozone/g barley) min, respectively.

4.5.2 Rapid quality control system for barley grain

Near infrared spectroscopy (NIRS) is an excellent candidate for a rapid and low cost method for the detection of aflatoxins in barley. Fernandez et al. (2007) studied the utility of near infra red spectroscopy (NIRS) for rapid detection of mycotoxigenic fungi as AFB_1. A total of 152 samples were analyzed for aflatoxin content and the results of spectroscopic models developed have demonstrated that NIRS technology is an excellent alternative for fast AFB_1 detection in barley. Fernandez Pierna et al. (2012) detected diverse impurities from different origins such as straw, broken grains, grains for mother crops, weed seeds, insects, plastic, stones, pieces of wood and paintings, animal faeces in barley in a case study. Images were acquired using the whiskbroom imaging system by creating discriminate models for the different categories of impurities based on the information obtained from the principal component analysis (PCA).

Balasubramanian et al. (2007) used conducting polymer-based electronic nose system to analyze the headspace from barley samples infected with *Fusarium* and stored at moisture contents of 13, 18, 20 and 25 g of water/100 g sample. Classification models using linear (LDA) and quadratic discriminate analyses (QDA) were developed. The performance of the developed models was validated using leave-1-out cross validation. The models classified the barley samples stored into two groups based on the ergosterol content, i.e., "acceptable" (ergosterol content < 3.0 mg/g) and "unacceptable" (ergosterol content >3.0 mg/g). The study proved there is potential in using an electronic nose system for indicating mold spoilage in stored grains, and necessitates future studies in this direction. The possibility of using an electronic nose or gas chromatography combined with mass spectrometry (GC–MS) to quantify ergosterol and colony forming units (CFU) of naturally contaminated barley samples was investigated by Olsson et al. (2000). The main volatile compounds of grain with normal odour were 2-hexenal, benzaldehyde and nonanal with off-odours. GC–MS and electronic nose data could be used to predict the ergosterol levels with high accuracy with the help of Regression models (partial least-squares, PLS) built by them. They further detected and quantified ochratoxin A, and deoxynivalenol (DON) using both an electronic nose and gas chromatography combined with mass spectrometry (2001). Samples with normal odour had no detectable ochratoxin A and average DON contents of 16 mg kg^{-1}, while samples with off-odour had average OA contents of 76 mg kg^{-1} and DON contents of 69 mg kg^{-1} according to the investigation.

4.6 Safe storage of green gram

Safe storage study of green gram seeds was aimed at maximum disinfestation of insect pests (*Callosobruchus* species) of stored green gram seed without affecting the nutritional profile of seed. The two control strategies *viz.*, microwave and fungal treatment (entomopathogenic fungi, *Beauveria bassiana*), were successfully applied for insect disinfestations without affecting the quality of green gram seed (Pande et al., 2013). The microwave treatment parameters were optimized using central composite rotatable design (CCRD) and response surface methodology (RSM) for maximum insect pest mortality. The effect of independent parameters, viz., microwave power level and time of exposure on the moisture content, insect mortality, color, and phytic acid was optimized using the RSM, with the optimized value for power of 808 W and time 79 S (Pande et al, 2012). The optimally treated green gram seed has 8.9 % moisture, 99.5 % insect mortality, 2.22 Δa^* (green color of seed), and 591.79 mg/100 g of phytic acid (Pande et al, 2012). *In vitro* protein digestibility (IVPD) of the control (untreated) sample was 83 ± 0.289 %, and that of the microwave and fungal treated sample using optimum conditions was 85 ± 0.296 % and 86 ± 0.320 %, respectively (Pande and Mishra, 2012). These values are significantly higher ($p<.05$) than those of control. The mineral elements studied were Zn, Fe, Mg, Mn, Cu, K, Ca and Na. The microwave treatment resulted in a non-significant ($p<.05$) effect for Mg, Mn, Cu, K, and Na but a significant ($p<.05$) effect for

Zn, Ca, and Fe (Pande et al., 2012). On the other hand, fungal treated seed did not had much affect in the mineral content. The GAB model fitted well with both the adsorption and desorption data of the untreated seed. For microwave treated seed Modified Chung-Pfost was found to be the best model for adsorption and GAB model was the best fit for desorption (Pande and Mishra, 2014). The shelf life of seeds normally for 1 months without any treatments but with the help of microwave and biological treatment at 90% RH, 30 °C it is increased approximately 4 month in metalized polyester package. A rapid and nondestructive FT-NIR spectroscopic method was developed for the determination of phytic acid in green gram seed (Pande and Mishra, 2015).

5. Summary

Although microwaves have potential for disinfestation applications in the grain industry, they have not been used widely due to their assumed adverse effects on various quality parameters. The major problems associated with microwave heating are the non-uniform temperature distribution and thereby incomplete killing. The non-uniform heating pattern of microwaves are one of the important factors for the quality degradation of products during microwave treatment. A hot spot can be defined as a local area of very high temperature that results from the temperature dependence of material properties. Although hot spots are desirable to increase insect mortality, detrimental effects on the medium are not. Researchers have also tried the microwave heating along with gamma irradiation to control insects in stored cereal sand cereals products. El-Naggar and Mikhaiel (2011) evaluated the biochemical analyses on the samples of wheat grain and flour subjected to combined microwave and gamma irradiation treatment where high mortality was obtained. There were no detectable changes in the quality of protein, fat, fiber, carbohydrates, or ash. Researchers have reported that microwave treatment is an attractive quarantine treatment. The major advantage of microwave heating is that it interacts directly with food grains and significantly reduces the amount of time required for food grain to reach the lethal temperature for insects as compared with conventional heating methods.

For NIR, FT-NIR spectrophotometric measurement, chemometric techniques have to be used to model data to extract relevant information. A large enough number of samples are required to encompass all variations in physical and/or chemical properties. For construction of a model, it requires prior knowledge of the value for the target parameter, which must be previously, determined using a reference method. Despite significant advances in recent years, no specific methodology for this has yet gained widespread acceptance.

A major obstacle to the development of commercially useful machine vision systems for insect detection in grain is the limited rate of sample throughput. Simple, fast and reliable algorithms are needed to allow a statistically significant portion of a grain sample to be inspected in the short time available. Furthermore, device complexity, particularly in terms of camera type, computer specification and sample delivery system, must be minimized to make this approach cost effective. Another disadvantage is that it can detect only external insects in the grain bulks whereas NIR spectroscopy and X-ray methods can detect internal insects.

Electronic nose technology is certainly the most promising among new analytical methodologies for objective odour evaluation, but it is not a complete substitute for reference methods that use sensory panels. The main problem is the lack of a system that fits the current needs of many food analysts, but it is expected that improvements will result in further applications in the future. Although MOS sensors respond to a broad range of volatiles, they have higher affinity for aldehydes, alcohols and ketones and they are less responsive to molecules like terpenes, aromatic compounds or organic acids. To solve the lack of selectivity, some approaches have been undertaken. One option is to increase the sensitivity and selectivity of the sensor by exploitation of the operating temperatures for the sensor.

References

Allen, B., Wu, J., Doan, H. 2007. Inactivation of Fungi Associated with Barley Grain by Gaseous Ozone. Journal of Environmental Science and Health, Part B: Pesticides, *Food Contaminants, and Agricultural Wastes*, 38 (5): 617-630.

Annis, P.C., Banks, H.J. 1993. Is hermetic storage of grains feasible in modern agricultural systems? In "Pest control and sustainable agriculture" In: SA Corey, DJ Dall, WM Milne. CSIRO, Australia. 479-482.

Annis, P.C. and Morton, R. 1997.The acute mortality effects of carbon dioxide on various life stages of Sitophilus oryzae. *Journal of Stored Products Research*, 33: 115-124.

Balasubramanian, S., Panigrahi, S., Kottapalli, B., Wolf-Hall, C.E. 2007. Evaluation of an artificial olfactory system for grain quality discrimination. *LWT Food Science and Technology*, 40: 1815–1825.

Banks, J., Fields, P. 1995. Physical methods for insect control in stored-grain ecosystems. In: Jayas DS, White NDG, Muir W E. Stored-Grain Ecosystems. Marcel Dekker, Inc., New York: p 353–410.

Banks, J., Fields, J. B. 1955. Physical methods for insect control in stored grain ecosystems. Stored-Grain Ecosystems, D. S.Jayas, N.D. G.White, W. E. Muir, Eds., pp. 353–410, Marcel Dekker, New York, NY, USA.

Banks, Sharpe, A.K. 1979. Trial use of CO_2 to control insects in exported, containerized wheat. *CSTRO Aust. Div. Entomol. Rep.*, No. 15: 111.

Basavaraj, Raviteja, G., Deshpande, S. 2015. Physical Properties of Rice for Puffing. *Journal of Latest Trends in Engineering and Technology*, 376-380 (5).

Beckett, S.J., Morton, R. 2003. The mortality of three species of Psocoptera, Liposcelis bostrychophila Bandonnel, Liposcelis decolor Pearman and Liposcelis paeta Pearman, at moderately elevated temperatures. *Journal of Stored Products Research,* 39:103-115.

Bedi, S.S., Singh, M. 1992. Microwaves for control of stored grain insects. *National Academy of Science Letter*, 15 (6): 195–197.

Birck, N.M., Lorini, I., Scussel, V.M. 2006. Fungus and mycotoxins in wheat grain at post harvest. In: Proceedings of the 9th International Working Conference on Stored Product Protection. Campinas, SP, Brazil: ABRAPOS, 198-205.

Blanco, M., Villarro, I. 2002.NIR spectroscopy: a rapid-response analytical tool. *Trends in analytical chemistry*, 21(4): 240-250.

Boulanger, R.J., Boerner, W.M., Hamid, M.A.K. 1971. Microwave and dielectric heating system. Milling. *Journal of Stored Products Research,* 153:18-21 and 24-28.

Boxall, R.A. 1998. Grains post-harvest loss assessment in Ethiopia. Final report NRI Report No 2377. Chatham, U K, Natural Resources Institute 44 p.

Cosic, J., Vrandecic, K., Alexa, E., Postic, J., Svitlica, B. 2012. Influence of disinfection methods on fungal population of barley grains. Conference precedings: Meðunarodni znanstveno-struccni skup, 14. Ruzzickini dani, Danas znanost-sutra industrija, Vukovar, Croatia, 13-15.

Caleb, O.J., Opara, U.L., Witthuhn, C.R. 2012. Modified Atmosphere Packaging of Pomegranate Fruit and Arils: A Review. *Food and Bioprocess Technology,* 5(1): 15-30.

Caswell, G.H. 1973. The storage of Cowpea. *Samaru Agric News ltt* 15:20-73.

Chakraverty, A., Singh, R.P. 2014. Food grain storage. In: Chakraverty A, Singh RP. Postharvest Technology and Food process Engineering. Boca Raton, FL: CRC Press Taylor & Francis Group. p 131-157.

Chen, C.H. 1999. Information processing for remote sensing. Hackensack, NJ: World Scientific.

Cogburn, R.R., Tilton, E.W., Brower, J.H. 1972. Bulk-grain gamma irradiation for control of insects infesting wheat. *Journal of Economic Entomology,* 65:818-821.

Decareau, R.V. 1985, Microwaves in the Food Processing Industry, Academic Press, Natick, Mass, USA.

Dendy, A., Elkington, H.D. 1920. Roy. Soc. Repts. Grain Pest War Committee No.7.

Dermott, T., Evans, D.E. 1977. An evaluation of fluidized-bed heating as a means of disinfesting wheat. *Journal of Stored Products Research,* 14:1-12.

Dowell, F.E., Ram, M.S., Zhang, N., Herrman, T.J, Seitz LM. 1998. Detection of scab in single wheat kernels using NIR spectroscopy. ASAE Annual International Meeting, Orlando, Florida, USA, 12–16 July, 1998, ASAE Paper no 983062, 1–11.

Du, C.J., Sun, D.W. 2006. Learning techniques used in computer vision for food quality evaluation: A review. *Journal of Food Engineering*, 72(1):39–55.

El-Naggar, S.M., Mikhaiel, A.A. 2011. Disinfestation of stored wheat grain and flour using gamma rays and microwave heating. *Journal of Stored Products Research,* 47: 191-196.

Enu, R., Enu, P. 2014. Sterilization of grains using ionizing radiation: the case in Ghana. *European Scientific Journal,* 10 (6):117–136.

FDA, United States Food and Drug Administration.1982. GRAS status of ozone. *Federal Register* 47: 50209-50210.

Federal Grain Inspection Service. 1997. Grain grading procedures. In: Grain inspection handbook II. Washington, DC: USDA-GIPSA-FGIS.

Fernandez, Pierna, J.A., Vermeulen, P, Amand O, Tossens A, Dardenne P, Baeten V. 2012. NIR hyperspectral imaging spectroscopy and chemometrics for the detection of undesirable substances in food and feed. *Chemometrics and Intelligent Laboratory Systems*, 117:233–239.

Fernandez-Ibanez V, Soldado A, Martinez-Fernandez A, Roza-Delgado B. 2007. Application of near infrared spectroscopy for rapid detection of aflatoxin B1 in maize and barley as analytical quality assessment. *Food Chemistry*, 113: 629–634.

Food and Agriculture Organization. 2015. FAO cereal supply and demand brief. Accessed from http://www.fao.org/worldfoodsituation/csdb/en/ on 11th March, 2015.

Frost, S.W., Dills, I.E., Nicholas, J.E. 1944. The effects of infrared radiation on certain insects, *Journal of Economic Entomology,* 37 (2):287.

Gardner JW, Bartlett PN. 1993. A Brief History of Electronic Noses. In: Sensor Actuators B 18 (1-3): 211-220.

General Mills India Private Limited, Malegaon Industrial Area, Malegaon, Nashik, Maharashtra.

Girolamo AD, Lippolis V, Nordkvist E, Visconti A. 2009. Rapid and non-invasive analysis of deoxynivalenol in durum and common wheat by Fourier-Transform Near Infrared (FT-NIR) spectroscopy. *Food Additives and Contaminants*, 26 (6): 907–917.

Giusti, A. M., Bignetti, E., & Cannella, C. (2008). Exploring new frontiers in total food quality definition and assessment: from chemical to neurochemical properties. *Food and Bioprocess Technology* 1:130-142.

Goettel MS, Hajek AE. 2000. Evaluation of non target effects of pathogens used for management of arthropods. In: Wajnberg E, Scott JK, Quimby PC. Evaluating indirect ecological effects of biological control. Wallingford: CABI Publishing. pp 81– 97.

Graham, D.M.1997.Use of ozone for food processing. *Food Technology* 51: 72–75.

Gunasekaran N, Rajendran S. 2005.Toxicity of carbon dioxide to drugstore beetle Stegobium paniceum and cigarette beetle Lasioderma serricorne. *Journal of Stored Products Research*, vol. 41(3):283-294.

Hajek AE, Leger RJ. 1994. Interactions between fungal pathogens and insect hosts. *Annual Review of Entomology,* 39:293-322.

Hall DW. 1980. Handling and storage of food grains in tropical and subtropical areas. FAO, Rome, Italy, p350.

Hall DW. 1980. Storage methods. In: Hall DW. Handling and storage of food grains in tropical and subtropical areas. Rome: Food and Agriculture organization of the United Nations. p 156-198.

Hamid MAK, Boulanger RJ. 1969. A new method for the control of moisture and insect infestations of grain by microwave power. *Journal of Microwave Power*. 4(1): 11–18.

Hashem MY, Ahmad SS, EI-Mohandes MA, Gharib MA. 2012. Susceptibility of different life stages of saw-toothed grain beetle Oryzaephilus surinamensis (L.) (Coleoptera: Silvanidae) to modified atmospheres enriched with carbon dioxide. *Journal of Stored Products Research,* 48: 46-51.

http://dir.indiamart.com/dewas/storage-tank.html. Accessed 2015 March 19.

http://www.bestapples.com/facts/facts_controlled.asp. Assessed 2015 March 19.

http://www.fowlerwestrup.com/galvanized-silo-storage-system.php. Accessed 2015 March 19.

Hunter-Fujita FR, Entwistle PF, Evans HF, Crook NE. 1998. Insect viruses and pest management, West Sussex: John Wiley and Sons.

Ignatowicz S. 2004. Irradiation as an alternative to methyl bromide fumigation of agricultural commodities infested with quarantine stored product pests. Irradiation as a phyto-sanitary treatment of food and agricultural commodities. IAEA-TEC-DOC-1427, 51-66.

Isikber AA, Oztekin S. 2009. Comparison of susceptibility of two stored-product insects, Ephestia kuehniella Zeller and Tribolium confusum du Val to gaseous ozone.*Journal of Stored Products Research,* 45: 159-164.

Jayas DS, Jeyamkondan S. 2002. Modified atmosphere storage of grains meats fruits and vegetables. *Biosys Engg* 82(3): 235-251.

Jayas DS, White NDG. 2003. Storage and drying of grain in Canada: Low cost approaches. *Food Con* 14:225–261.

Karunakaran C, Jayas DS, White NDG. 2003. Soft X-ray inspection of wheat kernels infested by Sitophilus oryzae *Transactions of the ASAE*, 46 (3): 739–745.

Kellsand, Mason LJ, Maier DE, Woloshuk CP. 2001. Efficacy and fumigation characteristics of ozone in stored maize. *Journal of Stored Products Research* Oct, 37(4):371-382.

Khamis, M., Subramanyam, B., Dogan, H., Flinn, P.W., Gwirtz, J.A. 2010. Effects of flameless catalytic infrared radiation on Sitophilus oryzae (L.) life stages. *Journal of Stored Products Research,* 47: 173-178.

Kim, J.G., Yousef, A.E., Dave, S. 1999. Application of ozone for enhancing the microbiological safety and quality of foods: a review. *Journal of Food Protection,* 62: 1071–1087.

Kirkpatric, R.L., Tilton, E.W. 1972. A comparison of microwave and infra red radiation to control adult stored product coleopteran. *Journal of Georgia Entomological Society,* 7:73-75

Kirkpatrick, R.L., Brower, J.H., Tilton, E.W. 1973. Gamma, infra-red and microwave radiation combinations for control of rhyzopertha dominica in wheat. *Journal of Stored Products Research,* 9: 19-23.

Lacey, L.A., Frutos, R., Kaya, H.K, Vail P. 2001. Insect pathogens as biological control agents: Do they have a future? *Biological Control,* 21:230-248.

Lakshmana, K. 2000. Documentation and analysis of indigenous farm practices on major crops. An Inventory. Unpub. M.Sc. (Ag) Thesis, TNAU, Coimbatore, Tamil Nadu.

Lancova, K., Hajslova, J., Poustka, J., Krplova, A., Zachariasova, M., Dostálek, P., Sachambula L. 2008. Transfer of Fusarium mycotoxins and 'masked' deoxynivalenol (deoxynivalenol-3-glucoside) from field barley through malt to beer. Food Additives and Contaminants: Part A, Chemistry, Analysis, Control, Exposure and Risk Assessment, 25:732–744.

Legeron, J.P. 1984. Ozone disinfection of drinking water. In: Rice, R.G., Netzer, A. (Eds.), Handbook of Ozone Technology and Applications, Vol. II. Butterworth, Boston, pp. 99–121.

Locatelli, D.P., Traversa, S. 1989. Microwaves in the control of rice infestants. *Italian Journal of Food Science,* 1(2): 62.

Mahroof, R., Subramanyam, B., Throne, J.E., Menon, A. 2003. Time-mortality relationships for Tribolium castaneum (Coleoptera: Tenebrionidae) life stages exposed to elevated temperatures. *Journal of Economic Entomology,* 96(4):1345-1351.

Mills, J.T., Woods, S.M. 1994. Factors affecting storage life of farm-stored field peas (Pisum sativum L.) and white beans (Phaseolus vulgaris L.). *Journal of Stored Products Research,* 30(3):215-226.

Mills, J.T. 1996.Quality of stored cereals. In Cereal Grain Quality, ed. R.J.Hendry and P.S. Kettlewell, London,UK: Chapman and Hall, 441-478.

Mohan, R.J., Sangeetha, A., Narasimha, H.V., Tiwari, B.K. 2011. Post-harvest Technology of Pulses. In: Tiwari BK, Gowen A, Mckenna B. Pulse Foods Processing, Quality and Nutraceutical Applications. Food Science and Technol International Series. p 171–192.

Mokobia, C.E., Anomohanran, O. 2005. The effect of gamma irradiation on the germination and growth of certain Nigerian agricultural crops. *Journal of Radiological Protection,* 25: 181–188.

Monro, 1967. Effect of phosphine fumigation on greenhouse plants. London, Ontario, Research Institute, Canada Department of Agriculture. (Unpublished report)

Mujumdar, A.S., Beke, J. 2003. Grain Drying: Basic principles. In: Chakraverty A, Mujumdar AS, Ramaswamy HS. Handbook of Postharvest technology: Cereals, Fruits, Vegetables, Tea, and Spices. CRC Press. p 119-212.

National Stored Product Research Institute (NISPRI) 25th Annual Reports, 1989.

Nelson, S.O. 1972. Possibilities for controlling stored grain insect with RF energy. *Journal of Microwave Power,* 7:231-239.

Olsson, J., Borjesson, T., Lundstedt, T., Schnurer, J. 2000. Volatiles for mycological quality grading of barley grains: determinations using gas chromatography-mass spectrometry and electronic nose. *International Journal of Food Microbiology,* 59 (3) 167–178.

Opadokun, J.S .1996. The use of ionizing radiation in reducing post-harvest losses in food crops: experience of the Nigerian Stored Products Research Institute and potentials for future. *Journal of Irradiation National Dev.,* 11:57–62

Pande, R., Misha, H.N. 2015. Fourier transform near-infrared spectroscopy for rapid and simple determination of phytic acid content in green gram seeds (*Vigna radiata*). *Food Chemistry,* 172:880-884.

Pande, R., Mishra, H.N., Singh, M.N. 2012. Microwave drying for safe storage and improved nutrition quality of green gram seed (Vigna radiata). *Journal of Agricultural and Food Chemistry,* 60:3809-3816.

Pande, R., Mishra, H.N. 2012. Studies on safe storage of green gram (Vigna radiata) seeds. Ph.D Thesis. Agricultural and Food Engineering Department. Indian Institute of Technology, Karagpur, West Bengal, India.

Pande, R, Mishra, H.N. 2014. Influence of microwave heating on moisture sorption isotherm of green gram (Vigna radiata) seeds. *Journal of Food Process Preservation,*doi:10.1111/jfpp.12251.

Pande, R., Mishra, S., Mishra, H.N., Singh, M.N. 2013. Role of Entomopathogenic Fungi (Beauveria Bassiana) in Stored Food Grain Safety. In: Tiwari SP, Sharma R, Singh R. Recent advances in microbiology. Volume 1. USA: Nova Science Publishers. p 575-588.

Paolesse, R., Alimelli, A., Martinelli, E., Natale, CD, D'Amico, A., D'Egidio, M., Aureli, G., Ricelli, A., Fanelli, C. 2006. Detection of fungal contamination of cereal grain samples by an electronic nose. *Sensors and Actuators*, 119: 425–430.

Pell, J.K., Eilenberg, J., Hajek, A.E., Steinkraus, D.C. 2001. Biology, ecology and pest management potential of Entompathorales, In: Butt TM, Jackson C, Magan N. Fungi as biocontrol agents: progress, problems and potential. Wallingford: CABI International. p 71-153.

Perkowski, J., Busko, M., Chmielewski, J., Goral, T., Tyrakowska, B. 2008. Content of trichodiene and analysis of fungal volatiles (electronic nose) in wheat and triticale grain naturally infected and inoculated with *Fusarium culmorum. International Journal of Food Microbiology*, 126: 127–134.

Pettersson, H., Aberg, L. 2002. Near infrared spectroscopy for determination of mycotoxins in cereals. *Food Control*, 14: 229–232.

Prakash, A., Pasalu, I.C., Mathur, K.C. 1985. Varietal resistance of stored paddy grains to *Stotroga celealella. Oliv Oryza,* 20: 133-137.

Rice, R.G., Farquhar, J.W., Bollyky, L.J., 1982. Review of the applications of ozone for increasing storage times of perishable foods. *Ozone Science and Engineering,* 4: 147–163

Ridgway, C., Davies, E.R., Chambers, J., Mason, D.R., Bateman, M. 2002. Rapid machine vision method for the detection of insects and other particulate bio-contaminants of bulk grain in transit. *Biosystems Engineering*, 83 (1): 21-30.

Sabbour, M.M., Shadia, E.A.E.A. 2007. Efficiency of some bio insectides against broad bean beetle Bruchus rufimanus (Coleoptera: Bruchdae). *Res J Agric Biological Sci* 3(2):67–72.

Shah PA, Pell JK. 2003. Mini Review Entomopathogenic fungi as biological control agents. *App Microbiol Biotech,* 61:413"423.

Shayesteh, N., Barthakur, N.N. 1996. Mortality and behavior of two stored-product insect species during microwave irradiation. *Journal of Stored Products Research*, 2(3): 239–246.

Shah, P. A., & Pell, J. K. 2003. Entomopathogenic fungi as biological control agents. *Applied Microbiology and Biotechnology*, 61(5-6), 413-423.

Sinha, R. N., Yaciuk, G., and Muir, W. E. 1973.Climate in relation to deterioration of stored grain-A multivariate study, *Oecologia*, 12:69–88.

Siuda, R., Balcerowska, G., Sadowski, C. 2006. Comparison of the usability of different spectral ranges within the near ultraviolet, visible and near infrared ranges (UV-VIS-NIR) region for the determination of the content of scab-damaged component in blended samples of ground wheat. *Food Additives & Contaminants*, 23(11): 1201–1207.

Sousa AH, Faroni LRDA, Guedes RNC, Totola MR, Urruchi MI. 2008. Ozone as a management alter-native against phosphine-resistant insect-pests of stored products. *Journal of Stored Products Research*, 44: 379-385.

Stetter, J.R. 1992. Chemical sensor arrays: Practical insights and examples in: Sensors and Sensory Systems for an Electronic Nose. J. W. Gardner and N. Bartlett, eds. Kluwer Academic Publishers; Dordrecht, 273-301.

Strait, C.A. 1998. Efficacy of ozone to control insects and fungi in stored grain. M.Sc. Thesis, Purdue University, West Lafayette, INFDA, United States Food and Drug Administration, 1982.GRAS status of ozone. *Federal Register,* 47:50209–50210.

Subramanya, S., Ranganna, B., Ramakumar, M.V., Babu, C.K. 1999. A report on the use of microwave radiation in the disinfestations of spices. PHT Scheme, UAS, Banglore. Report submitted to the project Co-ordinator, CIPHET, PAU, Ludhiana.

Suchita, M.G., Reddy, G.P.U., Murthy, M.M.K. 1989. Relative efficacy of pyrethroids against rice weevil (*Sitophilus oryzae* L.) infesting stored wheat. *Indian Journal of Plant Protection*, 17:243–246.

Sutar, P.P., Prasad, S. 2008. Microwave drying technology-recent developments and R&D needs in India. in proceedings of the 42nd ISAEAnnual Convention, pp. 1–3,Kolkata, India, February.

Tilton, E.W., Schroeder, H.W. 1963. Some effects of infrared irradiation on the mortality of immature insects in kernels of rough rice. *Journal of Economic Entomology,* 56: 727–730.

Tilton, E.W., Vardell, H.H. 1982. Combination of microwaves and partial vacuum for control of four stored-product insects in stored grain. *Georgia Entomological Society*. 17:96–106.

Thapa, R.B., Dhakal, D.D. 1997. Loss of maize grains due to rats and insect pests in storage. *Nepalese Journal of Agriculture,* 18: 73 76.

Tyler, P., Boxall, R.A. 1984. Post–harvest loss reduction programmes: A decade of activities: What consequences? *Tropical Stored products Information,* 50: 4-13.

USDA 2011.http://data.worldbank.org/indicator/AG.YLD.CREL.KG. Accessed on 21 june, 2016.

Vadivambal, R., Jayas, D.S., White NDG. 2007.Wheat disinfestation using microwave energy. *Journal of Stored Products Research*, 43(4):508-514.

Vadivambal, R., Jayas, D.S., White, N.D.G. 2008. Determination of mortality of different life stages of Tribolium castaneum (Coleoptera: Tenebrionidae) in stored barley using microwaves. *Journal of Economic Entomology,* 101(3):1011-1021.

Wallace, H.A.H., Sholberg, P.L., Sinha, R.N., Muir, W.E.1983. Biological, physical and chemical changes in stored wheat. *Mycopathologia*, 82:65-72.

Waters, F.L. 1976. Microwave radiation for control of Tribolium confusum in wheat and flour. *Journal of Stored Products Research,* 12:19-25.

Webber, H. H., Wangner, R. P., Pearson, A. G. 1946. High frequency electric fields as lethal agents for insects. *Journal of Economics Entomology*, 39:481-498.

White, N.D.G. 1995. Insects, mites, and insecticides in stored-grain ecosystem. In: Jayas DS, White NDG, Muir WE. Stored-grain ecosystems. Marcel Dekker., New York, USA: p 123-127.

White, N.D.G 1995. Insects, mites and insecticides in stored grain ecosytems. In Stored-Grain Ecosystems, ed. D.S. Jayas, N.D.G. White and W.E. Muir, 123-167. New York, NY: Marcel Decker Inc.

Wilson, A.D., Banks, H.J., Annis, P.C., Guiffre, V. 1980. Pilot commercial treatment of bulk wheat with CO2 for insect control: the need for recirculation. *Australian Journal of Experimental Agriculture,* 20:618-624.

Wilson, C.L. 2013. Establishment of a world food preservation center: concept, need, and opportunity. *Journal of Post-Harvest Technology*, 1(1): 1-7.

Yeomans, A.H. 1952. Radiant energy and insects. Insects: The Yearbook of Agriculture , USDA 91952-411.

Zayas, I.Y., Flinn, P.W. 1998. Detection of insects in bulk wheat samples with machine vision. *Transactions of the ASAE,* 41:883–888.

Zhang, H.M., Wang, J. 2007. Detection of age and insect damage incurred by wheat with an electronic nose. *Journal of Stored Products Research*, 43:489–495.

Zia-Ur-Rehman.2006. Storage effects on nutritional quality of commonly consumed cereals. *Food Chemistry* ,95:53-57.

7

Natural Antioxidants and Colors

R Upadhyay, S Halder, M Bhattacharya, H N Mishra

1. Introduction

Antioxidants are considered an important class of nutraceuticals which are associated with health improving properties. Antioxidants are compounds which when present in foods at concentrations lower than an oxidizable substrate, markedly delay or prevent the oxidation of substrate (Halliwell, 1999). They are increasingly applied in food systems to retard the oxidation process, particularly in lipid containing foods (Droge, 2002; Lee et al., 2004). The commercially available solutions to counter food oxidation are predominated by synthetic antioxidants such as butylated hydroxyanisole (BHA), butylated hydroxytoluene (BHT) and tertiary butyl hydroquinone (TBHQ), among others, which are increasingly rejected by consumers and legitimized by regulators due to long term health implications (Park et al., 2001). As a consequence, the development and utilization of natural antioxidants as alternatives to synthetic ones have attracted global interest among researchers.

The use of spices and herbs in foods was known since ancient times to modify or to improve the flavor. However, the antioxidant properties of only some of the spices have been recognized. Different spices and their extracts such as turmeric, ginger, fennel, rosemary, and sage, among others, are reported to possess very high antioxidant properties, owing to the content of biologically active compounds, which enables their food application (Madhavi et al., 1996). Exploitation of natural antioxidants in consumer goods is increasingly accepted and will gain momentum due to the rise of green consumerism. In this chapter, an attempt has been made to provide an insight into natural antioxidants and colors. Beginning with the source distribution of natural antioxidants, followed by review of extraction methods for their isolation from plant based matrices and bio-preservative properties in vegetable oils to delay rancidity. Lastly, a thorough collection of literature pertaining natural colors has been presented to potentiate the application of clean label ingredients in food systems. The application of natural antioxidants and colors in food products can vouch the food safety with value addition in terms of their nutraceutical properties.

2. Natural Antioxidants

Antioxidants are classified as exogenous (natural or synthetic) or endogenous compounds, both responsible for removal of free radicals, scavenging reactive oxygen species (ROS) or their precursors, inhibiting formation of ROS and binding metal ions needed for catalysis of ROS generation (Gilgun-Sherki et al., 2001). Natural antioxidant system is sorted in two major groups, enzymatic and non-enzymatic. Enzymatic antioxidants such as catalase, glutathione peroxidase as well as superoxide dismutase (SOD), whereas non-enzymatic antioxidants include direct acting antioxidants against oxidative stress (OS) such as ascorbic and lipoic acid, polyphenols and carotenoids and indirect acting antioxidants include chelating agents and bind to redox metals to prevent free radical generation. Antioxidants can be classified according to the mechanism of action to reduce lipid oxidation into two groups. First is the primary antioxidant that acts as hydrogen donors to the lipid free radical formed during the lipid oxidation and rearrange to a stable conformation. The primary antioxidants (AH) react with lipid peroxyl radicals (ROO•) and convert them to more stable, antioxidant radicals (A•). The primary antioxidant is able to scavenge lipid radicals (Eq. 1).

$$ROO\bullet + AH \rightarrow ROOH + A\bullet \quad (1)$$

Antioxidant radicals are stable due to delocalization of the unpaired electron around a phenol ring and cannot easily react with fatty acids. They are able to terminate radical chain process by reacting with radicals. These free radical interceptors react with peroxy radicals (ROO•) to stop chain propagation; thus, they inhibit the formation of peroxides (Eq. 2).

$$ROO\bullet + A\bullet \rightarrow ROOA \quad (2)$$

The most effective antioxidants interrupt the free radical chain reaction and usually contain aromatic rings capable of donating H• to the free radical formed during lipid oxidation. The formed antioxidant radical is stabilized by delocalization of the unpaired electron around the phenol ring to form a stable resonance hybrid. Second is the secondary antioxidant, in opposite to the primary antioxidants, do not break free radical chain or do not convert free radicals into stable molecules. Secondary antioxidant act through various mechanisms to slow the rate of oxidation reactions, as reducers and chelators of metal ions, provide H to primary antioxidants, decompose hydroperoxide to nonradical species, deactivate singlet oxygen, absorb ultraviolet radiation, or act as oxygen scavengers (Reische et al., 2002).

2.1 Classification of food antioxidants

An antioxidants which can inhibit or retard oxidation either by scavenging free radicals, as primary antioxidant, or by a mechanism that does not involve direct scavenging of free radicals, a secondary antioxidant (Figure 1). Examples of primary antioxidants include phenolic compounds and vitamin E. Secondary antioxidants acts by different mechanisms, including binding metal ions, scavenging O_2, converting lipid hydroperoxides to non-radical species, absorbing UV radiation or deactivating singlet O_2. Secondary antioxidants show antioxidant activity in the presence of a second minor component. This can be seen in the case of sequestering agents such as citric acid, which are effective only in the presence of metal ions, and reducing agents such as ascorbic acid, which are effective in the presence of tocopherols or other primary antioxidants.

In the recent past, considerable interest in defining the benefits of plant secondary metabolites as food preservatives and antioxidants has occurred. The major breakthroughs in this area were achieved in within past 20 years, when more structured analytical techniques were introduced in areas such as separation, isolation and characterization of bioactive components derived from plant sources. The reason behind using plant derived compounds for application in the food industry is the increasing concern over the safety of synthetic antioxidants, such as BHA, BHT and TBHQ (Wong et al., 1995; Cuvelier et al., 1994). Besides the reasons stated above, BHT and BHA are also quite volatile and easily decompose at high temperature, and consequently do not provide satisfactory effectiveness for some common food products such as French

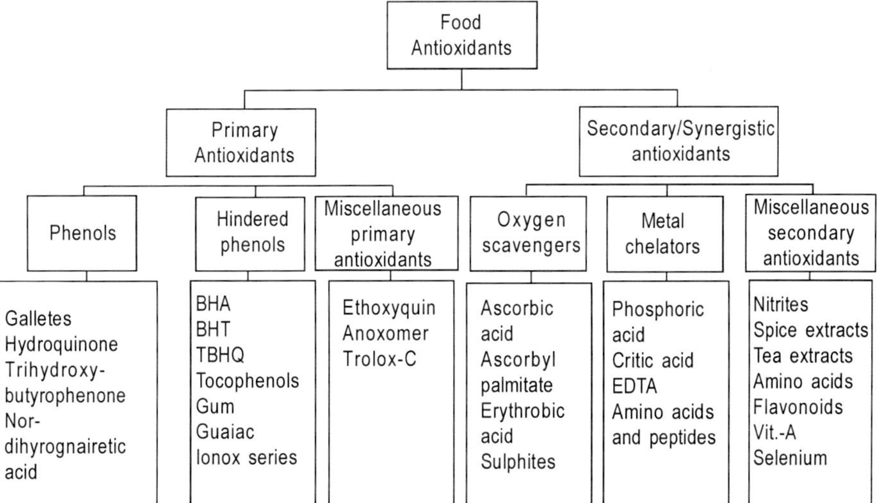

Fig. 1: Classification of antioxidants used in food

fries and potato chips that involve storage (Irwandi et al., 2000). Also, these antioxidants are not effective in vegetable oils and in preventing the development of off-flavors, though addition of TBHQ results in lower peroxide value, it also some time results in the development of objectionable flavors (Chang et al., 1977).

2.2 Sources of natural antioxidants

There has been a great interest in obtaining and utilizing the antioxidants from plant sources. The search for safe and effective natural antioxidants is now focused on extracts from natural sources such as edible plant, spices and herbs owing to their rich content of bioactive polyphenols (Madhavi et al., 1996). Today, the most important natural antioxidants used in food application includes ascorbic acid, citric acid and their salts and esters, tocopherols (α-isomer) and spice and herbal extracts (Weinreich, 1998). Relaying the interest in natural antioxidants, the greatest level of research attention has been focused on antioxidant potential of natural extracts from spices and herbs along with the identification of their chemical composition (Senorans et al., 2000). There are reports in the literature which indicate the antioxidant behavior of herbal extracts exceeding the synthetic antioxidants (Zhang et al., 2010). Researchers have evaluated the effectiveness of different natural antioxidants on the stability of variety of food systems (Hamied et al., 2009; Hras et al., 2000; Richheimer et al., 1996). There is a paradigm shift in the food industries in context with antioxidant selection for processed foods and herbal extracts are now widely accepted.

Oxidation is one of the principal causes of quality deterioration in fats and oils. It is mediated by free radical initiated chain reaction called autooxidation. Antioxidants are widely practiced in food products to prevent or retard the oxidation of fats and oils (Hopia et al., 1996). Antioxidants are class of compounds which when present at concentrations lower than the oxidizing substrate significantly reduces the oxidation of substrate. Synthetic antioxidants (BHA, BHT, and TBHQ) are one such class of food additives which are added judiciously to food products to increase their shelf life. However, the use of synthetic antioxidants is regulated and critically restricted in terms of application and usage level due to human health concerns (Kahl & Kappus, 1993; Chen et al., 1992). This greatly influences the acceptance criteria resulting in increased rejection of food containing synthetic antioxidants. Recently, natural antioxidants surged in food additives world not only for their usefulness as a preservation method but also because of beneficial effect on human health. The hunt for identification of natural sources to screen their effective application as food antioxidant is focused objective of today's food research. It has been presumed that market of natural antioxidants will grow and predominate in future due to consumer's preference.

2.2.1 Rosemary as a potential antioxidant

Of the natural antioxidant extracts, the studies have been reported on the Labiatae (*Lamiaceae*) family, which includes rosemary (*Romarinus officinalis* L.) along with sage (*Salvia officinalis* L.) (Upadhyay & Mishra, 2014). Rosemary is highly prized for its natural healing and flavoring properties (Aruoma et al., 1996; Altinier et al., 2007). More effort recently has been given to elucidate the antioxidant mechanisms of rosemary; possible reasons rely heavily on its strong antioxidant properties (Jaswir et al., 2000). Rosemary has always been known a versatile, aromatic herb and addition to being used as a food flavoring agent (Oluwatuyi et al., 2004). The studies on the antioxidant properties of rosemary cover a range of whole plant tissues to standardized isolates of single compounds. The dried leaves of rosemary were found to delay the development of warmed over flavor of cooked minced meatball (Huisman et al., 1994). The development of commercialized rosemary extract could be attributed to the work of Chang et al. (1977), who tried to use different solvents with various polarities to extract the antioxidant compounds from rosemary and evaluated the antioxidant activity in a lard based model system at 60 °C. The first identified antioxidant compound from rosemary was carnosol (Houlihan et al., 1984). The active compounds in rosemary with antioxidant properties include carnosol, carnosic acid, and rosmarinic acid (Upadhyay & Mishra, 2014) (Figure 2). Carnosol and rosmarinic acid are highly effective than tocopherols, BHT and BHA when assessed with active oxygen method (Nakatani & Inatani, 1981). Similar results on the purpose of comparing the antioxidant activity of different compounds are also described elsewhere (Thomsen et al., 1999; Richheimer et al., 1996; Chen et al., 1992).

The use of rosemary extract to replace synthetic antioxidants was initially proposed to overcome the disadvantage of BHT and BHA in thermal oxidation of edible oil (Chang et al., 1977). The rosemary used in these experiments was basically oleoresins, made by solvent extraction of herbs and spices followed by solvent removal. Oleoresin was originally considered to be a concentrated mixture of natural colors, aromas and flavors in vegetable oils, which has been commercially available for more than 40 years (Guzinski, 1996). The oleoresins from rosemary are USFDA and EU approved and listed GRAS for food application as natural flavoring/antioxidant. The production procedure for manufacturing the oleoresin consists of grinding, extraction, desolventization and solvent standardization. The oleoresin rosemary has been used as a form of natural antioxidant extracts in which carnosic acid and carnosol are present in high concentration. Generally, the commercial rosemary oleoresins are not flavorless, and thus the implication with use in lipid system may be the potential taste of oleoresin, which should be considered before deciding to use oleoresins as preservatives.

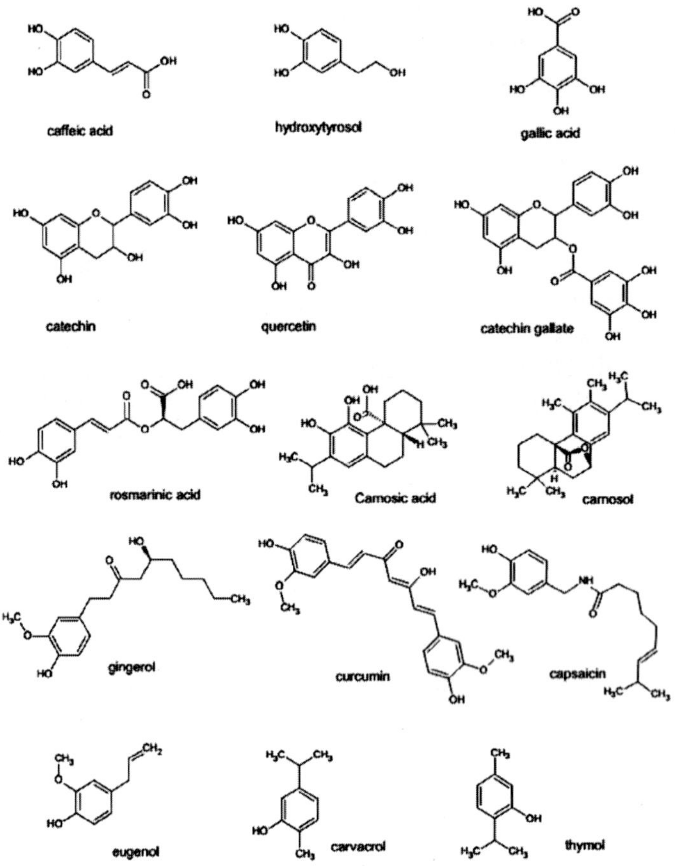

Fig. 2: Some representative bioactive antioxidant compounds from natural sources

Rosemary is widely used in its ground form or as an extract in many food applications. Rosemary is used in food processing for the preparation of poultry, lamb, shellfish, sausages, and salads as well as soups (Nissen et al., 2000). It is also used as seasoning in potato chips and french fries (CheMan & Tan, 1999). Besides used as flavoring and seasoning, the dried leaves of rosemary possess high antioxidant, anti-inflammatory, antitumorigenic and neuroprotective effects (Satoh et al., 2008; Chang et al., 2008; Danilenko et al., 2003). A number of phenolic compounds that vary in structure, polarity and mutual interactions have been identified in rosemary extracts (Terpinc et al., 2009; Thorsen & Hildebrandt, 2003; Basaga et al., 1997). The published data attributed the main antioxidant effect of rosemary extracts to phenolic diterpenes, flavonoids and phenolic acids (Erkan et al., 2008; Frankel, 1996; Economou et al., 1991). Many investigators have studied the free radical scavenging activity to empathize the antioxidative properties of rosemary extracts. Over the years, several reports have been appeared on the preparation of rosemary extracts for retarding lipid oxidation in sunflower oil (Upadhyay & Mishra, 2016; Upadhyay & Mishra,

2015a; Upadhyay & Mishra, 2015b; Upadhyay & Mishra, 2015c; Upadhyay & Mishra, 2015d; Upadhyay & Mishra, 2015e; Zhang et al., 2010; Hras et al., 2000). Many antioxidants in rosemary are non-volatile and exhibit preservative effect against oxidative changes in food product. There are plentiful extraction processes that have been developed to bring rosemary extracts and other potential sources of natural antioxidants in the market. In the next section, the conventional and non-conventional extraction of bioactive compounds from plant matrices has been thoroughly discussed.

3. Extraction of Natural Antioxidants

Extraction aids in identification and isolation of antioxidants/bioactive compounds and various extraction techniques have been formulated to facilitate optimal extraction of these compounds. Plant materials act as a source for the extraction of the bioactive compounds and numerous techniques have been formulated to facilitate optimal extraction of such compounds (Spigno et al., 2007). Each bioactive compound demands specific extraction technique depending on its chemical affinity, polarity and molecular structure based on number of hydroxyl groups and aliphatic or aromatic rings (Li et al., 2005). Extraction efficiency depends upon various factors such as pH, temperature, time, pressure, particle size and the nature and concentration of the solvent (Juntachote et al., 2006). Therefore, it is appropriate to choose optimal extraction conditions according to the chemical structure and properties of target compounds. The optimal extraction of the bioactive compounds depends on the substrate specific selection of the extraction technique ensuring maximum antioxidant concentration in the extract. Among the commonly used methods solvent, pressurized liquid, sub-critical fluid, ultrasound-assisted, microwave-assisted and supercritical fluid extraction are recognized as the most efficient techniques for extraction of antioxidants.

3.1 Solvent extraction

Solvent extraction is basically employed to obtain the desired analyte from different sources such as microorganisms, micro-algae and plant tissues. Maceration, heat reflux and mechanical agitation are well known traditional extraction methods of antioxidants from plant sources, where the mass transfer phenomenon is enhanced by agitation of the raw material which in turn facilitates improved solubility and hence optimum extraction of bioactive compounds. The nature of the solvent used for extraction along with the extraction time and temperature plays a dominant role in deciding the antioxidant yields and capacity. The solvents such as hexane, ethanol, benzene, chloroform are employed in the extraction selectively adhere to the compound of interest. To achieve optimum recovery of phenolic antioxidants basically polar solvents like ethanol and methanol are preferred over acetone and ethyl acetate. Combination of solvents like ethanol and methanol have been used to extract bioactive compounds from rosemary, sage, sumac, rice bran, wheat grain and bran, mango seed kernel, citrus peel, and many other fruit peels. Bonoli et al. (2004) found that a mixture of ethanol and acetone yielded maximum phenolic compounds from barley flour. Several studies involving the usage of polar organic solvents for the extraction of phenolic compounds suggest that extraction conducted between temperature 25–65°C, extraction time of 5–45 min and solvent to feed ratio of 5:1 to 20:1 yielded the optimum results (Upadhyay et al., 2015; Anwar et al., 2006). However, the solvent has to be efficiently removed prior to its application in food. As compared to other methods solvent extraction involves low processing cost and excels in ease of application. The major disadvantage being the consumption of large amount of solvent as the extraction involves longer duration. The conventional extraction method involves loss of bioactive compounds due to long extraction time and higher extraction temperatures (Chen et al., 2012). The extract obtained after centrifugation and filtration is generally used as a food additive or as an ingredient in functional foods (Starmans & Nijhuis, 1996).

3.2 Pressurized liquid extraction

Pressurized liquid extraction employs organic solvents and the process takes place at high temperature (50–200 °C) and pressure (1450–2175 psi) to facilitate higher extraction yield (Dunford et al., 2010). Extraction temperature controls the dielectric constant of the solvent which in turn is responsible for governing the polarity of the solvent. The temperature could be used to enhance the extraction yield, as the polarity of the solvent lowers with rise in temperature as it results in decreasing the dielectric constant of the solvent (Abboud & Notario, 1999). Thus, optimum yield could be ensured by regulating extraction temperature so that the polarity of solvent coincides with the analyte (Miron et al., 2010). Food related applications involve usage of GRAS solvents such as water and ethanol. In addition, the high pressure forces the solvent into solid matrix facilitating faster extraction. The yield obtained in pressurized liquid extraction is higher as compared to the conventional extraction. However, higher extraction temperature may lead to deterioration of thermolabile bioactive compounds (Ajila et al., 2011).

3.3 Sub-critical fluid extraction

Sub-critical water extraction employs extraction pressure and temperature below supercritical conditions. It is generally preferred over conventional extraction as the consumption of solvents can be greatly reduced. The other advantages over conventional methods are shorter extraction duration and higher extraction yield (Plaza et al., 2010a). Moreover, this extraction technique is environment-friendly as it employs water for extraction. Sub-critical water extraction is carried out under high temperature (100–374°C) and pressure (145–870 psi). As the extraction temperature is increased, the dielectric constant of the solvent increases leading to lowering its polarity. Thus, increasing the solubility index of the non-polar bioactive compounds, so this extraction can be preferred for the extraction of non-polar bioactive compounds. The sub-critical water extraction has been applied to vegetable and other related matrices. Highest total phenolic content and antioxidant capacities was observed at 160°C extraction temperature for canola seed meal using sub-critical water extraction (Hassas-Roudsari et al., 2009). Similarly, catechins and proanthocyanidins were extracted from wine-related products by employing sequential extraction at three different temperatures of 50, 100, and 150°C (Garcia-Marino et al., 2006). Bioactive compounds have been also extracted from other vegetable matrices such as citrus pomaces (Kim et al., 2009), oregano (Rodriguez-Meizoso et al., 2006) and rosemary (Plaza et al., 2010b), as well as some microalgae (Herrero et al., 2006). Nevertheless, the sub-critical water extraction leads to the formation of new bioactive compounds, as the disintegration of cells in vegetable matrix under high temperature and pressure leads to the exposure of antioxidant compounds to elevated extraction conditions. This may lead to structural modification of antioxidants which may be attributed to Maillard reaction, thermo-oxidation or caramelization. This phenomenon was studied in various microalgae, algae, rosemary, thyme and verbena extracts and it was observed that the overall antioxidant capacity of the extract was affected depending on the nature of the sample (Plaza et al., 2010b). In addition, sub-critical water extraction is an environmental friendly technique and ensures shorter extraction duration.

3.4 Ultrasound assisted extraction

Ultrasound assisted extraction (UAE) works on the principle of acoustic cavitation. The high intensity ultrasound waves create micro-bubbles in the solvent during its compression and rarefaction cycle. Once the bubble formation takes place the bubble expands during rarefaction cycle and recompresses back during compression cycle. Further, after a series of compression and rarefaction cycles the bubble implodes. The implosion of the micro-bubbles propels a fine jet of solvent into the plant cells resulting into its disintegration and the release of desired bioactive compound. Thus, UAE ensures better recovery of bioactive compounds along in a shorter time interval with lower solvent consumption as compared to the conventional solvent extraction (Upadhyay et al., 2015). UAE preserves the molecular and structural integrity of the bioactive compounds as the extraction takes place at lower temperatures. For these reasons ultrasound is employed for the extraction of antioxidants from plant sources, such as extraction of phenolic compounds has been specially dealt by optimizing their yield and antioxidant capacity by means of experimental design. The optimized yield of flavonoids from *Prunella vulgaris* L. was obtained at 79 °C extraction temperature, 41% ethanol concentration and 30.5 min of ultrasonication time (Zhang et al., 2011a,b). The flavonoids associated with the lowering of coronary heart disease was extracted from the Hawthorn seeds using an ethanol concentration of 72%, at 65°C of extraction temperature for 37 min, in a 40 W ultrasonic water bath. It was observed that mostly the extraction was carried out within 40–80% of ethanol concentration, 20:1-40:1 ratio of solvent to raw material and 15-40 min of ultrasonication time using 40 W ultrasonic water-baths.

These studies have showed that the antioxidant capacity is affected independently and/or interactively by many factors such as extraction temperature, frequency, time, pH, solvent to solid ratio, particle size, and type of solvent and solvent concentration (Pinelo et al., 2005). The most common solvents used for the extraction of bioactive compounds include ethanol, methanol, mixture of ethanol: water and methanol: water, and acetone. As compared to other extraction methods it was found that ultrasound extracted phenolic compounds retained most of its bioactivity, in some cases even no degradation of phenolic compound was claimed (Dobias et al., 2010).

3.5 Microwave assisted extraction

Microwaves are electromagnetic radiation ranging from 0.001–1 m in wavelength. The principle that governs microwave-extraction is dipole-rotation, which is responsible for heating the moisture inside the plant cells. The moisture evaporates and builds pressure inside the plant cell. This may lead to rupture of the plant matrix enhancing the solubilization of bioactive compounds in solvent (Routray & Orsat, 2011). Microwave-assisted extraction has proved to be one of the most advanced extraction methods as it heats the matrix internally as well as externally. In addition, it also ensures efficient extraction using lesser amount of solvent under shorter duration. The extraction of antioxidants *viz.,* phenolic compounds (Upadhyay et al., 2012; Li et al., 2011), carotenoids (Zhao et al., 2006) and saponins (Zhang et al., 2011b) were successfully performed using microwave-assisted extraction giving higher extraction yields. Polar solvents with high dielectric constant such as water prove to be better solvents over non-polar solvents (Wang & Weller, 2006). The extraction of phenolic compounds were reportedly higher with ethanol: water mixtures than pure water and ethanol which could be attributed to higher dissipation factor of mixture resulting in faster generation of heat, which in turn results in building up pressure inside cells (Ajila et al., 2011). Finally, it leads to rupture of the cellular matrix resulting in better solubilization of bioactive compounds in the ethanol-water mixture. Thus, a mixture of should be preferred to obtain an optimum extraction yield (Simsek et al., 2012).

3.6 Super critical fluid extraction

Bioactive compounds are generally extracted using conventional techniques. However, conventional extractions, such as heating, boiling, or reflux results in loss of antioxidant properties due to ionization, oxidation and hydrolysis, as comparatively long extraction times are involved (Li et al., 2005). In order to preserve the structural integrity of target compounds, a milder extraction method was preferred, thus supercritical fluid extraction seemed to be a most preferable alternative to the conventional methods. Moreover, this extraction technique advocates high selectivity, shorter extraction periods and environmentally benign extraction (Wang & Weller, 2006). Several solvents can be used as a supercritical fluid such as carbon dioxide (CO_2), ammonia, water, propane, ethane, hexane, pentane and butane; however, the supercritical behavior, technical viability, toxicity, cost, and solvation power determines the suitability of the solvent for a particular application. Among them, CO_2 is the most suitable for supercritical fluid extraction (SFE) based on the fact that it is GRAS, nonflammable, noncorrosive and inexpensive (Rizhvi, 1994). The CO_2 when raised above its critical temperature (31.1 °C) and pressure (7.5 MPa) attains the supercritical state. The low critical temperature of CO_2 ensures minimal alterations in the chemical structure of the thermolabile bioactive compounds leading to the preservation of curative properties during extraction (Moyler & Heath, 1988). The extraction efficiency of supercritical fluids are remarkably higher than conventional extraction techniques as the CO_2 in supercritical state has a higher diffusion coefficient (0.07×10^6 $m^2.s^{-1}$), lower viscosity (0.01 - 0.03 mPa.s) than liquids and no surface tension ensuring almost complete penetration into the plant matrix. Extraction of non-polar organic compounds can be performed efficiently using supercritical CO_2 because of its non-polar nature. The extraction of polar bioactive compound requires addition of co-solvent along with the supercritical CO_2 as it alters the polarity of the supercritical fluid resulting in enhanced extraction yield. The fluid density and vapor pressure of supercritical fluid, along with molecular weight and polarity of solute plays a key role in the solubility of the bioactive compound in the solvent.

Clearly, the different extraction techniques have advantages and dis-advantages associated with them. Nevertheless, the ease of extraction, better extraction efficiency and economical solvent and energy consumption drives the attention of scientific community towards the exploitation of non-conventional extraction techniques. In the next section, the bio-preservative potential of bioactive extracts/ antioxidant rich concentrates in vegetable oils is discussed with particular inclination towards shelf life extension and frying application.

4. Bio Preservation of Vegetable Oils using Natural Antioxidants

This section elucidates the recent compilation of investigation pertaining bio-preservative properties and frying suitability of natural antioxidants in vegetable oils. Starting with thorough understanding of concept of lipid oxidation, its causes and mechanism, followed by the application of natural antioxidants to control the oxidation has been covered. Finally, the literature relevant to the application of natural antioxidants in stabilizing the oils during high temperature treatment is also discussed.

4.1 Lipid oxidation

Lipid oxidation is one of the major chemical changes that occur during processing, storage, shipment, and final preparation of food stuffs containing lipids. Lipids are categorized as triglycerides, phospholipids, and sterols, all of which can be subjected by oxidation while there are oxygen molecules in the environment. Unsaturated double bonds of lipid molecules make them prone to be oxidized. Food stuffs containing polyunsaturated fatty acids are particularly highly susceptible to lipid oxidation. The greater the degree of unsaturation of fatty acids, generally speaking, the more susceptible is the food. Lipid oxidation is propagated by the removal of hydrogen atoms and the subsequent addition of oxygen at α-positions to a fatty acid double bond, producing free radical species R and peroxyl radicals (ROO) (Figure 3). Lipid oxidation widely occurs in food systems and is mainly mediated by oxygen free radicals, or more widely by reactive oxygen species. A free radical is defined as any chemical species capable of independent existence and containing one or more unpaired electrons. A free radical species containing an oxygen atom is called as an oxygen free radical. A typical example of oxygen free radicals is $O^{\cdot 2-}$.

Reactive oxygen species (ROS) are collective name of non-free radical molecules which causes the lipid oxidation including hydrogen peroxide (H_2O_2), hypocholorous acid (HOCl), ozone (O_3), and singlet oxygen ($^1O_2$). The reactive oxygen species belonging to free radicals include superoxide negative ion radical ($O^{\cdot 2-}$), hydroxyl free radical (OH·, a very powerful oxidization agent), peroxyl (ROO·), and alkoxyl (RO·). The OH· is extremely reactive; whereas $O^{\cdot 2-}$ and H_2O_2 are much more selective type (Min & Lee, 1999; Halliwell et al., 1995). A ground or triplet oxygen molecule is a diatomic molecule possessing two unpaired valence electrons with the lowest energy status (an unpaired electron refers to being one alone in an atomic or molecular orbital). When a ground state O_2 absorbs sufficient energy via photosensitization, one of the unpaired ground state electrons shifts to higher energy state called singlet O_2. It is an unstable energy state of O_2 molecule which readily reacts with electron-rich double bonds of unsaturated fatty acids to release its excess energy. Oxidation reaction also occurs in ground state O_2; however, the singlet O_2 has greater oxidation reactivity. Singlet O_2 reacts with linoleic acid at least 1450 times faster than triplet oxygen (Min & Lee, 1999) and it is major agent causing rancidity of edible oils during oil storage, shipping, and processing.

It has widely been accepted that a chain reaction of free radicals causing lipid oxidation mediated by reactive oxygen species is conveniently illustrated in the following three phases:

Initiation $RH + \text{initiator} \rightarrow R\cdot$ (3)

$ROOH + \text{initiator} \rightarrow ROO\cdot$

Propagation $R\cdot + O_2 \rightarrow ROO\cdot$ (4)

$ROO\cdot + RH \rightarrow ROOH + R\cdot$

Termination $R\cdot + R\cdot \rightarrow R\text{-}R$ (5)

$ROO\cdot + R\cdot \rightarrow ROOR$

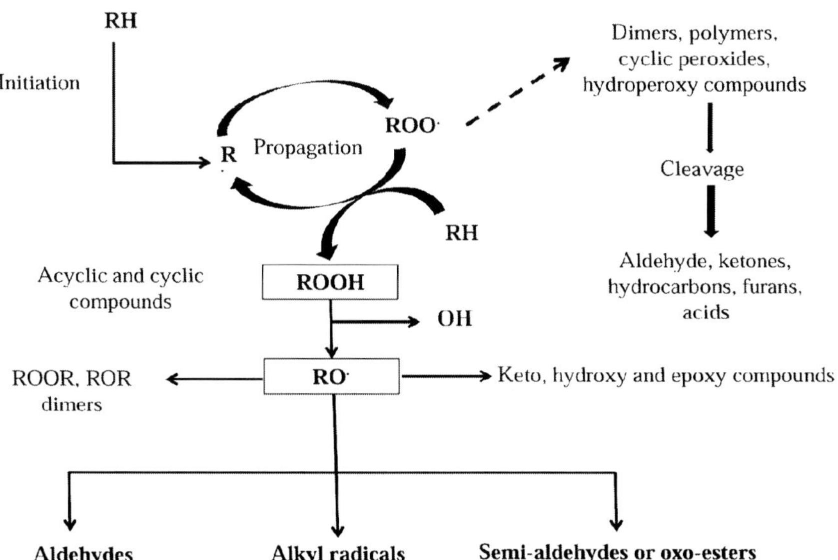

Fig. 3: Schematic diagram of lipid oxidation reaction

The above mechanism is initialized by the various physical or chemical factors (initiators). The factors include temperature, reactive oxygen species, and photosensitizers such as chlorophyll, transition metal ions, heating, or radiation. It should be pointed out that the controversy concerning the generation of the initial radicals remains. The initiators trigger free radical chain reactions by mediating oxidization of the substrate molecules, causing the substrate molecule to lose a hydrogen atom to form a free radical R•. The initiation often occurs at an allytic methylene group of an unsaturated fatty acid (RH) or a lipid-hydroperoxide (ROOH), shown in the initiation step at Eq. (3). The produced free radical R• then reacts with oxygen to form a peroxy radical (ROO•). This immediate product further reacts with another lipid molecule to generate a lipid hydroperoxide (ROOH) and another lipid free radical (R•), causing a cascade mode of chain reactions until the free radical is neutralized by other free radicals, the stage of which is shown in Eq. (4). In the termination phase, two radicals combine into non-free radical products and stop the cascade mode of chain reaction, shown in Eq. (5). In addition, the reaction chain is also terminated by some antioxidants or free radical scavengers. Transition metal ions, especially iron and copper, are powerful catalysts to initialize the process of the reaction. Additionally, lipoxygenase also acts as an initiator to induce the oxidation to generate the peroxides in food material containing lipid. It should be mentioned that hydroperoxide, a primary product of the oxidation, is very unstable and readily decomposes into secondary products including hundreds of compounds. The compounds belong to a few chemical groups such on aldehyde, ketone, alcohols, acids, or hydrocarbons. These compounds adversely affect the quality, appearance, and edibility of a food product by producing objectionable off-odors and/or off-flavors on foods and hence affect the nutritional values, wholesomeness, safety, color, flavor, and texture of foods (Min & Lee, 1999; Angelo et al., 1996).

4.1.1 Factors affecting lipid oxidation

Light and O_2 are very important factors leading to oxidative changes. Another mechanism of oxidation occurs in the presence of sensitizer and UV-light that is photosensitized oxidation. Photo-oxidation pathway is an alternative route leading to the formation of hydroperoxides instead of the free radical mechanism. The excitation of unsaturated fatty acid or oxygen may occur in the presence of light and a sensitizer. There are two types of photo-oxidation (Gordon, 1991). (I) – an electron or a hydrogen atom transfers between an excited triplet sensitizer and a substrate (PUFA), producing free radicals or radical ions; and (II) – triplet oxygen ($^3O_2$) can be excited by light to singlet oxygen ($^1O_2$), which reacts with the double bond of unsaturated fatty acids, producing an allylic hydroperoxide (Frankel, 1995). Photosensitized oxidation is a direct reaction of light-activated, singlet oxygen with unsaturated fatty acids, and subsequently hydroperoxides are formed. Photosensitized oxidation happens because of the presence of molecules that can absorb visible or near UV light to become electronically excited (sensitizers). Transition metal ions (Fe^{2+} and Fe^{3+}) can directly braking down unsaturated lipids into lipid oxidation products such as hydroperoxides (LOOH) into alkoxyl (LO•) and peroxyl (LOO•) reactive radicals by the following mechanisms (Eq. 6 and 7).

$$Fe^{2+} + LOOH \rightarrow Fe^{3+} + LO^{\cdot} + OH^{-} \quad (6)$$

$$Fe^{3+} + LOOH \rightarrow Fe^{2+} + LOO^{\cdot} + H^{+} \quad (7)$$

4.1.2 Measurement of lipid oxidation

There are different methods which are used for the evaluation of oxidative stability and shelf life of fats and oils. These methods are based on the measurement of specific compounds generated during the oxidative, hydrolytic and thermo-oxidative degradation of fats and oils. Broadly, the assessment of oxidative stability depends upon the measurement of primary and secondary oxidation products. An exhaustive list of different method for the measurement of primary and secondary oxidation products of lipid degradation are summarized in Table 1.

4.2 Thermo-oxidative stability of oils

Food manufacturers are becoming aware of the health benefits of vegetable oils. This reflects in selection of vegetable oils for frying purposes in conditions with extremely high cooking temperatures. A number of common snack foods currently contain vegetable oils *viz.,* New York fries, French fries, Mix healthy snacks, Kettle Chips, Sun Chips, Sunflower Chips, Ruffles, Walkers and Lay's potato chips etc. Because sunflower oil is primarily composed of healthier but less stable PUFA and MUFA, it can be particularly susceptible to damage by high heat. Considerable interest in the use of natural antioxidants for frying purposes has been expressed (Kitts, 1996). Rosemary has been studied intensively and proven effective for stabilizing the frying oils. A study by Man and Jaswir (2000) revealed a synergistic interaction among rosemary extract, sage extract, and citric acid that resulted in a significant stability of palmolein and improved sensory scores of the fried food during extended frying of potato at 180°C for 6 days. Therefore, it would be interesting to develop a synergistic blend of rosemary with secondary antioxidants which can provide better thermo-oxidative stability and frying performance to sunflower oil.

Whereas reports abound on the antioxidant activity of extracts and components from various spices and herbs in edible oils under ambient and storage conditions, investigations on their effectiveness during frying have received relatively less

Table 1: Summary of methods to analyze primary and secondary oxidation products

Method	Principle	Measurement
Primary oxidation products		
Iodometric titration	Reduction of ROOH with KI and measurement of I_2	Titration with $Na_2S_2O_3$
Ferric ion complexes	Reduction of ROOH with $Fe^{2+}$and formation of Fe^{3+}complexes	Absorption at 500–510 nm of the red complex with SCN
FTIR	Reduction of ROOH with TPP	Absorption at 542 cm^{-1} of TPPO
Chemiluminescence	Reaction with luminol in thepresence of heme catalyst	Chemiluminescence emission of oxidized luminol
GC-MS	Reduction of ROOH to ROH and quantitation of ROH derivatives	ROH derivatives
UV spectrometry	Estimation of conjugated dienes and trienes	Absorption at 234 and 268 nm
Secondary oxidation products		
TBA	Measurement of TBARS, mainly malonaldehyde	UV/VIS spectrometry (532 nm)
p-Anisidine	Measurement of aldehydes, mainly alkenals	UV/VIS spectrometry (350 nm)
Carbonyls	Measurement of total carbonyls or specific carbonyl compounds	UV/VIS spectrometry and HPLC for total or specific carbonyl compounds
Total volatile acids	Induction period/ oil stability index (C_4-C_6 acids)	Monitoring change inelectrical conductivity
Total volatile carbonyls and hydrocarbons	Non-volatiles polymers	Direct headspace rapidanalysis (hexanal, trans-2-hexenal, nona-2,6-dienal, pent-2-enal) Dimers and polymers linkedether, carbon, and/ or peroxybridges Headspace GC-MS, SPME-LC-MS

attentions. In general, extracts from different parts of spices, herbs, and fruits studied so far indicated their efficacy in inhibiting thermo-oxidative degradation and extending the fry life of vegetable oils. An exhaustive list of frying application of phenolic extracts from plant sources has been presented in Table 2. Commercially obtained oleoresin rosemary and sage extracts at 0.04% markedly improved the frying performance of refined, bleached, and deodorized palm olein, based on wet and FTIR spectrometry measurements of peroxide value, iodine value, and free fatty acid content of the oils after extended frying at 180°C for 5 consecutive days (Jaswir & Man, 1999). According to Yanishlieva et al. (1997), thermal degradation of sunflower oil was significantly inhibited in the presence of summer savory ethanolic extract; the time to reach 73% unchanged triacylglycerol (27% TPC) increased by up to 14 h in the presence of the extract during heating in an oven at 180°C for 8 h per day for 7 days. In a recent study, Patel et al. (2013) reported a significant improvement in stability of clarified butterfat (Indian ghee) fortified with 0.5% commercial steam distilled coriander extract and oleoresin in a model frying of wet cotton balls at 180°C. Apart from the more frequently studied common spices and herbs, a number of vegetal sources including fruits and berries also present viable sources for frying antioxidants. Freeze dried, expelled juice from pompozia fruit (*Syzygium cumini*) added at 0.12% offered significantly better protection than BHT against thermo-oxidative deterioration of sunflower oil during continuous frying of frozen French fries at 180°C for 12 h (Ali, 2010).

5. Natural Colors

Coloration of foods, either with natural or synthetic food additives indicates good quality, assist marketing, satisfy consumers and restores desired natural food colors lost by exposure to air, light, temperature, moisture, or improper storage conditions (Goyle & Gupta, 1998; Clydesdale, 1993, Francis, 1985). Inappropriate color is often associated with "not fit for edible purposes". Color enhances acceptability of food by making them more appealing and appetizing. Food colors also elicit cephalic phase of gastric secretion thereby promoting digestion and absorption of nutrients from food (Babu & Shenolikar, 1995). Awareness regarding to health issues is increasing day by day and interests of consumers are shifting towards natural and safe food products. Strict regulations have been imposed on the use of synthetic colorants by USFDA. Only eight synthetic pigments are accepted in India *viz.*, Ponceau 4R, Carmoisine, Erythrosine, Tartazine, Sunset Yellow FCF, Indigo Carmine, Brilliant Blue FCF and Fast Green FCF. The prolonged use of synthetic colorants causes various health affects like hyper acidity, thyroid, tumors, urticaria (hives) dermatitis, asthma, nasal congestion, allergies, abdominal pain, nausea, eczema, liver and kidney damage and cancer (Kapoor, 2006). This has caused vociferous public

Table 2: Frying performance of vegetable oils supplemented with plant extracts

Sources	Fat / oil substrate	Conditions	Reference
Olive leaf, hazelnut leaf, hazelnut green leafy cover	Canola oil	Frying of dough patties (180°C)	Aydeniz & Yilmaz (2012)
Andea mashua (*Tropaeolum tuberosum*)	Soybean oil	Frying of potatoes (175°C)	Bettalleulz-Pallardel et al. (2012)
Oregano	Soybean oil	Heating of oil at 180°C	Pereira & Jorge (2012)
Grape seed	Sunflower oil	Heating at 180°C	Poiana (2012)
Inca munã (*clinopodium bolivianum*) leaves	Soybean oil	Frying of potatoes (180°C)	Chirinos et al. (2011)
Rosemary	Sunflower oil	Heating of oil (180°C)	Filip et al. (2011)
Majorana syriaca	Corn oil	Potato frying (185°C)	Al-Bandak & Oreopoulou (2011)
Thyme flower	Corn oil	Potato frying (180°C)	Karoui et al. (2011)
Pompozia fruits	Sunflower oil	Frying of French fries (180°C)	Ali (2010)
Olive waste cake	Sunflower oil	Heating of oil at 180°C	Abd-ElGhany et al. (2010)
Mulberry leaves	Rice bran oil	Heating at 180°C	Roy et al. (2010)
Apple, blueberry, mangosteen, dragon fruits	Peanut oil	Potatoes immersed in fruit extractsand then fried at 170°C	Cheng et al. (2009)
Denatured carob fiber	Sunflower oil	Heating at 180°C	Botega et al. (2009)
Curcuma longa (turmeric) leaf	Palm oil	Frying of French fries at 180°C	Nor et al. (2009)
Murraya koenigii leaf	Palm oil	Frying of French fries at 180°C	Nor et al. (2011)
Rooibos tea	Soybean oil	Rancimat at 120, 140, 160 and 180°C	Fukasawa et al. (2009)
Olive leaf	Sunflower oil, olive oil	Pan frying at 175°C for 6 min	Chiou et al. (2009)
Pandanus amaryllifolius leaf	Palm oil	Frying of French fries at 180°C	Nor et al. (2008)
Bamboo, green tea leaves	Vegetable oil	Mixed with wheat flour prior to dough preparation, bread stickdeep fried at 180°C	Zhang & Zhang (2007)
Greek sage and summer savory	Olive oil, sunflower oil	Heating at 180°C	Kalantzakis & Blekas (2006)
Black tea, garlic bulb, onion skin	Corn oil	Heating at 140°C	Navas et al. (2006)
Black tea leaves	Corn oil	Frying of potato at 180°C	Naz et al. (2005)
Oregano	Cottonseed oil	Frying of potato at 185°C	Houhoula et al. (2004)

Contd.

Table 2: contd.

Sources	Fat / oil substrate	Conditions	Reference
Oleoresin rosemary, sage	Palm oil	Frying of potato at 180°C	Man & Jaswir (2000)
Citrus hystrix peel	Palm oil	Frying of fish crackers at 180°C	Jamilah et al. (1998)
Rosemary	Rapeseed oil	Frying of potato chips at 180°C	Gordon & Kourimska (1995)
Mung bean hull	Soybean oil	Heating at 180°C	Duh et al. (1997)
Cassia essential oil	Rapeseed oil, soybean oil	Frying of precooked beef at 130, 150,and 190°C	Du & Li (2008)
Corinander oleoresin, essential oil	Butter fat	Frying of wet cotton at 180°C	Patel et al. (2013)
Rosemary	Olive and sunflower oil	Frying of potatoes at 180°C	Gamel & Kiritsakis (1999)
Rosemary	Soybean oil	Heating of oil at 180°C	Ramalho & Jorge (2008)
Spinach	Soybean oil	Flour dough fortified with spinach powder and deep fried at 160°C	Termentzi et al. (2008)
Old tea leaves	Rapeseed oil	Frying of potatoes at 180°C	Zandi & Gordon (1999)
Green tea, green coffee	Lard	Frying of fortified donuts at 180°C	Budryn et al. (2013)

demand for the development of natural colors. This section is attempted to provide a holistic insight into the varied sources of natural colors distributed in nature, their stability, and finally the legal status related to their food application.

5.1 Sources of natural food colors

Natural colors are extracted from fruits, vegetables, seeds, roots and microorganisms and they are often termed as "bio-colors" due to their biological origin. Algae, bacteria and plants are three important sources of bio colors (Joshi et al., 2009).

5.1.1 Quinones

Quinones cover a wide range of colors, from pale yellow, to orange, red, purple, and brown and are widely distributed in higher plants but also present in fungi, lichens, and some invertebrate animals. Depending on their structures, quinones are classified into benzoquinones, naphthoquinones, anthaquinones, or extended quinones. Quinone pigments are derivatives of 1, 4-naphthoquinone and 9, 10-anthaquinone. In higher plants, quinones can be found in leaves, flowers, fruits, roots, bark and heartwood (Pintea, 2007). The best example of napththoquinone is henna, a natural dye obtained by crushing leaves of *Lawsonia alba*. Another important naphthoquinone is juglone, derived from buds, nut hulls, and roots of the walnut tree (*Juglans regia*). Plumbagin (5-hydroxy-2-methyl-1, 4-naphthoquinone) is a yellow pigment identified in walnut roots, leaves, and bark, in the roots of Plumbago zeylanica, and in Droseraceae, Ancestraladaceae, and other plants (Duroux et al., 1998) . The yellow pigment lapachol (another naphthoquinone) and its α-lapachone and β-lapachone derivatives are isolated from the stems and seeds of the South American Bignoneaceae tree.

The madder roots of *Rubia tinctorium* are used as red dye which contains anthaquinone alizarin, along with purpurin, rubiadin and lucidin-primeveroside. Another orange anthaquinone, emodin is found in Rhubarb (genus *Rheum*), aloe vera, *Rumex acetosa*, and in fungi and lichens. St. John's Wort (*Hypericum performatum* L.) plant produces hypericin, a naphthodiantrone (Britton 1983). The fungus *Mollisia fallens* produces mollisin - a yellow substituted 1, 4-naphthoquinone. The strain *Penicilium oxalicum* var. Armeniaca CCM 8242 produces an anthaquinone type pigment related to carmine and patented as Arpink Red (Dufosse, 2006). The most important anthaquinones for the food industry are the pigments carminic acid produced by female cochineal insects (*Dactylopius coccus Costa*) and carmesic acid. Another insect, *Laccifer lacca* Kerr produces lac (or lake) which contains laccaic acids (A, B, C, and D) that are closely related to carmesic acid, a yellow to red pigment produced by *Kermes ilices* (Thomson, 1962). Extended quinones like protoaphins are found in various aphids and are converted after the deaths of the insects to red erythroaphin (Britton, 1983).

5.1.2 Chlorophylls

Chlorophylls are photosynthetic pigments found in the chloroplasts of higher plants, algae, and some bacteria. The chlorophylls are derivatives of dihydroporphyrin chelated with a centrally located magnesium atom. Chlorophyll a and chlorophyll b are important as food colorants. These pigments are obtained from land plants and differ only by a $-CH_3$ and $-CHO$ group, respectively, on carbon 7 (Hendry, 1996). Chlorophylls a and b are found in superior plants, ferns, mosses, green algae, and the prokaryotic organism prochloron; the other types have been found in other groups such as algae and bacteria (Delgado-Vargas et al., 2000; Rüdiger et al., 1998). Chlorophyll colorant for food use are mainly obtained only from terrestrial plants like alfalfa (*Medicago sativa*), nettles (*Urtica dioica*), and several pasture grasses (Frick, 2003).

5.1.3 Monascus

Monascus pigments are widely used in Asian countries for centuries as food colorants and spices and in traditional medicine (Mudgett, 2000). These pigments are produced by cultivation of *Monascus* species on carbohydrate-rich substrates such as rice, wheat, corn, potatoes, and soybeans. Three species of *Monascus* identified are pilosus, purpureus, and ruber. Several strains were isolated and characterized from various food sources. The main source of *Monascus* pigment commonly called anka pigments is fermented red rice called koji or angkak. Certain region of Asia such as Japan, China, and Taiwan are the biggest producers of *Monascus* colorants. *Monascus* pigments belong to the azaphylone class of pigments and various secondary metabolites of typical polyketide structures are from it. Six main pigments are produced by this fungus: yellows (ankaflavin and monascin) (Manchand & Whalley, 1973; Inouye et al., 1962; Fielding et al., 1961), oranges (rubropunctatin and monascorubrin) and red-purples (rubropunctamine and monascorubramine) (Teng & Feldheim, 1998). The major source of red pigments in Asia is *Monascus purpureus*.

5.1.4 Iridoids

Iridoids are found in saffron (*Crocus sativus* L.) and Cape jasmine fruit (*Gardenia jasminoids* Ellis). Iridoids belong to monoterpenoids having methylcyclopentane skeletons. Iridoids are not considered as a colorant in the USA and EU countries. However, their range of colors goes from green to yellow, red and blue colors. Their stability has attracted the attention of food formulators in Japan where traditional applications include noodles, beverages and pickles. Iridoid compounds are also identified from different sources like aucubin from *Aucuba japonica* and *Plantago* sp., catalposide from *Catalpa*, jasminine from *Jasminum*, nepetalactone iridoid from catmint, *Nepeta cataria* etc. Iridoids on reaction with primary amines, amino acids and proteins produces different colored products (Jensen et al., 2002; Francis, 1999).

5.1.5 Carotenoids

Carotenoids are compounds of eight isoprenoid units whose order is inverted at the molecule center (Britton, 1996). The smallest linear carotenoid is lycopene which forms the starting material for all other carotenoids by reactions involving: (1) hydrogenation, (2) dehydrogenation, (3) cyclization, (4) oxygen insertion, (5) double bond migration, (6) methyl migration, (7) chain elongation and (8) chain shortening. Carotenoids are classified into two groups according to their chemical structure, *viz*., (1) carotenes and that are constituted by carbon and hydrogen; (2) xanthophylls that have carbon, hydrogen, and, additionally, oxygen. Carotenoids are also classified as primary or secondary by their functionality. Primary carotenoids group are those compounds required by plants in photosynthesis (β-carotene, violaxanthin, and neoxanthin), whereas secondary carotenoids are localized in fruits and flowers (α-carotene, β-cryptoxanthin, zeaxanthin, antheraxanthin, capsanthin, capsorubin) (Delgado-Vargas et al., 2000).

Carotenoids are widely distributed among photosynthetic and non-photosynthetic organisms such as in higher plants, algae, fungi, bacteria, and at least in one species of each form of animal life. Carotenoids are responsible for many of the brilliant red, orange, and yellow colors of fruits, vegetables, fungi, flowers, and also of birds, insects, crustaceans, and trout. Plants and microorganisms synthesize carotenoids de novo and animals get carotenoids from these two sources. Nature produces about 10^8 tons/year, most of which is found in four carotenoids, viz., fucoxanthin, in marine algae; and lutein, violaxanthin, and neoxanthin, in green leaves (Haugan et al., 1992).

More than 40 carotenoids have been identified in flowers and more than 70 in fruits. In addition, carotenoids have been identified in wood (oak, *Quercus* sp., chestnut, *Castanea sativa* and beech, *Fagus silvatica* L.) with $\beta_2$carotene and lutein as the principle components. Carrots and sweet potatoes also contain high amount of carotenoids (Masson et al., 1997). The Carotenoid present in algae is called fucoxanthin. In bacteria, different structural elements are attached to the carotenoid structure: sulfate groups and aromatic rings, shorter and longer than 40 carbon chains, among others. The most common fungal carotenoids are carotenes, mono- and bi-cyclic carotenoids but without ε-rings. Canthaxanthin is the most important fungal carotenoid (Goodwin, 1992). One of the major carotenoids in human diet is lycopene which is also the main pigment in tomatoes. It also occurs in water melon, guava, pink grape fruit (MacDougall and Francis, 2002). The spice paprika which belongs to Capsicmum annum species consists of several different carotenoid pigments that develop during ripening. The most important are capsaxanthin, capsorubin and beta-carotene, accounting for about 90% of total pigments. Annatto, the dried seed of *Bixa orellana* L., is mainly used as a food color. The orange red color is due to carotenoids which constitute more than 80% in the annatto seed coat (Satyanarayana et al., 2003).

5.1.6 Anthocyanins

Anthocyanins are combination of anthocyanidins with one or more sugar molecules (Harborne & Grayer, 1988). They are water soluble pigments and are widely distributed in the plant kingdom but are absent in some lower plants such as liverworts and algae. They occur as different glycoside combinations that produce red, blue or purple coloration in various fruits and vegetables. Most of the rich red, pink, blue and violet ornaments of the plant kingdom owe their color to one of the anthocyanins. Anthocyanins are consumed naturally in fruits and vegetables such as grapes, strawberries, raspberries, apples, radish, red cabbage, etc. Less familiar source of anthocyanins include purple corn, black carrot and passion fruit and host of other exotic fruits. Grape color extract is used in nonbeverage food; grape skin extract is used in still and carbonated drinks and ades, beverage bases and other alcoholic beverage. Other examples of anthocyanin use include cherries in yogurt and ice cream, fruit filling, candy and confections.

5.1.7 Betalains

Betalains are immoniun derivatives of betalamic acid. They are found in the seeds of amaranthus spp., in the roots of red beet (*B. vulgaris*), in the flowers of *Mirabilis jalapa* and *Bougainvillea* spp., and in the fruits of Prickly pear (Delgado-Vargas et al., 2000). Betalains only from *B. vulgaris* and prickly pear (*Opuntiaficus indica*) are approved to be used in food (Jackman & Smith, 1996). They are also present in the higher fungi Amanita, Hygrocybe, and Hygrosporus (Strack et al., 1993).

5.1.8 Caramel

The USFDA defines caramel as "the color additive caramel is the dark brown liquid or solid resulting from carefully controlled heat treatment of the following food grade carbohydrates: dextrose, invert sugar, lactose, malt syrup, molasses, starch hydrolysates and fractions thereof, sucrose". Caramels are been widely used as food colorant from ancient times. Four classes of caramels are recognized. viz., burnt sugar, caustic, ammonia, and ammonium sulfite. The first is used mainly as a flavoring additive, whereas the other three classes are regarded as coloring agents. Ammonia caramels are the most common coloring agents in foods and drinks (Francis, 1999)

5.1.9 Turmeric

Turmeric is important both as a spice and a coloring agent. The pigments responsible for the bright yellow-orange color turmeric are called curcuminoids. Curcuminoids are a group of polyphenolic bioactive components including curcumin, demethoxycurcumin, and bis-demethoxycurcumin as active molecules (Priyadarsini, 2014). Curcuminoids are extracted using ethanol, acetone, and ethyl acetate and purified by crystallization (Madsen et al., 2003). It is a yellowish crystalline, odorless powder, heat stable (mp 184-186 °C) and poorly soluble in water, petroleum ether, and hexane (Strimpakos & Sharma, 2008).

5.2 Regulatory perspective

In most countries, the use of food additives (including colorants) is governed by strict regulation. The legislation specifies which colorants may be used; the source(s) of the colorant, the purity of the colorant, to which foods the colorant may be added, and at what level the colorant may be added to a specific food. Major differences as to which colorants are allowed between the EU and US legislation lies when it comes to which sources are allowed and which foods may be colored. For instance, in the USA, only source of sodium copper chlorophyllin is alfalfa (*Medicago sativa*) and only be used in citrus-based dry beverage mixes, whereas in the EU, allowed sources are alfalfa, grass, nettle, and edible plant material, and a long list of foods may be colored. Local, traditional usage of coloring matter also influences Legislation. Thus, *lac*, *monascus*, *gardenia*, and *spirulina* are important colorants in some parts of Asia, but none of them are allowed in the EU or in the USA, where there is no traditional use of the raw materials. Table 3 shows the list of natural colorants considered in the EU and the USA (exempt colorants). According to the European 94/36/ EC directive, restrictions on utilization of colorants are claimed for different food categories based on their safety profile. In general, the toxicological studies are performed by animal testing although alternative in vitro methods are increasingly used. ADI values are determined by chronic toxicity studies, not acute toxicity measurements. ADI, expressed as milligrams of test substance per kilogram of body weight (ppm), with the recommendation not to eat more than the ADI per day. The FDA, EU, and WHO agree on the ADI principle (Socaciu C, 2007).

Table 3: List of natural food colorants and permitted leaves for use in food

Natural food colorants	EU Code
Curcumin, turmeric, turmeric oleoresin	E 100
Riboflavin (ii) riboflavin-5-phosphate	E 101
Carminic acid and carmine, cochineal extract	E 120
Chlorophyll grass	E 140
Chlorophyllins (copper complex of chlorophyll)	E 141
Caramel colors (I, II-sulfite, III-ammonia; IV-ammonia sulfite)	E 150a–d
Carotenes (mixtures including β-carotene), β-carotenE	E 160a
Bixin, norbixin annato extracts	E 160b
Capsanthin, capsorubin, paprika, paprika oleoresin	E 160c
Lycopene	E 160d
Canthaxanthin	E 161a
Lutein	E 161b
Betanin, dehydrated beets	E 162
Anthocyanins, grape color extract, grape skin extract	E 163

Food product	Colorant allowed	Maximum level
Malt bread	E 150a-d	NMM
Breakfast cereals	E 120, E 150c, E 160a-c, E 162, E 163	25-200
Chips (potato)	E 100	NMM
Butter	E 160a	NMM
Margarine, emulsions of fats in water	E 100E 160a,b,	10 for E 160b
Cheese	E 120, E 140, E 141, E 160a–c, E 163	1.5–50 for E 160b
Vinegars and aromatic wines	E 150a–d, E 163	NMM
Bitter beverages	E 100, E 101, E 120, , E 129	100 mg/l
Fruit and vegetable juices	E 160a, E 160d, E 160e	NMM
Beer and cider	E 150a–d	NMM
Vegetables conserved in vinegar, salt or oil	E 101, E 140, E 150a–d, E 141, E 160, E 163	NMM
Jams and marmalades	E 100, E 140, E 141, E 150a–d, E 160c,E 161b, E 162, E 163, E 120,	NMM
Sausages, meat pastes, fish products	E 100, E 101, E 120, E 150a–d, E 160a, E 160c, E 162	20-100
Hamburger	E 120, E 150a–d	100 only for E 120

NMM = No maximum level mentioned (*Source*: Delgado-Vargas *et al*., 2000; Francis, 1999)

5.3 Stability of natural colors

Synthetic colorants have achieved better results than natural or nature-identical colorants in the present market because of greater stability and higher ratios of coloring yield. The existing technologies used for the extraction, concentration, and purification of natural plant pigments to be used as food colorants still produce lower yields and the final products are still expensive. Also, the processing or storage of colored foods is sensitive to treatments. Carotenoids are unstable when they're exposed to light or oxygen because of their properties as highly conjugated and intensely colored isoprenoid plant compounds (Yadav & Sehgal, 1997). Degradation and isomerization are observed during processing or storage and the patterns of isomerization are similar to those observed in model systems. Degradation of chlorophylls proceeds by oxidation of the ring structure to chlorins and ultimately by formation of colorless end products. It has been observed that high temperatures and exposure to light are the most deleterious conditions during processing of natural colorant-containing foods (Schwartz & Lorenzo, 1990). Different methodologies have been used to preserve the green color of fruits and vegetables. Techniques employed are pH control, microencapsulation, use of salts, control of the thermal treatment, use of modified atmospheres, and combinations of these, among others. Other factors are also important in the stability of natural colorants (taken from anthocyaninss) and in general they must be processed and stored at low temperatures, with low availability of oxygen, and out of light. Some colors are highly affected by pH. For example, curcuminoids are degraded at alkaline pH to feruloyl methane and ferulic acid; uncolored forms of anthocyanins are favored as the pH goes toward neutral values. Thus, acidic pH favors the appearance of the colored forms of most anthocyanins ($pH < 4$) (Strimpakos & Sharma, 2008; Bridle & Timberlake, 1997). Sometimes water activity may hasten the rate of degradation of some pigments. Betalains have high stability at low water activity (Cohen & Saguy, 1983).Natural colors such as annatto, β-carotene, and turmeric present solubility problems during their use and may create dust clouds. To overcome these problems, extensive research and technological development are invested by scientists for the extraction and formulation of natural colorants extracted from plants or produced by biotechnology *de novo* or via bioconversions of colorant precursors *in vitro*. Knowing the limitations of each pigment means that a specific pigment can be avoided for certain applications, in which the conditions are unfavorable for the pigment, and that alternatives can be sought, or that attempts can be made to increase the stability of the pigments by formulation.

6. Summary

This chapter presented an overview of natural antioxidants and colors with literature supported application in food system. The sources of bioactive antioxidants are highlighted along with necessary methods describing their extraction from different matrices. The bio-preservative properties of antioxidants are presented in support of frying suitability. Lastly, an insight into natural colors, sources and stability are presented to accentuate their food application leading to the concept of clean label ingredients.

References

Abboud, J.L.M., Notario, R. 1999. Critical compilation of scales of solvent parameters. Part I. Pure, non-hydrogen bond donor solvents. *Pure and Applied Chemistry,* 71(4):645–718.

Abd-ElGhany, M.E., Ammar, M.S., Hegazy, A.E. 2010. Use of olive waste cake extract as a natural antioxidant for improving the stability of heated sunflower oil. *World Applied Sciences Journal,* 11(1):106–113.

Ajila, C.M., Brar, S.K., Verma, M., Tyagi, R.D., Godbout, S., Valero, J.R. 2011. Extraction and analysis of polyphenols: recent trends. *Critical Reviews in Biotechnology,* 31(3):227–49.

Al Bandak, G., Oreopoulou, V. 2011. Inhibition of lipid oxidation in fried chips and cookies by *Majorana syriaca. International Journal of Food Science & Technology*, 46(2):290–296.

Ali, R.F.M. 2010. Improvement the stability of fried sunflower oil by using different levels of pomposia (*Syzyygium cumini*). *Electronic Journal of Environmental, Agricultural & Food Chemistry*, 9:396–403.

Altinier, G., Sosa, S., Aquino, R.P., Mencherini, T., Loggia, R.D., Tubaro, A. 2007. Characterization of topical antiinflammatory compounds in *Rosmarinus officinalis* L. *Journal of Agricultural & Food Chemistry*, 55:1718–1723.

Angelo, S.t., Vercellotti, A.J., Jacks, J., Legendre, M .1996. Lipid oxidation in foods. *Critical Reviews in Food Science & Nutrition*, 36(3):175–224.

Anwar, F., Jamil, A., Iqbal, S., Sheikh, M.A. 2006. Antioxidant activity of various plant extracts under ambient and accelerated storage of sunflower oil. *Grasas y Aceites, 57*:189-197.

Aruoma, O.I., Spencer, J.P.E., Rossi, R., Aeschbach, R., Khan, A., Mahmood, N., Munoz, A., Murcia, A., John, Butler, Halliwell, B. 1996. An evaluation of the antioxidant and antiviral action of extracts of rosemary and Provencal herbs. *Food & Chemical Toxicology*, 34(5): 449–456.

Aydeniz, B., Yilmaz, E. 2012. Enrichment of frying oils with plant phenolic extracts to extend the usage life. *European Journal of Lipid Science & Technology*, 114(8):933–941.

Babu, S. and Shenolikar, I.S. 1995. Health and nutritional implications of food colors. *Ind J Med Res*, 102: 245-249.

Basaga, H,. Tekkaya, C., Acikel, F. 1997. Antioxidative & free radical scavenging properties of rosemary extract. *LWT-Food Science & Technology*, 30:105–108.

Betalleluz-Pallardel, I., Chirinos, R., Rogez, H., Pedreschi, R., Campos, D. 2012. Phenolic compounds from Andean mashua (*Tropaeolum tuberosum*) tubers display protection against soybean oil oxidation. *Food Science & Technology International*, 18(3):271–280.

Bonoli, M., Verardo, V., Marconi, E., Caboni, M.F. 2004. Antioxidant phenols in barley (*Hordeum vulgare* L.) flour: comparative spectrophotometric study among extraction methods of free and bound phenolic compounds. *Journal of Agricultural and Food Chemistry,* 52:5195-5200.

Botega, D.Z., Bastida, S., Marmesat, S., Pérez-Olleros, L., Ruiz-Roso, B., Sánchez-Muniz, F.J. 2009. Carob fruit polyphenols reduce tocopherol loss, triacylglycerol polymerization and oxidation in heated sunflower oil. *Journal of the American Oil Chemists' Society*, 86(5):419–425.

Bridle, P., and Timberlake, C.F. 1997. Anthocyanins as natural food colour-selected aspects. *Food Chemistry*, 58: 103–109.

Britton, G. 1983. The Biochemistry of Natural Pigments. In: Bannister WH, editor. Biochemical Education. Cambridge: Cambridge University Press. p 366.

Britton, G. 1996. Carotenoids. In: Hendry GAF,and Houghton JD Editors, Natural Food Colorants. New York: Chapman & Hall. p 197–243

Budryn, G., Zyzelewicz, D., Nebesny, E., Oracz, J., Krysiak, W. 2013. Influence of addition of green tea and green coffee extracts on the properties of fine yeast pastry fried products. *Food Research International*, 50(1): 149–160.

Chang CH, Chyau CC, Hsieh CL, Wu YY, Ker YB, Tsen HY, et al 2008. Relevance of phenolic diterpene constituents to antioxidant activity of supercritical CO_2 extract from the leaves of rosemary. *Natural Products Research*, 22:76–90.

Chang SS, Ostric Matijasevic BISERKA, Hsieh OA, Huang CL 1977. Natural antioxidants from rosemary and sage. *Journal of Food Science*, 42(4):1102–1106.

CheMan YB, Tan CP 1999. Effects of natural & synthetic antioxidants on changes in refined, bleached, & deodorized palm-olein during deep-fat frying of potato chips. *Journal of the American Oil Chemists' Society*, 76:331–339.

Chen Q, Shi H, Ho CT 1992. Effects of rosemary extracts & major constituents on lipid oxidation & soybean lipoxygenase activity. *Journal of American Oil Chemists Society,* 69:999-1002.

Chen R, Li S, Liu C, Yang S, Li X 2012. Ultrasound complex enzymes assisted extraction and biochemical activities of polysaccharides from *Epimedium* leaves. *Process Biochemistry* 47(12): 2040-2050.

Cheng KW, Shi JJ, Ou SY, Wang M, Jiang Y 2009. Effects of fruit extracts on the formation of acrylamide in model reactions and fried potato crisps. *Journal of Agricultural & Food Chemistry*, 58(1):309–312.

Chiou A, Kalogeropoulos N, Salta FN, Efstathiou P, Andrikopoulos NK 2009. Pan frying of French fries in three different edible oils enriched with olive leaf extract: Oxidative stability and fate of micro constituents. *LWT-Food Science & Technology*, 42(6):1090–1097.

Chirinos R, Huamán M, Betalleluz-Pallardel I, Pedreschi R, Campos D 2011. Characterization of phenolic compounds of Inca muna (*Clinopodium bolivianum*) leaves and the feasibility of their application to improve the oxidative stability of soybean oil during frying. *Food Chemistry*, 128(3):711–716.

Clydesdale FM. 1993. Color as a factor in food choice. *Critical Reviews in Food Science and Nutrition,* 3:83-101.

Cohen E, and Saguy I. 1983. Effect of water activity and moisture content on the stability of beet powder pigments.*Journal of Food Science,* 48: 703–707.

Cuvelier ME, Berset C, Richard H 1994. Antioxidant constituents in sage. *Journal of Agricultural & Food Chemistry*, 42:665–669.

Danilenko M, Wang Q, Wang X, Levy J, Sharoni Y, Studzinski GP 2003. Carnosic acid potentiates the antioxidant & prodifferentiation effects of 1alpha 25-dihydroxyvitamin D (3) in leukemia cells but does not promote elevation of basal levels of intracellular calcium. *Cancer Research*, 63:1325–1332.

Delgado-Vargas F, Jiménez AR, and Paredes-López O. 2000. Natural pigments: carotenoids, anthocyanins, and betalains - characteristics, biosynthesis, processing and stability. *Critical Reviews in Food Science and Nutrition*, 40: 173–289.

Dobias P, Pavlikova P, Adam M, Eisner A, Benova B, Ventura K 2010. Comparison of pressurised fluid and ultrasonic extraction methods for analysis of plant antioxidants and their antioxidant capacity. *Central European Journal of Chemistry* 8(1):87–95.

Droge W 2002. Free radicals in the physiological control of cell function. *Physiological Reviews*, 82(1):47-95.

Du, H., & Li, H. 2008. Antioxidant effect of Cassia essential oil on deep-fried beef during the frying process. *Meat Science*, 78(4): 461–468.

Dufosse L. 2006. Microbial Production of Food Grade Pigments. *Food Technol Biotechnol* 44 (3): 313–321.

Duh, P. D., Yen, W. J., Du, P. C. & Yen, G. C. (1997). Antioxidant activity of mung bean hulls. *Journal of the American Oil Chemists' Society*, 74(9):1059–1063.

Dunford N, Irmak S, Jonnala R 2010. Pressurised solvent extraction of policosanol from wheat straw, germ and bran. *Food Chemistry* 119(3): 1246–9.

Duroux L, Delmotte FM, Lancelin JM, Kéravis G and Jay-Allemand C. 1998. Insight into napthoquinone metabolism: β-glucosidase-catalysed hydrolysis of hydrojuglone β-pyranoside. *Biochem J.* 333: 275–283.

Economou, K.D., Oreopoulou, V., Thomopoulos, C.D. 1991. Antioxidant activity of some plant extracts of the family *Labiatae*. *Journal of the American Oil Chemists' Society*, 68: 109–113.

Erkan, N., Ayranci, G., Ayranci, E. 2008. Antioxidant activities of rosemary (*Rosmarinus Officinalis*) extract, black seed (*Nigella sativa*) essential oil, carnosic acid, rosmarinic acid and sesamol. *Food Chemistry*, 110(1):76–82.

Fielding, B.C., Holker, J.S.E., Jones, D.F., Powell, A.D.G., Richmond, K.W., Robertson, A., Whalley, W.B. 1961. The chemistry of fungi. XXXIX. The structure of monascin. *J Chem Soc,* 4579–4589. DOI: 10.1039/JR9610004579.

Filip, S., Hribar, J., Vidrih, R. 2011. Influence of natural antioxidants on the formation of trans fatty acid isomers during heat treatment of sunflower oil. *European Journal of Lipid Science & Technology*, 113(2):224–230.

Francis, F.J. 1985. Pigments and other colorants. In: Fennema OR, editor, 2nd ed. Food Chemistry. New York: Marcel Dekker. pp 545-584.

Francis, F.J. 1999. Anthocyanins and betalains. In: Francis FJ editor, Colorants. St Paul: MN-Eagan Press. p 55–66.

Frankel, E.N. 1995. Chemistry of autoxidation: Mechanism, products and flavor significance. In D. B. Min, & T. H. Smouse (Eds.), *Flavor chemistry of fats and oils* (pp. 1–38). AOCS Press: Urbana-Champaign.

Frankel, E.N. 1996. Antioxidants in lipid foods and their impact on food quality. *Food Chemistry*, 57:51–55.

Frick, D. 2003. The coloration of food. Review of Progress in Coloration and Related Topics. Color Technol33: 15–32. doi: 10.1111/j.1478-4408.2003.tb00141.x.

Fukasawa, R., Kanda, A., Hara, S. 2009. Anti-oxidative effects of rooibos tea extract on auto-oxidation and thermal oxidation of lipids. *Journal of Oleo Science*, 58(6):275–283.

Gamel, T. H., & Kiritsakis, A. 1999. Effect of methanol extracts of rosemary and olive vegetable water on the stability of olive oil and sunflower oil. *Grasas y Aceites*, 50(5): 345–350.

Garcia-Marino, M., Rivas-Gonzalo, J., Iba´nez, E., Garc´ýa-Moreno, C. 2006. Recovery of catechins and proanthocyanidins from winery by-products using subcritical water extraction. *Analytica Chimica Acta* ,563(1):44–50.

Gilgun-Sherki, Y., Melamed, E., Offen, D 2001. Oxidative stress induced-neurodegenerative diseases: the need for antioxidants that penetrate the blood brain barrier. *Neuropharmacology*, 40(8):959–975.

Goodwin TW. 1992. Biosynthesis of carotenoids: an overview. Methods Enzymol 214: 330–340.

Gordon, M.H. 1991. Oils & fats: taint or flavor? *Chemistry in Britain*, November, 1020–1022.

Gordon, M. H., & Kourkimskå, L. 1995. The effects of antioxidants on changes in oils during heating and deep frying. *Journal of the Science of Food & Agriculture*, 68(3): 347–353.

Goyle, A., Gupta, R.G. 1998. Use of synthetic colors in food. *Sci Cult* 64: 241.

Guzinski, J. 1996. Oleoresins and essential oils. In Y. H. Hui (Ed.), *Bailey's industrial oils and fat products,* (pp. 145–158). Wiley-Interscience Publication: New York.

Halliwell, B. 1999. Establishing the significance and optimal intake of dietary antioxidants: the biomarker concept. *Nutritional Reviews*, 57:104–113.

Halliwell, B., Murcia, M.A., Chirico, S., Aruoma, O.I. 1995. Free radicals and antioxidants in food and *in vivo*: what they do and how they work. *Critical Reviews in Food Science & Nutrition*, 35:7–20.

Hamied, A.A., Nassar, A.G., Badry, N.E. 2009. Investigations on antioxidant & antibacterial activities of some natural extracts. *World Journal of Dairy & Food Sciences,* 4(1):1–7.

Harborne, J.B., and Grayer, RJ. 1988. The anthocyanin. In: Harborne JB editor. The flavonoids.

New York: Chapman & Hall. p 1–20.

Hassas-Roudsari, M., Chang, P., Pegg, R., Tyler, R. 2009. Antioxidant capacity of bioactives extracted from canola meal by subcritical water, ethanolic and hot water extraction. *Food Chemistry,* 114(2):717–26.

Haugan, J.A., Akermann, T., and Liaaen-Jensen, S. 1992. Isolation of fucoxanthin and peridinin. *Methods Enzymol*, 213: 231–245.

Hendry, G.A.F. 1996. Chlorophylls and chlorophyll derivatives. In: Hendry GAF and Houghton JD editors, Natural food colorants. New York : Chapman & Hall. p 131–156.

Herrero, M., Cifuentes, A., Ibaez, E. 2006. Sub and supercritical fluid extraction of functional ingredients from different natural sources: plants, food-by-products, algae and microalgae: a review. *Food Chemistry,* 98(1):136–48.

Hopia AI, Huang SW, Schwarz K, German JB, Frankel EN 1996. Effect of different lipid systems on antioxidant activity of rosemary constituents carnosol and carnosic acid with and without á-tocopherol. *Journal of Agricultural & Food Chemistry*, 44:2030–2036.

Houhoula, D. P., Oreopoulou, V., & Tzia, C. 2004. Antioxidant efficiency of oregano in frying and storage of fried products. *European Journal of Lipid Science &Technology*, 106(11), 746–751.

Houlihan, C.M., Ho, C.T., Chang, S.S. 1984. Elucidation of the chemical structure of a novel antioxidant, rosmaridiphenol, isolated from rosemary. *Journal of the American Oil Chemists Society*, 61(6):1036–1039.

Hras, A.R., Hadolin, M., Knez, Z, Bauman D 2000. Comparison of antioxidative and synergistic effects of rosemary extract with α-tocopherol, ascorbyl palmitate and citric acid in sunflower oil. *Food Chemistry*, 71(2):229–233.

Huisman, M., Madsen, H.L., Skibsted, L.H., Bertelsen, G. 1994. The combined effect of rosemary (*Rosmarinus officinalis* L.) and modified atmosphere packaging as protection against warmed over flavor in cooked minced pork meat. *Zeitschrift für Lebensmittel-Untersuchung und Forschung*, 198(1):57–59.

Inouye, Y., Nakanishi, K., Nishikawa, H., Ohashi, M., Terahara, A., Yamamura, S. 1962. Structure of monascoflavin. Tetrahedron, 18: 1195–1203.

Irwandi, J., Man, Y.C, Kitts DD, Bakar J, Jinap S 2000. Synergies between plant antioxidant blends in preventing peroxidation reactions in model and food oil systems. *Journal of the American Oil Chemists' Society*, *77*(9):945–951.

Jackman, R.L. and Smith, J.L. 1996. Anthocyanins and betalains. In: Hendry GAF and Houghton JD editors, Natural food colorants. New York : Chapman & Hall. p 244–310.

Jamilah, B., Man, Y. B., & Ching, T. L. 1998. Antioxidant activity of *citrus hystrix* peel extract in RBD palm oleen during frying of fish crackers. *Journal of Food Lipids*, *5*(2): 149–157.

Jaswir, I., Man, Y.B.C. 1999. Use optimization of natural antioxidants in refined, bleached, and deodorized palm olein during repeated deep-fat frying using response surface methodology. *Journal of the American Oil Chemists' Society,* 76(3):341–348.

Jaswir, I., Man, Y.B.C., Kitts, D.D. 2000. Use of natural antioxidants in refined palm olein during repeated deep-fat frying. *Food Research International*, 33(6):501–508.

Jensen, S.R., Franzyk, H., and Wallander, E. 2002. Chemotaxonomy of the Oleaceae: iridoids as taxonomic markers. *Phytochemistry,* 60: 213-231.

Joshi, P., Jain, S., Sharma, V. 2009. Turmeric (*Curcuma longa*) a natural source of edible yellow colour. *International Journal of Food Science and Technology,* 44 (12): 2402–2406.

Juntachote, T., Berghofer, E, Siebenhandl, S., Bauer, F. 2006. The antioxidative properties of holy basil and galangal in cooked ground pork. *Meat Science,* 72: 446– 456.

Kahl, R., Kappus, H. 1993. Antioxidantien, BHA & BHT im Vergleich mit dem natuÈrlichen Antioxidans Vitamin E. *Zeitschrift fuÈr Lebensmittel- Untersuchung und Forschung*, 196:329–338.

Kalantzakis, G., Blekas, G. 2006. Effect of Greek sage and summer savory extracts on vegetable oil thermal stability. *European Journal of Lipid Science & Technology*, 108(10):842–847.

Kapoor, V.P. 2006. Food Colors: Concern regarding their safety and toxicity. EnviroNews, Newsletter of ISEB India, April 2006. International Society of Environment Botanists.

Karoui, I.J., Dhifi, W., Ben, Jemia, M., Marzouk, B. 2011. Thermal stability of corn oil flavoured with Thymus capitatus under heating and deep frying conditions. *Journal of the Science of Food & Agriculture*, 91(5):927–933.

Kim, J., Nagaoka, T., Ishida, Y., Hasegawa, T., Kitagawa, K., Lee, S. 2009. Subcritical water extraction of nutraceutical compounds from citrus pomaces. *Separation Science and Technology* 44(11):2598-608.

Kitts, D.D. 1996. Toxicity and safety of fats and oils. *Bailey's Industrial Oil and Fat Products.*

Lee J, Koo N, Min DB 2004. Reactive oxygen species, aging, and antioxidative nutraceuticals. *Comprehensive Reviews in Food Science and Food Safety*, 3(1):21-33.

Li, H., Chen, B., Yao, S. 2005. Application of ultrasonic technique for extracting chlorogenic acid from *Eucommia ulmodies* Oliv. (*E. ulmodies*). *Ultrasonics Sonochemistry,* 12: 295-300.

Li, Y., Skouroumounis, G.K., Elsey, G.M., Taylor, D.K. 2011. Microwave-assistance provides very rapid and efficient extraction of grape seed polyphenols. *Food Chemistry,* 129(2): 570–576.

MacDougall, D.B., and Francis, F.J. 2002. Food colorings. In: Watson DH editor, Colour in Food: Improving Quality. Cambridge, UK: Woodhead Publishing Co. p 297-330.

Madhavi, D.L., Despande, S.S., Salunkhe, D.K. 1996. *Food Antioxidants*. Marcel Dekker: New York.

Madsen. B., Hidalgo, G.V., Hernandez, V.L. 2003. Purification process for improving total yield of curcuminoid colouring agent. United States Patent 6,576, 273, June 10, 2003.

Man, Y.B.C., Jaswir, I. 2000. Effect of rosemary and sage extracts on frying performance of refined, bleached and deodorized (RBD) palm olein during deep-fat frying. *Food Chemistry*, 69(3):301-307.

Manchand, P.S., Whalley, W.B. 1973. Isolation and structure of ankaflavin: a new pigment from Monascus anka. *Phytochemistry,* 12: 2531–2532.

Masson, R.G., Baumes, Puech, J.L., and Razungles, A. 1997. Demonstration of the presence of carotenids in wood: quantitative study of cooperage oak. *Journal of Agricultural and Food Chemistry*, 45: 1649–1652.

Min, D.B., Lee, H.O. 1999. *Flavor chemistry: 30 years of progress*. In Teranishi (Ed.), (pp. 175–187). Kluwer Academic/Plenum Publishers: New York.

Miron, T., Plaza, M., Bahrim, G., Iba´nez, E., Herrero, M. 2010. Chemical composition of bioactive pressurized extracts of Romanian aromatic plants. *Journal of Chromatography,* 1218(30):4918–27.

Moyler, D.A., Heath, H.B. 1988. Flavors and fragrances world perspective, in *Developments in Food Science*, Vol. 18, eds. B.M. Lawrence, B.D. Mookherjee and B.J. Willis, Elsevier, Amsterdam pp. 41–64.

Mudgett, R.E. 2000. Monascus. In: Lauro GJ, Francis FJ editors, Natural food colorants science and technology. New York: Marcel Dekker. p 31-86.

Nakatani, N., Inatani, R. 1981. Structure of rosmanol, a new antioxidant from rosemary (*Rosmarinus officinalis*). *Agricultural& Biological Chemistry*, 45(10):2385–2386.

Navas, P. B., Carrasquero Durán, A., & Flores, I. 2006. Effect of black tea, garlic and onion on corn oil stability and fatty acid composition under accelerated oxidation. *International Journal of Food Science & Technology*, 41(3): 243–247.

Naz, S., Siddiqi, R., Sheikh, H., & Sayeed, S.A. 2005. Deterioration of olive, corn and soybean oils due to air, light, heat and deep-frying. *Food Research International*, 38(2): 127–134.

Nissen, L.R., Mansson, L., Bertelsen, G., Huynh-Ba, T., Skibsted, L.H. 2000. Protection of dehydrated chicken meat by natural antioxidants as evaluated by electron spin resonance spectrometry. *Journal of Agricultural & Food Chemistry*, 48:5548–5556.

Nor, F., Suhaila, M., Nor Aini, I., Razali, I. 2011. Antioxidative properties of *Murraya koenigii* leaf extracts in accelerated oxidation and deep-frying studies. *International Journal of Food Sciences &Nutrition*, 60(S2):1–11.

Nor, F.M., Mohamed, S., Idris, N.A, Ismail R 2008. Antioxidative properties of *Pandanus amaryllifolius* leaf extracts in accelerated oxidation and deep frying studies. *Food Chemistry*, 110(2): 319–327.

Nor, F.M., Mohamed, S., Idris, N.A., Ismail, R. 2009. Antioxidative properties of *Curcuma longa* leaf extract in accelerated oxidation and deep frying studies. *Journal of the American Oil Chemists' Society*, 86(2):141–147.

Oluwatuyi, M., Kaatza, G.W., Gibbons, S. 2004. Antibacterial and resistance modifying activity of *Rosmarinus officinalis*. *Phytochemistry*, 65:3249–3254.

Park, P.J., Jung, W.K., Nam, K.S., Shahidi, F., Kim, S.K. 2001. Purification and characterization of antioxidative peptides from protein hydrolysate of lecithin-free egg yolk. *Journal of the American Oil Chemists' Society*, 78(6):651-656.

Patel, S., Shende, S., Arora, S., Singh, A.K. 2013. An assessment of the antioxidant potential of coriander extracts in ghee when stored at high temperature and during deepfat frying. *International Journal of Dairy Technology*, 66:207–213.

Pereira, M.A.C., Jorge, N. 2012. Antioxidant potential of oregano extract (*Origanum vulgare* L.). *British Food Journal*, 114(7):954–965.

Pinelo, M., Del Fabbro, P., Manzocco, L., Nuñez, M.J., Nicoli, M.C. 2005. Optimization of continuous phenol extraction from *Vitis vinifera* byproducts. *Food Chemistry*, 92: 109-117.

Pintea, A.M. 2007. Other Natural Pigments. In: Socaciu C, editor. Food Colorants: Chemical and functional properties. Boca Raton: CRC Press. p 25-49.

Plaza, M., Amigo-Benavent, M., del Castillo, M., Iba´nez, E., Herrero, M. 2010a. Facts about the formation of new antioxidants in natural samples after subcritical water extraction. *Food Research International*, 43(10):2341–48.

Plaza, M., Santoyo, S., Jaime, L., Garc´ýa-Blairsy Reina, G., Herrero, M., Senorans,´ F., Iba´nez E. 2010b. Screening for bioactive compounds from algae. *Journal of Pharmaceutical and Biomedical Analysis*, 51(2):450–5.

Poiana, M.A. 2012. Enhancing oxidative stability of sunflower oil during convective and microwave heating using grape seed extract. *International Journal of Molecular Sciences*, 13(7): 9240–9259.

Priyadarsini, K.V. 2014. The Chemistry of Curcumin: From Extraction to Therapeutic Agent. Molecules, 19: 20091-20112; doi:10.3390/molecules191220091

Ramalho, V. C., & Jorge, N. 2008. Antioxidant action of Rosemary extract in soybean oil submitted to thermoxidation. *Grasas y aceites*, 59(2), 128–131.

Reische, D.W., Lillard, D.A., Eitenmiller, R.R. 2002. Antioxidants. In C.C. Akoh, & D. B. Min (Eds.), *Food lipids: Chemistry, nutrition and biotechnology* (pp. 489–516), Marcel Dekker: New York.

Richheimer, S.L., Bernart, M.W., King, G.A., Kent, M.C., Beiley, D.T. 1996. Antioxidant activity of lipid-soluble phenolic diterpenes from rosemary. *Journal of the American Oil Chemists' Society*, 73(4):507–514.

Rizvi, S.S.H. 1994. Supercritical Fluid Processing of Food and Biomaterials, First ed., Blackie Academic and Professional, Glassgow, UK pp. 36-38.

Rodriguez-Meizoso, I., Marin, F., Herrero, M., Senorans, F., Reglero, G.,˜ Cifuentes, A., Iba´nez E 2006. Subcritical water extraction of nutraceuticals˜ with antioxidant activity from oregano. Chemical and functional characterization. *Journal of Pharmaceutical and Biomedical Analysis* 41(5):1560–5.

Routray, W., Orsat, V. 2011. Microwave-assisted extraction of flavonoids: a review. *Food Bioprocess Technology,* 5:409–24.

Roy, L.G., Arabshahi-Delouee, S., Urooj, A. 2010. Antioxidant efficacy of mulberry (*Morus indica* L.) leaves extract and powder in edible oil. *International Journal of Food Properties*, 13(1): 1-9.

Rüdiger, W. and Schoch, S. 1988. Chlorophylls. In: Plant Pigments, Vol. 1. Goodwin TW editor, Plant pigments. New York: Academic Press. p 1–60.

Satoh, T., Kosaka, K., Itoh, K., Kobayashi, A., Yamamoto, M., Shimojo, Y., et al 2008. Carnosic acid, a catechol-type electrophilic compound, protects neurons both *in vitro & in vivo* through activation of the Keap1/Nrf2 pathway via Salkylation of targeted cysteines on Keap1. *Journal of Neurochemistry*, 104:1116–1131.

Satyanarayana, A., Probhakara Rao, P.G., and Rao, D.G. 2003. Chemistry, processing and toxicology of annatto (Bixa *orellana* L.). *Journal of Food Science and Technology,* 40: 131-141.

Schwartz, S.J., and Lorenzo, T.V. 1990. Chlorophylls in foods. *Critical Reviews in Food Science and Nutrition*, 29: 1–17.

Senorans, F.J., Ibanez, E., Cavero, S., Tabera, J., Reglero, G. 2000. Liquid chromatographic–mass spectrometric analysis of supercritical-fluid extracts of rosemary plants. *Journal of Chromatography A*, 870:491–499.

Simsek, M., Sumnu, G., Sahin, S. 2012. Microwave-assisted extraction of phenolic compounds from sour cherry pomace. *Separation Science and Technology,* 47(8):1248–54.

Socaciu, C. 2007. Natural Pigments as Food Colorants. In: Socaciu C, editor. Food olorants. Boca Raton: CRC Press. p 551-616.

Spigno, G., Tramelli, L., De Faveri, D.M. 2007. Effects of extraction time, temperature and solvent on concentration and antioxidant activity of grape marc phenolics. *Journal of Food Engineering,* 81: 200-208.

Starmans, D., Nijhuis, H. 1996. Extraction of secondary metabolites from plant material: a review. *Trends in Food Science and Technology,* 7(6):191–7.

Strack, D., Steglich, W., and Wray, V. 1993. Betalains. In: Conn EE editor, Methods in Plant Biochemistry, Vol. 8. Orlando, FL: Academic Press. p. 421–450.

Strimpakos, A.S., and Sharma, R.A. 2008. Curcumin: preventive and therapeutic properties in laboratory studies and clinical trials. *Antioxid Redox Signal,* 10(3): 511-545.

Teng, S.S., Feldheim, W. 1998. Analysis of Anka pigments by liquid chromatography with diode array detection and tandem mass spectrometry. *Chromatographia,* 47: 529–536.

Termentzi, A., Kefalas, P., & Kokkalou, E. 2008. LC–DAD–MS (ESI+) analysis of the phenolic content of Sorbus domestica fruits in relation to their maturity stage. *Food Chemistry*, 106(3):1234–1245.

Terpinc, P., Bezjak, M., Abramovic, H. 2009. A kinetic model for evaluation of the antioxidant activity of several rosemary extracts. *Food Chemistry*, 115:740–744.

Thomsen, M.K., Vedstesen, H., Skibsted. L.H. 1999. Quantification of radical formation in oil in water food emulsions by electron spin resonance spectroscopy. *Journal of Food Lipids*, 6(2):149–158.

Thomson, R.H. 1962. Quinones: structure and distribution. In: Florkin AM and Mason HS editors, Comparative Biochemistry Vol. III: Constituents of Life. Part A. New York: Academic Press. p 631–725.

Thorsen, M.A., Hildebrandt, K.S. 2003. Quantitative determination of phenolic diterpenes in rosemary extracts: Aspects of accurate quantification. *Journal of Chromatography A*, 995:119–125.

U.S. Food & Drug Administration, 1999. Summary of Color Additives Listed for Use in the United States in Food, Drugs, Cosmetics and Medical Devices, Washington, D.C.

Upadhyay, R., Jha, A., Singh, S.P., Kumar, A., Singh, M. 2015. Appropriate solvents for extracting total phenolics, flavonoids and ascorbic acid from different kinds of millets. *Journal of Food Science & Technology*, 52:472–478.

Upadhyay, R., Mishra, H.N. 2014. Antioxidant activity measurement of oleoresin from rosemary and sage. *Industrial Crops and Products*, 61:453–459.

Upadhyay, R., Mishra, H.N. 2015a. Predictive modeling for shelf life estimation of sunflower oil blended with oleoresin rosemary (*Rosmarinus officinalis* L.) and ascorbyl palmitate at low and high temperatures. *LWT-Food Science and Technology*, 60:42–49.

Upadhyay, R., Mishra, H.N. 2015b. Multivariate analysis for kinetic modeling of oxidative stability and shelf life estimation of sunflower oil blended with sage (*Salvia officinalis*) extract under Rancimat conditions. *Food and Bioprocess Technology*, 8:801-810.

Upadhyay, R., Mishra, H.N. 2015c. A multivariate approach to optimise the synergistic blend of oleoresin rosemary (*Rosmarinus officinalis* L.) and ascorbyl palmitate added into sunflower oil. *International Journal of Food Science and Technology*, 50:974–981.

Upadhyay, R., Mishra, H.N. 2015d. Classification of sunflower oil blends stabilized by oleoresin rosemary (*Rosmarinus officinalis* L.) using multivariate kinetic approach. *Journal of Food Science*, 80: E1746–E1754.

Upadhyay, R., Mishra, H.N. 2015e. Effect of relative humidity and light conditions on the oxidative stability of sunflower oil blends stabilised with synthetic and natural antioxidants. *International Journal of Food Science & Technology*, 51:293–299.

Upadhyay, R., Mishra, H.N. 2016. Multivariate optimization of a synergistic blend of oleoresin sage (*Salvia officinalis* L.) and ascorbyl palmitate to stabilize sunflower oil. *Journal of Food Science and Technology* (DOI 10.1007/s13197-015-2157-9).

Upadhyay, R., Nachiappan, G., Mishra, H.N. 2015. Ultrasound-assisted extraction of flavonoid and phenolic compounds from *Ocimum tenuiflorum* leaves. *Food Science & Biotechnology*, 24:1951-1958.

Upadhyay, R., Ramalakshmi, K., Rao, L.J.M. 2012. Microwave-assisted extraction of chlorogenic acids from green coffee beans. *Food Chemistry*, 130:184–188.

Upadhyay, R., Sehwag, S., Singh, S.P. 2015. Antioxidant activity and polyphenol content of *Brassica oleracea* varieties. *International Journal of Vegetable Sciences* (DOI:10.1080/19315260.2015.1048403).

Wang, L., Weller, C. 2006. Recent advances in extraction of nutraceuticals from plants. *Trends in Food Science and Technology,* 17(6):300–12.

Weinreich, B. 1998. Frischeschutz mit "weisser Weste". *Die Zeitschrift für die Lebensmittelwirtschaft*, 49:24–26.

Wong, J.W., Hashimoto, K., Shibamoto, T. 1995. Antioxidant activities of rosemary and sage extracts and vitamin E in a model meat system. *Journal of Agricultural & Food Chemistry*, 43(10):2707–2712.

Yadav, S.K., and Sehgal, S. 1997. Effect of home processing and storage on ascorbic acid and β-carotene content of bathua (Chenopodium album) and fenugreek (Trigonella foenum graecum) leaves. *Plant Foods for Human Nutrition,* 50: 239–247.

Yanishlieva NV, Marinova EM, Marekov IN 1997. Gordon, M. H., Effect of an ethanol extract from summer savory (*Saturejae hortensis* L.) on the stability of sunflower oil at frying temperature. *Journal of Science of Food & Agriculture*, 74:524–530.

Zandi, P., Gordon, M.H. 1999. Antioxidant activity of extracts from old tea leaves. *Food Chemistry*, 64(3):285–288.

Zhang, G., He, L., Hu, M. 2011. Optimized ultrasonic-assisted extraction of flavonoids from *Prunella vulgaris* L. and evaluation of antioxidant activities in vitro. *Innovative Food Science & Emerging Technologies,* 12(1):18-25.

Zhang, H.F., Yang, X.H., Wang, Y. 2011b. Microwave assisted extraction of secondary metabolites from plants: current status and future directions. *Trends in Food Science and Technology,* 22(12):672–88.

Zhang, Y., Yang, L., Zu, Y., Chen, X., Liu, F. 2010. Oxidative stability of sunflower oil supplemented with carnosic acid compared with synthetic antioxidants during accelerated storage. *Food Chemistry*, 118: 656–662.

Zhang, Y., Zhang, Y. 2007. Study on reduction of acrylamide in fried bread sticks by addition of antioxidant of bamboo leaves and extract of green tea. *Asia Pacific Journal of Clinical Nutrition*, 16(1):131–136.

Zhang, Y.L., Yin, C.P., Kong, L.C., Jiang, D.H. 2011a. Extraction optimization, purification and major antioxidant component of red pigments extracted from *Camellia japonica*. *Food Chemistry,* 129: 660-664.

Zhao, L., Zhao, G., Chen, F., Wang, Z., Wu, J., Hu, X. 2006. Different effects of microwave and ultrasound on the stability of (all-E)-astaxanthin. *Journal of Agricultural and Food Chemistry,* 54(21):8346–51.

8

Extension of Shelf Life of Fruits and Vegetables

S Billoria, S Biswas, C Sen, H N Mishra

1. Introduction

India is a land of large varieties of fruits and vegetables due to its vast soil and climatic diversity. India is the second largest producer of fruits and vegetables in the world, with an annual production of fruits 45.5 Million tonnes and vegetables 90.8 Million tonnes. It contributes 10.23% and 14.45% of the total world production of fruits and vegetables (F&V) respectively (NHB, 2010). Though fruits and vegetables are grown only on 7–8% of gross cultivated area, these contribute more than 18.8% of the gross value of agricultural output and 52% export earnings from total agricultural produce. These have a high export potential of 20-30 times more foreign exchange per unit area than cereals due to higher yields and higher price available in the international market (Sen et al., 2012). According to estimates, the per capita consumption of fruits and vegetables in India is only around 46g and 130g which are far below the stipulation of a minimum of 92g and 300g respectively as recommended by the Indian Council of Medical Research and National Institute of Nutrition. The low availability is mainly due to considerably high post-harvest losses. About 15-20% of the fruits and vegetables are wasted every year in the country, due to poor post-harvest management and lack of infrastructural facilities for processing and storage, amounting to a revenue loss of over Rs. 625 billion.

The sensorial, nutritional, and organoleptic quality of fresh produce starts deteriorating after harvest as a result of soon after harvest the altered plant metabolism and microbial growth. The quality deterioration is the result of produce respiration and transpiration, senescence, ripening-associated processes, wound-initiated reactions, development of postharvest disorders, microbial proliferation leading to postharvest quality loss. Therefore, it is necessary to understand the respiration and ripening processes and the physiological and biochemical changes during ripening. The best way to reduce respiratory metabolism is to reduce the storage temperature. All biological processes proceed more slowly at lower temperature (Yahia, 2007). In general, O_2 levels must be reduced to less than 10 % to see a reduction in the rate of respiration. Respiration rate is temperature dependent; so, as the temperature reduces the amount of oxygen required also drops. Therefore, oxygen levels must be substantially reduced at low temperature for it to affect the rate of respiration. It must however, be ensured that O_2 levels do not fall below extinction point as anaerobic respiration and decay may occur. Tolerance to low O_2 varies between species and cultivars and is also time dependent. Commodities may be able to tolerate low O_2 for short periods of time. Low O_2 can also be tolerated when CO_2 is low.

Fruits can be divided into two groups according to the regulatory mechanisms underlying their ripening process which majorly govern the shelf life of fresh produce and forms the basis for formulation of strategies to control the deteriorating processes occurring within the commodity and its surroundings.

1.1 Climacteric

After harvest, fruit starts ripening with increase in respiration rate at outset of its post-harvest life just after a small decrease followed by a sudden decline in respiration rate which causes fruit senescence. This is known as climacteric rise. The phyto hormone ethylene being the major trigger and coordinator of the ripening process. Ethylene production very promptly reaches the highest during this time, then subsides during the fruit senescence. Ethylene is catalytic

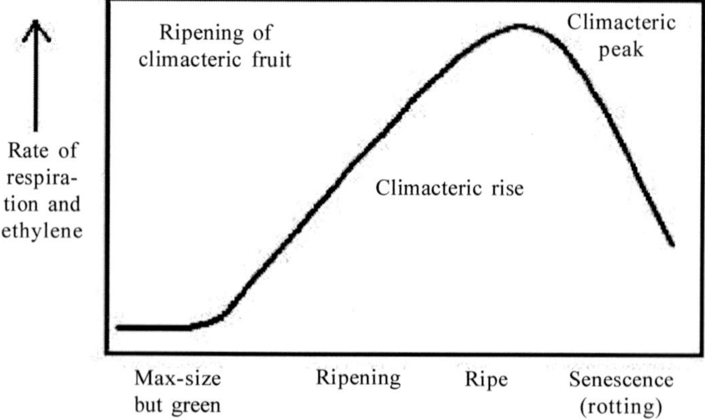

Fig. 1: Respiratory behavior of the horticultural produce after harvest

in action, typically, fruit generates barely detectable amounts of ethylene until ripening when there is sudden a burst of production (Figure 1). The respiration rate and ethylene production occur side by side, therefore, major rise in ethylene production may take place before, just after or close to the respiratory peak. Examples include apple, banana, fig, mango, passion fruit, papaya, pear, avocado, and tomato whereas banana, pear and avocado show strong respiratory rise. Climacteric fruit ripen after harvest, and can be harvested when they are unripe but mature. Ethylene is responsible for various changes in the fruits and vegetables during ripening like color change, aroma production and softening of the produce.

1.2 Non-climacteric

In these type of fruits, respiration rate either remains almost unchanged or shows a steady decline until senescence intervenes, with little or no increase in ethylene production; these are called 'non-climacteric' fruits. The fruit ripen only if they remain attached to the parent plant e.g. blueberry, cherry, citrus, cucumber, grape, pineapple and strawberry. Non-climacteric fruits, are characterized by the lack of ethylene-associated respiratory peak and the signaling pathways that drive the ripening process remain elusive.

Extension of shelf-life of perishables becomes the prime importance of post-harvest research owing to apparent losses in the present scenario and this can be achieved by retarding deteriorative processes. Temperature, relative humidity and atmospheric compositions are the environmental factors, which can be manipulated to lower the respiration rate and minimize the microbial spoilage if proper sanitation procedures are adopted. These factors form the basis of technologies such as controlled/modified atmosphere storage/packaging, edible coating and active packaging, which are of great potential for the extension of shelf life of perishable commodities like whole fruits and vegetables. These technologies work within the aerobic range of surrounding atmosphere of the targeted commodity for manipulation of the environmental factors based on species and cultivars.

2. Factors Affecting Shelf Life

2.1 Role of respiration

Respiration is a metabolic process that provides the energy for plant biochemical processes. Various substrates used in important synthetic metabolic pathways in the plant are formed during respiration. Aerobic respiration consists of oxidative breakdown of organic reserves to simpler molecules, including CO_2 and water, with release of energy. The organic substrates broken down in this process may include carbohydrates, lipids, and organic acids. The process consumes O_2 in a series of enzymatic reactions. Glycolysis, the tricarboxylic acid cycle, and the electron transport system are the metabolic pathways of aerobic respiration (Fonseca et al., 2002). The energy and organic molecules produced during respiration are used by other metabolic processes to maintain the health of the commodity. Heat produced during respiration is called vital heat and contributes to the refrigeration load that must be considered in designing storage rooms.

There are various factors that affect the respiration rate of the horticultural produce including temperature; commodity type and variety, maturity, climacteric behavior; availability of oxygen, carbon dioxide, ethylene, growth regulators et al., (stress and injuries, etc.).Vegetables include a great diversity of plant organs (roots, tubers, seeds, bulbs, fruits, sprouts, stems and leaves) that have different metabolic activities and consequently different respiration rates. Even different varieties of the same product can exhibit different respiration rates. In general, non-climacteric commodities have higher respiration rates in the early stages of development that steadily decline during maturation. Respiration rates of climacteric commodities also are high early in development stage and decline until a rise occurs that coincides with ripening or senescence. Climacteric products exhibit a peak of respiration and ethylene (C_2H_4) production associated with senescence or ripening (Fonseca et al., 2002).

2.2 Changes during ripening

Ripening is a phase in maturation process of fruits and vegetables. As the fruits mature, their tissue undergoes innumerable changes ranging from texture, color, aroma, composition and/or nutritional changes. The terms ripening and maturity are sometimes used alternatively and are regarded as the state when the fresh produce becomes appropriately suitable for harvesting as well as for eating. Though the degree of ripeness has remarkable clarity in terms of visible changes including color, it is sometimes ambiguous to define the right stage of desired ripeness of fresh produce, depending on the human choice. As in the case of a fruit like tomato, ripeness may depend on the degree of sweetness or acidity an individual may find attractive. In contrast to the fruits, vegetables do not possess any obvious changes and maturity is exceedingly difficult to define. However, vegetables during the maturation period exhibit changes in the chemical and physical structure. Changes that occur during ripening are as:

(i) Changes in color

(ii) Softening and associated alteration in texture

(iii) Production of volatiles and flavor compounds

(iv) Changes in sugar and organic acid metabolism

(v) Changes in membrane permeability which releases compartmentalized enzymes

(vi) Increase in enzyme synthesis

(vii) Other changes associated with the ripening may include changes in carbohydrates, pectins, organic acids.

Generally, the changes in the carbohydrates can be attributed to starches and sugars. In most of fruits starch gets converted to sugars making them sweeter after ripening whereas in some other plant products sugars decrease and starch content increase after harvesting. Example of such plant products is ripe sweet corn which results in loss of quality parameters such as flavor and texture. Storage temperature also plays a crucial role in postharvest changes during ripening. Potatoes, if stored below 10 °C result in increased sugar level, whereas same potatoes stored above 10 °C does not mark any such change.

More pronounced changes during ripening are pectin changes wherein water insoluble pectic substances are converted to soluble pectin, resulting in gradual softening of fruits and vegetables. Further, breakdown of water-soluble pectin by pectin methyl esterase also occurs. The organic acids of fruit generally decreases during storage and ripening. This occurs in apples and pears and is especially important in the case of oranges. Oranges have a long ripening period on the tree and time of picking is largely determined by degree of acidity and sugar content which has major effects upon juice quality. It is important to note that the reduction of acid content on ripening influences more than just the tartness of fruit. Since many of the plant pigments are sensitive to acid, fruit color would be expected to change.

2.3 Moisture loss

Most of the fresh fruits and vegetables contain around 65-95% moisture content depending on the type and cultivar of the produce. In growing plants, there is constant flow of water which maintains the high moisture content and the pressure inside the plant helps to support it. Fresh produce after harvesting starts losing moisture due to transpiration and respiration but unlike the growing plant it can no longer get water from soil, thereby causing dearth of water supply. This lack of supply and continuous loss of water results in wilting and shriveling of the fruits and vegetables. Severe desiccation results in huge losses e.g. wilted leafy vegetables require excessive trimming to make them marketable. The grapes start to detach from the bunch if the stem becomes excessively dried. The relative humidity, temperature, surrounding atmosphere and velocity of the surrounding air are few factors that are required to be controlled in order to prevent moisture losses.

2.4 Microbiological spoilage

Microbiological deteriorations usually take place in foods with high water activity at temperatures above the foods' freezing point. As the temperature increases, the rate of microbiological growth and hence spoilage increases geometrically, usually up to the temperature at which the microorganisms is thermally disrupted. Most spoilage microorganisms are aerobic and so consume oxygen in their respiratory actions. The reduction or removal of oxygen from the environment of aerobic microorganism reduces their rate of growth and thus the rate of spoilage. This reduced rate of microbiological growth under reduced oxygen conditions is among the reasons for the effectiveness of vacuum and modified atmosphere packaging for non-sterile foods. On the other hand, complete removal of oxygen can result in conditions under which anaerobic spore forming microorganisms and particularly pathogenic types are capable of growing and producing toxins. Thus, oxygen reduction to retard microbiological spoilage could lead to an undesirable result in food products whose water activity is sufficiently high to permit microbial growth.

The commonly encountered microflora of fruits and vegetables are *Pseudomonas* spp., *Erwiniaherbicola, Flavobacterium, Xanthomonas, Entero-bacteragglomerans*, lactic acid bacteria such as *Leuconostoc mesenteroides* and *Lactobacillus* spp., and molds and yeasts (Nguyen and Carlin, 1994). Temperature is the main variant that affect the growth of these microflora in the food. Researchers have reported a significant reduction in the microbial load at lower temperatures. King et al., 1976 reported that although the cabbage stored at 7 °C spoiled in the same time as the one stored at 14°C but it had significantly lower microbial load than the one stored at 14°C.

3. Quality Issues and Challenges

3.1 Effect on chilling injury

The physiological disorder associated with low but not freezing temperatures is commonly known as chilling injury (CI). This sensitivity constitutes a limitation for storage and transport for long journeys at low temperatures. For banana, CI symptoms are generally characterized by soaked appearance and abnormal ripening. PPO activity is considered as one of the major reasons for browning reaction. PPO enzyme has often been found to localize to the chloroplasts, where they are associated with the internal thylakoid membranes (Nguyen et al., 2004). If chilling injury induces membrane damage of organelles such as vacuoles, the phenolics may come in contact with PPO resulting in browning.

Storage of mature green 'Cavendish' bananas in low-density polyethylene (0.05 mm thickness) bags for up to 30 days at 8, 11, and 14°C developed an in-package atmosphere of 3 to 11 kPa O_2 and 3 to 5 kPa CO_2 (Hewage et al., 1995). However, these authors reported that these storage conditions did not affect ripening and sensory quality, nor did they alleviate CI symptoms developed at 8 and 11°C. In contradiction to their result, Nguyen et al., (2004) reported that MAP (about 12 kPa O_2 and 4 kPa CO_2) resulted in less visible CI in 'KluaiKhai' bananas at 10°C. Total free phenolics in the peel of MAP bananas decreased slowly and the fruits had low phenylalanine ammonia lyase (PAL) and polyphenol oxidase (PPO) activities. Pulp softness, sweetness and flavor of MAP fruit were good. Sen et al., (2014) reported that active packaging was found to be beneficial in lowering the PPO activity which in turn can lower the chilling injury symptoms. However, the authors did not explain the underneath mechanisms.

Although respiration is normally reduced at low, but non-freezing temperatures, certain commodities, chiefly those originating in the tropics and subtropics, exhibit abnormal respiration when their temperature falls below 10 to 12 °C (50 to 53.6 ºF). Typically the Q_{10} is much higher at these low temperatures for chilling sensitive crops than it would be for chilling tolerant ones. Respiration may increase dramatically at the chilling temperatures or when the commodity is returned to non-chilling temperatures. This enhanced respiration presumably reflects the cells' efforts to detoxify metabolic intermediates that accumulated during chilling, as well as to repair damage to membranes and other sub-cellular structures.

3.2 Heat stress

After the obvious increase in respiration rate with rise in temperature, the respiration tend to fall when the temperature reaches beyond the physiological range. It becomes negative as the tissue nears its thermal death point, due to entropy in the metabolism and denaturation of enzyme proteins. Normally tissues can withstand the severity of high temperatures for short periods of time (in minutes), which is why they can be exposed to high temperatures for short time for killing surface fungi on some fruits. However, continued exposure to high temperatures causes phytotoxic symptoms, and then complete tissue collapse.

The repeated but short exposures to high temperatures, often modifies the tissue's response to subsequent harmful stresses.

3.3 CO_2 injury

Many fruits and vegetables are susceptible to the injuries caused by the rapid exposure of increased levels of CO_2 and low levels of O_2. In addition, there are few other factors including type and cultivar, maturity, and free moisture on the surface that make the fruit susceptible to the CO_2 injury. This injury affects the respiratory metabolism and leads to the accumulation of acetaldehyde, ethanol and succinic acid (Kader 1986; Trujillo et al., 2001). It renders the produce unacceptable by the production of off odors. In this condition, the fruit appears brown, roughened to pebbly, irregularly shaped with partly defined edges and also develop some partially sunken lesions in the skin. The injury begins from hypodermal cortex cells, cuticle and epidermis remaining unaffected. The radial walls of the affected area collapse and pleat that makes the skin surface near normal regions to sink. Loss of cytoplasmic integrity, coagulation of protoplast and cell wall disruption are some of the other associated features of CO_2 injury (Watkins et al., 1997). These injuries have resulted in economic losses for many apple cultivars such as 'Braeburn', 'Empire', 'Fuji', and 'McIntosh' under commercial conditions.

Also, variable sensitivity of single cultivars from different growing regions has restricted utilization of high CO_2 pretreatments to maintain firmness. This problem can be alleviated with the use of CO_2 scrubbers such as soda lime. Recently, it has been found that diphenyl alanine (DPA) treatments used for scald control can inhibit development of CO_2 injury on apples (Trujillo et al., 2001).

3.4 Anoxic conditions

The underlying principle in controlled/ modified atmosphere packaging is the reduction in respiration rate by reducing the O_2 content inside the package. Though this reduction in O_2 content often has favorable effect but too low oxygen concentration usually leads to switching over to the anaerobic respiration which can result in the production of aldehydes and ketones producing off-odors. Also, there are chances of growth of anaerobic pathogenic microbes with the zero-oxygen conditions. Thus, although the reduction in O_2 concentration is required for prolonging the shelf life of produce, the complete extinction of O_2 may be hazardous.

3.5 Temperature fluctuations

The effect of low temperature in the storage and packaging technologies have a greater importance than the reduced O_2 and increased CO_2 levels in extending the shelf life and maintaining the quality of the horticultural produce. Any fluctuations, mishandling or deficiency in the cold chain can lead to the serious issues by not only deteriorating the product but also by triggering safety problems. High temperature or fluctuating temperature results in increased relative humidity and imbalance in water content inside the package causing the accumulation of water, giving way to the proliferation of spoilage and pathogenic microorganisms.

At higher temperatures, the respiration of the fresh produce increases, resulting in decreased O_2 and CO_2 concentration. Generally, the increase in respiration rate with respect to temperature is much higher than the corresponding increase in permeability of the packaging films, which leads to occurrence of anaerobic conditions inside the package. Therefore, the packages optimized at low temperature may face these problems and the only solution to this problem is to exercise proper control of temperature. However, in most of the developing countries the uniform temperature control is not so easy. A valid alternative, therefore, is to develop the packages that can function at the highest temperatures that are encountered during life of packaged produce. Also, with the increased respiration rate, there is increased accumulation of water which is produced as one of the end product of the respiration process.

Aharoni et al., (2007) suggested a silicon based coating on the inner surface of packaging material as the possible solution to the water condensation without disturbing the RH inside the package. The silicon coating does not prevent condensation but reduces the size of water droplets and disperses them over larger surface area.

4. Methods for Shelf Life Extension of Fresh Produce

The above mentioned quality issues and the factors affecting the shelf life of fruits and vegetables form the basis of the storage and packaging technologies that aim to extend the shelf life of fresh produce. The storage and packaging technologies that can be used for the enhancement of shelf life are cold storage, modified / controlled atmosphere storage and packaging, active packaging, edible coatings, etc.

4.1 Modified atmosphere packaging (MAP)

MAP is the enclosure of fresh produces in a package in which the atmosphere inside the package is modified or altered to provide an optimum atmosphere for increasing shelf life while maintaining quality. Modification of the atmosphere may be achieved either actively or passively. Active modification involves displacing the air with a controlled, desired mixture of gases, and is generally referred to as gas flushing. Passive modification occurs as a consequence of the respiration/ metabolism of the enclosed commodity; the package structure normally incorporates a polymeric film and so the permeation of gases through the film (which varies depending on the nature of the film and the storage temperature) also influences the composition that develops.

The normal composition of air by volume is 78.08% nitrogen, 20.95% oxygen, 0.93% argon, 0.03% carbon dioxide, and traces of other nine gases in very low concentrations. The three main gases used in active MAP are O_2, CO_2 and N_2, either singly or in combination. The choices of suitable packaging materials for the MAP of respiring produce such as fruits and vegetables is much more complex due to the dynamic nature of the product. The main characteristics to be considered when selecting packaging materials for MAP are the package permeability to gases and the respiration characteristics of the commodity.

The exchange of gases between a plant organ and its environment can be considered in four steps, viz., (1) diffusion of the gas phase through the dermal system; (2) diffusion of the gas phase through the intercellular system; (3) exchange of gases between the intercellular atmosphere and the cellular solution (cell sap), and (4) diffusion of solution within the cell to centers of oxygen consumption or from centers of carbon dioxide and ethylene production, Kader (1986).

Gas diffusion within a fruit or vegetable is determined by the respiration rate, maturity stage, physiological age, commodity mass and volume, pathways and barriers for diffusion, properties of the gas molecule, concentration of gases in the atmosphere surrounding the commodity, magnitude of gas concentration gradient across barriers, and temperature. In turn, the respiration rate of a commodity inside a polymeric film package will depend upon the kind of commodity, maturity stage, physical condition, concentrations of O_2, CO_2 and C_2H_4 within the package, commodity quantity in the package, and storage temperature.

In MAP, the O_2 partial pressure is typically reduced while that of CO_2 is often increased. The purpose of this is to lower the respiration rate and slow down the metabolic pathways that negatively affect the quality of the stored product. The respiration rate is a good indicator of the physiological stage of the fruit and its storage potential; efforts have been put to model respiration or gas exchange of fruit and vegetables (Nicolai et al., 2009). Temperature is generally recognized as the most important external factor influencing respiration. Biological reactions generally increase two to three folds for every 10 °C rise in temperature. The respiration rate is usually maximal at moderate temperatures between 20 and 30°C, but decreases considerably to almost zero at around 0°C, depending on the genus, species and even cultivar. Respiration rate (RO_2 and RCO_2) is generally calculated using Eqs. 1 and 2. Respiratory Quotient (RQ)is calculated from Eq. 3 (Sen et al., 2012).

$$RCO_2 = ([CO_2]_{t+1} - [CO_2]_t / \Delta t)\ V_{fr} /W \tag{1}$$

$$RO_2 = ([O_2]_t - [O_2]_{t+1} / \Delta t)\ V_{fr} /W \tag{2}$$

$$RQ = RCO_2 / RO_2 \tag{3}$$

Where, RO_2 is the respiration rate, mL O_2 $kg^{-1}h^{-1}$; R_{CO2} is the respiration rate, mL CO_2 $kg^{-1}h^{-1}$; $[O_2]$ and $[CO_2]$ are the gas concentrations for O_2 and CO_2, respectively; t is the storage time in hours; Δt is the time difference between two gas measurements; V_{fr} is the free volume of the respiration chamber in mL and W is the mass of the fruit in kg.

At high temperatures over 30°C, enzymatic denaturation may occur resulting in reduced respiration rates. The respiration rate is affected by the development stage and the respiration pattern (climacteric/non climacteric) of the fruit as well (Nicolai et al., 2009). The more recent approach for modelling respiration rate by employing Michaelis–Menten type equations (Eq. 4-8) for prediction of respiration rate in terms of CO_2 evolution and O_2 consumption, respectively, is based on enzyme kinetics (Lee et al., 1991). McLaughlin and O'Beirne (1999) described the respiration rate in three ways as no inhibition, competitive inhibition and uncompetitive inhibition.

$$R_g = \frac{V_{m,g} C_{O_2}}{K_{m,g} + C_{O_2}} \tag{4}$$

$$R_g = \frac{V_{m,g} C_{O_2}}{K_{m,g}\left[1 + \frac{C_{CO_2}}{K_{i,g}}\right] + C_{O_2}} \tag{5}$$

$$R_g = \frac{V_{m,g} C_{O_2}}{(K_{m,g} + C_{O_2})\left[1 + \frac{C_{CO_2}}{K_{i,g}}\right]} \tag{6}$$

$$R_g = \frac{V_{m,g} C_{O_2}}{K_{m,g} + C_{O_2}\left[1 + \frac{C_{CO_2}}{K_{i,g}}\right]} \tag{7}$$

$$R_g = \frac{V_{m,g} C_{O_2}}{K_{m,g}\left[1+\frac{C_{CO_2}}{K_{i,g}}\right]+C_{O_2}\left[1+\frac{C_{CO_2}}{K_{i,g}}\right]} \quad (8)$$

Where, R_g is respiration rate in terms of gas (O_2/ CO_2); $V_{m,g}$ is the maximum rate of O_2 consumption or CO_2 production and $K_{m,g}$ is the dissociation constant of the enzyme–substrate complex or the concentration corresponding to the half-maximal respiration rate, and $K_{i,g}$ are inhibition constant. Linear dependence of respiration rate on oxygen concentration is presented by Eq. (4). The role of CO_2 in respiration was suggested to be mediated via inhibition mechanisms of the Michaelis–Menten equation and to be (i) competitive (Eq. 5), (ii) uncompetitive (Eq. 6), (iii) non-competitive (Eq. 7), and (iv) a combination of competitive and uncompetitive types of inhibition (Eq. 8).

4.1.1 Permeability

The mechanism by which substances travel through an intact plastic film is known as permeation. It involves dissolution of the penetrating substance, the permeant in the plastic, followed by diffusion of the permeant through the film, and finally by evaporation of the permeant on the other side of the film, all driven by a partial pressure differential for the permeant between the two sides of the film (Mangaraj et al., 2009).

The barrier performance of the film is generally expressed in terms of its permeability coefficient or permeability. For one-dimensional steady-state mass transfer, the permeability coefficient is related to the quantity of permeant, which is transferred through the film as represented by the Eq. (9)

$$P = \frac{Qx}{At\Delta p} \quad (9)$$

Where, P is the permeability coefficient, Q is the amount of permeant passing through the material, x is the thickness of the plastic film, A is the surface area available for mass transfer, t is the time, and Δp is the change in permeant partial pressure across the film.

Hence, the permeability coefficient (P) is the proportionality constant between the flow of the penetrant gas per unit film area per unit time and the driving force (partial pressure difference) per unit film thickness. The amount of gas penetrating through the film is expressed in terms of either moles per unit time (flux) or weight or volume of the gas at STP. Commonly, it is expressed in terms of volume.

The relationship between the rate of permeation and the concentration gradient is one of direct proportionality and is embodied in Fick's first law (Eq. 10)

$$J = -D\frac{\Delta c}{\Delta x} \tag{10}$$

Where, J is the flux per unit area of permeant through the polymer, D is the diffusion coefficient, c is the concentration of the permeant, Δx is the concentration gradient of the permeant across a thickness Δx.

It can be shown that the permeability coefficient (P), as defined by Eq. 9, is equal to the product of the Fick's law's diffusion coefficient, D, and the Henry's law's solubility coefficient, S (P = DS), in situations where these laws adequately represent mass transfer (ideally in dilute solutions, diffusion is independent of concentration). The permeability coefficient, under these circumstances, is a function of temperature, but is not a function of film thickness or permeant concentration.

4.1.2 Temperature quotient for permeability

The influence of temperature on permeability of polymeric films was quantified with the Q_{10}^{P} value, which is the permeability increase for a 10°C rise in temperature and is given by the following equation (Eq. 11).

$$Q_{10}^{P} = \left(\frac{P2}{P1}\right)^{10/(T2-T1)} \tag{11}$$

Where, Q^{P}_{10} is the temperature quotient for permeability; and P_1 and P_2 are the permeabilities at temperatures T_1 and T_2, respectively.

4.1.3 Effect of perforations in MAP

The polymeric films used for MAP are of three types is (i) polymeric films without perforations or microperforated; (ii) macro perforated polymeric films, and (iii) perforation mediated packaging systems.

Microperforated or non-perforated polymeric films yield low O_2 and low CO_2 concentrations because the CO_2 permeability of these materials is generally 3–6 times that of O_2 permeability. These materials are suitable for less CO_2 tolerant commodities such as mango, banana, grapes and apples. The gas permeability in microperforated polymeric films is temperature dependent and this dependence is commonly described by Arrhenius-type equations (Eq. 12)

$$P = P^{ref}\left[\frac{-E_a^P}{R}\left(\frac{1}{T}-\frac{1}{T_{ref}}\right)\right] \tag{12}$$

Where, P is the permeability at temperature T, E^p_a is the activation energy for permeation, R is the ideal gas constant, T is the absolute temperature, T_{ref} is the reference temperature, and P^{ref} is the permeability at the reference temperature.

Perforated films have higher permeability rate but the ratio of CO_2 to O_2 permeability is much lower, approaching unity. Such films are, therefore, of great interest for commodities tolerating simultaneously low O_2 and high CO_2 levels such as fresh-cut products, strawberry and mushroom that is commodities having high respiration rate.

In the perforation-mediated packaging, tubes are inserted in an airtight package. This system is also adequate for products requiring high CO_2/low O_2 concentrations and minimizes water accumulation inside the package. The perforation-mediated packaging system is a rigid one, which is suitable for bulk products and for products sensitive to mechanical damage (Mangaraj et al., 2009).

4.2 Controlled atmosphere (CA) storage

In controlled atmosphere storage gas composition is continuously maintained different from normal air with respect to CO_2 and O_2 levels within air tight compartments. This is done by intermediate monitoring and adjustment to the predetermined set value of CO_2 and O_2 levels using high end technologies. There are different types of controlled atmosphere storage depending mainly on the method or degree of control of the gases. Some researchers prefer to use the terms "static controlled atmosphere storage" and "flushed controlled atmosphere storage" to define the two most commonly used systems. "Static" is where the product generates the atmosphere and "flushed" is where the atmosphere is supplied from a flowing gas stream, which purges the store continuously. Systems may be designed which utilize flushing initially to reduce the O_2 content then either injecting CO_2 or allowing it to build up through respiration, and then maintenance of this atmosphere by ventilation and scrubbing. The main components of controlled atmosphere storage system are described below.

4.2.1 Gas-tight cells

A gas-tight cell is the most important component of a CA system mostly made up of an insulation core and on both sides powder coated galvanized iron sheets (PCGI) as they offer excellent anti corrosion properties. The chamber is closed with a gas-tight door which enables samples to be taken in and out of the chamber for loading, unloading or for analysis. Sometimes a window is fitted into the door to view the products without opening the door. Three types of commonly used gas-tight cells are briefly described below.

CA storage compartments

The CA storage compartment may be a small chamber or large room. Appropriately spaced perforated or non-perforated trays may be present for holding the commodity. Each chamber must have the following components

- Refrigeration unit
- Humidification system
- Nitrogen and carbon dioxide purging facility
- Venting facility (gas and condensate)
- Gas, temperature and humidity sensors
- Heating element

Palliflex storage unit

The Palliflex unit comprises a special plastic pallet and a plastic cover for gas-tight sealing, the later should remain supple at low temperatures (Figure 2). Temperature cannot be regulated per Palliflex unit as all the units in a particular cool cell will maintain the same target temperature set for the cell. This system is beneficial as different gas conditions can be set per pallet, i.e. individually. However, if different temperatures are required for the various products, then more than one cool cell must be used. In the Palliflex units, the following items are required for measuring and regulating the CA conditions

- Temperature sensor
- Bottles of nitrogen (N_2) or O_2 scrubber (oxygen reduction)
- Bottles of CO_2
- Gas inlet and outlet hoses
- An overpressure protection device, to be fitted in the cover
- Compressed air or an air pump
- A fully automatic measuring and regulation system with a built-in O_2/CO_2 meter.

Fig. 2: A typical Palliflex Storage System

Advantages of Palliflex unit over compartment storage are

- Individual setting of gas conditions per unit is possible
- Fungal spores and flavors/smells cannot migrate from unit to unit.
- No loss of CO_2 is these during when connecting and disconnecting units
- There is no disturbance to the conditions in other units when connecting and disconnecting.
- Easy to trace any leakages.

Bin storage unit

The bin units are manufactured of plastic with a special cover to make a gas tight connection and can store up to 140 kg.

Determining the capacity of the storage cell

The capacity of the storage cell depends upon the following factors

- Total tonnage of the commodity to be stored
- Bulk density of the commodity
- Volume of storage accessories
- Free volume inside storage required for material handling

According to the standard for commercial CA storage rooms, 65% of the total volume should be kept free.

4.2.2 Insulation

To prevent heat infiltration, insulated panels are supplied with a material having low thermal conductivity (Table 1) like poly urethane foam (PUF) or polyisocyanurate (PIR) on the wall, top, floor, partitions and doors. Internal and external surface are engineered to suit all-weather requirements.

Table 1 : Properties of poly urethane foam (PUF) and polyisocyanurate (PIR)

Properties	PIR	PUF
Density	32± 2 kg/m^3	36 ± 2 kg/m^3
Compression strength, kgf/cm^2(in direction of rise)	1.75	1.75
Thermal conductivity, w/m-k	0.021	0.023
Temperature range, °C	150 to – 200	110 to – 180
Ignitibility (bs- 476, part-5, 1965)	Class 'P'Not easily ignitable	Class 'P'Not easily ignitable
Mean extent of burn, mm	<25	< 125

The Fourier's law of heat transfer states that heat flux (Q) is given by Eq. 13, where, A is the area perpendicular to the direction of heat transfer; ΔT is temperature difference between inside and outside the chamber, and the overall heat transfer co-efficient (U) is calculated based on insulation and wall thickness (Δx) using Eq. 14; where K_w and K_{in} are the thermal conductivity of wall and insulation; h_0 and h_i are outside and inside convective heat transfer coefficient respectively. The overall heat transfer co-efficient (U) for PUF of different thickness in given in Table 2.

$$Q = U \times A \times \Delta T \tag{13}$$

$$\frac{1}{U} = \frac{1}{h_0} + \frac{dx_{in}}{K_{in}} + \frac{dx_w}{K_w} + \frac{1}{h_i} \tag{14}$$

Table 2: U values of PUF of different thickness

Thickness (mm)	60	80	100	120	150
'U' values (W/m^{2}K)	0.36	0.26	0.21	0.19	0.14

4.2.3 Refrigeration system

The refrigeration system contains the following four key components

- The evaporator unit is usually made up of stainless steel and mounted inside the cold controlled atmosphere room. Here the refrigerant liquid changes the state from liquid to gas absorbing heat from the products stored in controlled chamber.
- Compressor to compress the refrigerant gas.
- The condensing unit to condense the refrigerant gas into liquid by rejecting heat to the surrounding.
- Thermostat switches the compressor on or off to maintain a relatively constant set temperature.

Determination of capacity of refrigeration system

The heat load added by the products to be stored is important in designing refrigeration capacity of controlled atmosphere storage. The following three types of heat load should be considered.

Chilling load, Qc, kJ/h calculated from Eq (15)

$$Q_C\left(\frac{kJ}{h}\right) = m \times C_p \times \frac{T_1 - T_2}{t} \tag{15}$$

Where, m = is mass of product, C_p= mean specific heat of product, $T_1 - T_2$= initial and final temperature difference and t = chilling times, hours.

Biological products generate heat even when they are stored in cold atmosphere due to heat of respiration by living tissue and heat of reaction of chemical components of food materials.

The amount of heat generated by food materials is Q_R (kJ/h) = $\left(\frac{m \times h_g}{t}\right)$

Where, h_g = Heat generated per kg of food per hour.

The steady state heat flow by conduction from the walls and ceiling Q_H is given by Eq. 13 and 14. The total heat load is sum of all the above three heat loads.

4.2.4 Humidification

Fruit, that is stored in a controlled atmosphere cell, always loses moisture as cooling takes water out of the atmosphere, thus limiting the shelf life and quality of the stored commodity. This moisture loss reduces if the relative humidity levels (RH) are kept above 90%. A humidification system maintains relative humidity upto desired level. The RH in the storage room can be controlled by fine water spray or steam. Both the types are briefly discussed below.

Humidifiers with circulating water

Spray humidifier (Air washers) : Humidity is maintained by spraying of water in a humidification chamber. Due to the relatively large drops, only a small part of the injected water evaporates. The non-evaporated water is collected in a trough and conveyed to the nozzles. This system is having disadvantages such as necessity of high pump action and poor control possibility.

Contact humidifiers, trickling humidifiers: Humidity is maintained by spraying water on a porous surface passed by the air flow which is to be humidified. Evaporation takes place on the large and humid surface. The water which has not evaporated is conveyed back to the contact surface in circulation process. Classification of the contact surface is a major problem which must be addressed.

Disc atomizer : Water is injected on to a fast rotating disc and, due to the centrifugal power, spun off this as a fine mist. Calcification of the disc should be taken care of.

Humidifiers without circulating water

Ultrasonic atomizer : To maintain humidity in the compartment a membrane is initiated to vibrate through a high frequency signal. It transmits the vibrations to the water to be evaporated so that the water is atomized as fine mist. Fine atomization can be achieved but the unit is expensive and only small construction sizes are feasible.

Direct steam injection type : Steam is pure water vapor; it needs only to be mixed with air to satisfy the demands of the controlled system. In addition, direct steam humidifiers can meter output by means of a modulating control valve. As the system responds to control, it can position the valve anywhere from closed to fully open. As a result, response is quick and precise to fluctuating humidity demand. Assuming boiler makeup water is of satisfactory quality and there is no condensation, dripping or spitting in the ducts and no bacteria the major advantages of this humidifier are as follows.

- The high temperatures make it virtually a sterile medium and corrosion is rare.
- Scale and sediment whether formed in the unit or entrained in the supply steam are drained from the humidifier through the steam trap.
- The steam supply itself acts as a cleaning agent to keep system components free of mineral deposits that can clog many forms of water spray and evaporative pan systems.
- Response to control and pinpoint control of output.

Steam-to-steam humidifiers : In these humidifiers a heat exchanger is used and the heat of treated steam is used to create a secondary steam for humidification from untreated water. The secondary steam is typically at atmospheric pressure, placing increased importance on equipment location. Maintenance of steam-to-steam humidifiers is dependent on water quality. Impurities such as calcium, magnesium and iron can deposit as scale, requiring frequent cleaning. Response to control is slower than with direct steam because of the time required to boil the water.

Determination of capacity of humidifier: Humidity of the resultant stream is based on the principle of mass and energy balance. Let m_1, h_1, H_1 and m_2, h_2, H_2 be the flow rate, humidity ratio and enthalpy of two dry air streams, respectively.

Flow rate of resulting air stream, $m_3 = m_1 + m_2$ (16)

Humidity ratio of the resulting air stream, h_3 is given by

$$m_3h_3 = m_1h_1 + m_2h_2 \quad (17)$$

Enthalpy of the resulting air stream, H_3 is given by

$$m_3H_3 = m_1H_1 + m_2H_2 \quad (18)$$

4.2.5 CO_2 *maintenance*

Air contains 0.03% CO_2 and some amount of the gas is accumulated in the controlled atmosphere storage with time due to product CO_2 evolution during respiration. CO_2 cylinders with pressure regulators are provided with the storage chamber to purge the excess amount of the gas, if required to maintain the target level of CO_2 inside the chamber. Rising CO_2 level above the target level inside the controlled chamber is maintained by CO_2 scrubber which is a device that absorbs carbon dioxide. Many technologies are available for removing CO_2 but the most economical one used for controlled atmosphere storage is described below.

Active carbon filter

Air from the storage area is removed by the CO_2 scrubber, and the CO_2 purified air is then fed back in. The CO_2 scrubber has a cycle of two activities, viz.

- Absorption - removal of the carbon dioxide by using an active carbon filter, and
- Regeneration - cleaning the active carbon filter.

The scrubber's active carbon filter takes up carbon dioxide up to its saturation point during absorption. The purified air goes back into the cold store. During the last phase of the CO_2 take-up, a small part of the purified air is stored in a flexible buffer 'lung' otherwise under-pressure condition is created in the storage cell. After cleansing with ambient air, and removing the carbon dioxide from the active carbon, the 'lung air' is used to reduce the oxygen level in the tank with the active carbon filter to the level in the storage cell. This system, guarantees that the undesirable inflow of oxygen into cold store is minimized. The most important components of the CO_2 scrubber are enlisted below.

- Steel tank with active carbon which binds carbon dioxide,
- Air distribution system,
- Cabinet for internal process control,
- Flexible buffer made of PU film ('lung system') and
- Pump

In order to determine the capacity of a CO_2 scrubber for a particular CA unit, the following must be considered:

- Number of cool cells,
- Cool cell dimensions,
- Type & quantity of product in the cool cell (how much CO_2 does it produce) and
- Preferred storage conditions for CO_2 in the cell.

4.2.6 Nitrogen generators (O_2 scrubbers)

The concentration of oxygen in the controlled atmosphere can be decreased by purging in inert gas nitrogen. Nitrogen generators serve the purpose of separation of nitrogen and oxygen from air. The most commonly used technologies are discussed below.

Pressure swing adsorption

The PSA process cycle consists of the following key mechanisms:

- Pressurization / adsorption
- Pressure equalization, and
- Depressurization / desorption.

The separation takes place in an adsorber vessel filled with carbon molecular sieve due to faster kinetic diffusion of oxygen molecules into the pore structure of the carbon molecular sieve than for nitrogen molecules (Figure 3). Two such adsorber beds A and B are used. Sequential operation of each bed is discussed below.

Adsorber bed A : Oxygen, carbon dioxide and moisture are adsorbed from the entering compressed air after the operating pressure is reached in the bed, nitrogen product flows into a product receiver prior to entering the product piping. Then the bed undergoes depressurization (Figure 3b) and the oxygen enriched waste gas is vented to the atmosphere (Figure 3a). Depressurization permits the release of oxygen, carbon dioxide, and water vapour previously adsorbed during nitrogen production from adsorber bed A.

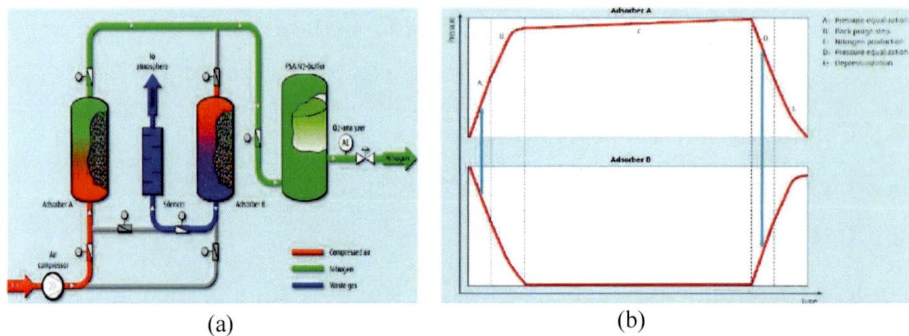

(a) (b)

Fig. 3: Typical process flow diagram and pressure time diagram for and Nitrogen PSA.

Adsorber bed B: Simultaneous to the nitrogen generation in bed A, adsorber bed B is depressurized to atmospheric pressure. Bed B (atmospheric pressure) is then pressurized to an intermediate pressure as the gas remaining in adsorber bed A after completion of nitrogen production (at operating pressure) flows into adsorber bed B. This step is known as equalization step and no air is consumed nor is product gas generated in this step. After that adsorber bed B is brought to operating pressure, and begins its nitrogen production portion of the cycle. Following nitrogen production, adsorber bed B undergoes equalization and subsequent depressurization. The cycle continues at the point where adsorber bed A undergoes pressurization and adsorber bed B is depressurized.

Therefore a nitrogen receiver is applied to allow for a constant flow, purity and pressure of the nitrogen product throughout the PSA cycle.

Vacuum pressure swing absorption

VPSA technology is based on an energy-efficient, low pressure technique. A VPSA nitrogen generator can both circulate and inject. During circulation, the residual oxygen level reduces along with the oxygen in the cold store. Circulation is the most energy efficient mode for lowering the oxygen level in a storage area. Pull-down time is an indication of the time necessary for circulation to bring the storage cells to the required oxygen levels. During circulation, the VPSA nitrogen generator reduces the oxygen concentration in the cell up to 30% quicker than when injecting.

Determination of capacity of nitrogen generator

The amount of N_2 flush depends on purity level and flow rate of N_2, O_2 concentration set point, and free volume of the CA storage.

The following assumptions were made for the development of the relationship between amount of N_2 required and different operating parameters of the CA chamber.

- The chamber is leak proof and perfectly well mixed.
- The change in gas composition due to respiratory metabolism of stored produce during flushing is negligible.
- Concentration of O_2 present in N_2 is the same throughout the flushing.

A mass balance between the inflow and outflow of O_2 in the CA chamber was made as follows (Eqs 19 and 20).

$$\text{Moles of } O_2 \text{ inflow} = V_{N_2} \times (O_2)_N \times \rho_{O_2} \times \frac{dt}{M_{O_2}} \tag{19}$$

$$\text{Moles of } O_2 \text{ outflow} = V_{N_2} \times (O_2)_t \times \rho_{O_2} \times \frac{dt}{M_{O_2}} \tag{20}$$

Where, V_f is the free volume in CA chamber in m^3; V_{N_2} is the flow rate of N_2 gas, m_3/s; $(O_2)N$ is concentration of O_2 present in N_2 gas, %; $(O_2)_t$ is the concentration of O_2 present in the CA chamber at time *t*, %; *t* is flushing time, s; ρ_{O_2} is the density, kg/m^3 and M_{O2} is the molecular weight of O_2, kg/mol. Combining Equations 19 and 20 leads to Eq. 21.

$$\text{Total moles of } O_2 \text{ accumulation} = \{(O_2)_t - (O_2)_N\} \times \rho_{O_2} \times V_{N_2} \times \frac{dt}{M_{O_2}} \tag{21}$$

Total moles of O_2 accumulated in the CA chamber during flushing time of *t* is also equal to

$$-V_f \times d(O_2)_t \times \frac{\rho_{O_2}}{M_{O_2}} \tag{22}$$

This can be equated with Eq. 19. The -ve sign indicates that the O_2 concentration of the CA chamber decreases with flushing of N_2 for time *t*.

$$-V_f \times d(O_2)_t = \{(O_2)_t - (O_2)_N\} \times V_{N_2} \times dt \tag{23}$$

It was assumed that $x = (O_2)_t\ (O_2)_N$. Where $(O_2)_N$ is constant and does not change with time. Differentiating this equation leads to d $(O_2)_t = dx$. Hence, Equation 22 can be presented as Eq. 24.

$$V_f \times dx = V_{N_2} \times x \times dt \tag{24}$$

Rearranging Eq. 24 with operational limits yield as $x_0 = (O_2)_a - (O_2)_N$ and $x = (O_2)_t - (O_2)_N$, where $(O_2)_a$ is O_2 concentration of ambient air (average 20.95%)

$$\int_{x_0}^{x} \frac{1}{x} dx = \int_{0}^{t} \frac{V_{N_2}}{V_f} \tag{25}$$

and integrating

$$V_{N_2} = \frac{V_f}{t} \times \ln \frac{(O_2)_a - (O_2)_N}{(O_2)_t - (O_2)_N} \tag{26}$$

Liquid nitrogen CA chamber : Liquid nitrogen is a cryogenic fluid which boils at a very low temperature at atmospheric pressure. Use of LN_2 for CA storage is favorable to attain low O_2 atmosphere and also act as a means of cooling. The required volume of liquid nitrogen in a controlled atmosphere chamber is

$$V_{N_2 liquid} = \frac{V_f}{t \times ER} \times \ln \frac{(O_2)_a - (O_2)_N}{(O_2)_t - (O_2)_N} \tag{27}$$

Where, $V_{N_2 liquid}$ is the amount of liquid N_2, m^3/s; and ER is expansion ratio which is defined as the ratio of density of liquid N_2 and density of N_2 vapor (ρ_{N_2}) at any temperature T. The $\rho_{N_2 liquid}$ is density of liquid N_2 which is calculated as 808 $kg/m^3$is calculated by ideal gas laws equation.

$$\rho_{N_2} = \frac{P}{R \times T} \tag{28}$$

Where P is the pressure inside the CA chamber during flushing. R is the gas constant and T is the absolute temperature.

Membrane nitrogen generator

The permeation of gases through polymer membranes is often understood in terms of a solution-diffusion model, for which the permeability coefficient is the product of a solubility coefficient (S_A) and a diffusion coefficient (D_A).

$$P_A = S_A \times D_A \tag{29}$$

Selectivity may then arise either because of differences in the solubility coefficient (solubility selectivity, S_A/S_B) or because of differences in the diffusion coefficient (mobility selectivity, D_A/D_B)

$$aA/B = \frac{P_A}{P_B} = \frac{S_A}{S_B}\frac{D_A}{D_B} \tag{30}$$

Materials that perform close to the upper bound in the separation of simple gases are generally glassy polymers. These typically show mobility selectivity, with smaller gas molecules diffusing more rapidly than larger ones. In contrast, rubbery polymers are more often used to separate large organic molecules from smaller gas molecules on the basis of high solubility selectivity.

The diffusion of gases through polymers is an activated energy process, thus, the relationships for *D*, *S*, and *P* as a function of temperature can be expressed as Eq. 32 to 33

$$D = D_o \exp(- E_d / RT) \tag{31}$$

$$S = S_o \exp(- \Delta H_s / RT) \tag{32}$$

$$P = P_o \exp(-E_p / RT) \tag{33}$$

Where, E_d is the activation energy of diffusion, ΔH_s is the heat of sorption and $E_p = E_d + \Delta H_s$ is the activation energy of permeation. The permeability of gases through polymers is dependent upon the gas–polymer combination. For specific gases, the range of permeability through known organic polymers can range from 6 to 10 orders of magnitude.

4.2.7 Ethylene decomposers

Ethylene decomposer removes ethylene from cold stores based on catalytic combustion. Using oxygen the ethylene decomposer combusts ethylene to form CO_2 and water, both inert substances. This enables ethylene to be kept at any level that may be required, both in the ppm and the ppb range. An ethylene decomposer is dimensioned based on the following parameters.

- F Output of the decomposer (F, m^3/h)
- The total tonnage of fruit
- Ethylene production of fruit (R, µL kg-1 h-1)
- Required ethylene concentration (c, ppm)
- D = Efficiency ratio (Standard is 95% 0.95)

Dimensioning is subsequently calculated using the formula given below in Eq. 34

$$F = \frac{R \times M}{C \times D} \tag{34}$$

Potassium permanganate: Potassium permanganate grains bind ethylene excellently, but large amounts are necessary to bind ethylene in large cells.

Ozone: Additionally ozone is often mentioned as a substance for breaking down ethylene but it must be produced on location because it is unstable and therefore not transportable.

4.2.8 Electrical control panel compartment

It receives the main electrical supply and distributes it through motor control center (MCC) to nitrogen generation section, refrigeration and relative humidity control section, electrical control panel and controlled atmosphere chamber section. The electrical control panel must have the following main compartments.

- Mains "ON/OFF" Switch.
- Three phase indicators.
- Ammeters /Voltmeters for individual phases.
- CO_2 indicator.
- Ethylene indicator.
- Refrigeration temperature indicator.
- Relative humidity indicator.
- Operating touch screen.
- Emergency switch (Push type).

4.2.9 Sensors for O_2 and CO_2 analysis equipment

This meter is used to measure O_2 and CO_2 composition inside the storage unit. They can be of two types:

- Inbuilt in the CA chamber: They comprise printed circuit boards. The O_2 meter is equipped with printed circuit boards and CO_2 meter are equipped with dual wave infrared sensor.
- Meter for manual measurement of O_2 and CO_2. They are used for manually controlling gas conditions in storage cells. Extremely suitable electronic cell or paramagnetic sensor is used for measurement of oxygen.

Some other type of meter use zirconium sensor for the same purpose. It should be ensured that the read outs should not be effected by the position of the meter.

4.2.10 Control system

A control system measures and registers O_2 and CO_2 and operates the CO_2 and O_2 scrubbers. In addition, the same control system can regulate cooling (switching on/off, defrosting and machine room regulation), carry out ethylene measurements and operate the ethylene converter.

4.2.11 CA storage of banana

Based on the above design considerations, a multi-product controlled / modified atmosphere storage facility (Figure 4) for the storage of fresh fruits and vegetables was designed, and commissioned at the Agricultural and Food Engineering Department, IIT Kharagpur. This unit has four chambers 250 kg fruit storage cap.. each and each chamber has individual provisions to control the parameters like O_2 (0-21%), CO_2 (0.03-25%), C_2H_4 (0-10,000 ppm), temperature (0-50 °C) and relative humidity (10-95%) separately. The CA/MA unit delays the ripening of the fresh produces by reducing its respiration rate achieved by maintaining the temperature, relative humidity and concentration of gases in combination as needed for the respective produce. Currently, the trialruns on the CA storage of tomatoes, guava, mushroom and banana are done. The synergistic effects of CA storage with the already developed process technologies of active packaging (extended shelf life 55 days) and edible coating (extended shelf life 48 days) for tomatoes are also evaluated.

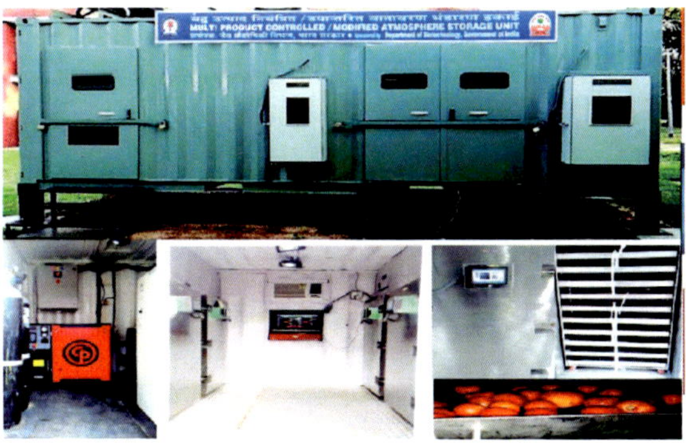

Fig. 4: Multiproduct CA/MA storage facility at IIT Kharagpur (See colour version on page 336)

4.3 Active packaging

4.3.1 Scavengers / emitters used in active packaging

Moisture absorbers

Moisture is one of the most critical factors that need special attention while designing a package for any food stuff. The packaging material with water vapor barrier properties is a prerequisite for the moisture sensitive food products. The packaging materials should be carefully chosen considering the water vapor permeability of the packaging material and the desired relative humidity requirements of the food stuff in the package headspace. Despite of this, a certain amount of moisture can be trapped in the packaging or develop during distribution. If not removed, this moisture will be absorbed by the product or condensate will be formed, causing microbial spoilage and/or low consumer appeal. Moisture control is required to prevent condensation when fresh horticultural produce respires so that water activity (a_w) is lowered and by the consequence, growth of molds, yeast and spoilage bacteria on foods with high a_w, is lowered.

The moisture control should be very precisely regulated to attain the optimal moisture content as even the excessive moisture loss can lead to the wilting (shrinkage) and toughening of texture and may sometimes cause lipid oxidation. In MAP these optimal moisture conditions are attained by the natural interplay between the permeability of the packaging material and the packaging contents which is a time taking process. This process can be hastened by using moisture controlling sachets or films containing moisture adsorber to achieve these conditions faster. Desiccants are successfully being used for a wide range of foods, such as cheeses, meats, chips, nuts, popcorn, candies, gums and spices (Anon 1995). The shelf-life of packaged tomatoes at 20 °C was extended from 5 to 15 days with a pouch containing NaCl, mainly by retardation of surface mold growth (Vermeiren et al., 1999).

Oxygen absorbers

The oxygen can be detrimental to many foods as it causes the food spoilage by oxidation of food constituents or spoilage by molds and aerobic bacteria. In case of fresh fruits and vegetables, the presence of oxygen may result in high respiration rates which make food susceptible to deterioration. Though O_2 concentration can be lowered with the modified atmosphere packaging or vacuum packaging, these technologies are often slow and imprecise. Moreover, there are chances of O_2 permeating through the packaging materials. The process of eliminating oxygen from the package can be enhanced by the addition of oxygen scavengers.The mechanism of action of these scavengers is based on the oxidation of one or more of the materials such as iron powder, ascorbic acid, photo-sensitive dyes, enzymes (such as glucose oxidase and ethanol oxidase), unsaturated fatty acids (such as oleic, linoleic and linolenic acids), rice extract. These scavengers are usually embedded in sachets that are highly permeable to oxygen and are labelled as “DO NOT EAT” to prevent possible accidental ingestion of the scavenger. These scavengers can reduce the oxygen content to as low as 0.01% as against the MAP’s usual achievable concentration of 0.3-3.0% (Day and Potter 2011).The recent developments in commercialization of oxygen scavengers involve the incorporation of these active materials directly into the packaging material increasing the surface area for their action.

Nowadays, some of the edible materials having the capacity to scavenge oxygen are being coated on the fruits and vegetables, especially cut fruits and vegetables, which are more prone to oxidation.

Carbon dioxide absorbers/ releasers

The produced CO_2 has to be removed from the package to avoid deterioration of food i.e. CO_2 injury and/or package destruction. In some cases, however, high concentration of CO_2 levels (10-80%) is desirable such as meat and poultry because high levels inhibit surface microbial growth and thereby extend shelf life. Removal of oxygen from the package by the use of oxygen scavenger creates a partial vacuum, which may result in a collapse of flexible packaging. In addition, when a package is flushed with a mixture of gases including CO_2 that dissolves partly in the product and create partial vacuum. The O_2 absorber/ CO_2 generator are mainly used in products where package volume and package appearance are critical e.g. peanuts or potato crisps. Such systems are based on either ferrous carbonate or a mixture of ascorbic acid and sodium bicarbonate.

Ethylene absorbers

Ethylene is a chemically simple, ubiquitous chemical that has diverse and profound effects on the physiology of plants. Ethylene has so many different effects on plants, is effective in very low concentrations, and its effects are so dose-dependent that it has been identified as a plant hormone. Ethylene has long been recognized as a problem in postharvest handling of horticultural produces. It is now recognized that ethylene, in very low amounts, can be responsible for a wide array of undesirable effects in plants and plant parts. In the case of climacteric fruit, ethylene can induce a rapid and irreversible elevation in respiration leading directly to maturity and senescence. In non-climacteric plant organs, ethylene induces a reversible increase in respiration. In most cases, exposure to a few parts per million (ppm) of ethylene leads to increased respiration and increased perishability.

Although some effects of ethylene are positive such as degreening of citrus fruit, ethylene is often detrimental to the quality and shelf-life of fruits and vegetables. To prolong shelf-life and maintain an acceptable visual and organoleptic quality, accumulation of ethylene in the packaging should be avoided. A lot of C_2H_4-adsorbing substances are described. Most of these are supplied as sachets or integrated into films. Many suppliers offer C_2H_4-scavengers based on potassium permanganate ($KMnO_4$), which oxidizes ethylene to acetate and ethanol. In this process, color changes from purple to brown indicating the remaining C_2H_4-scavenging capacity. Products based on $KMnO_4$ cannot be integrated into food-contact materials, but are only supplied in the form of sachets because $KMnO_4$ is toxic and has a purple color. The technology of C_2H_4-scrubbing has also been transferred to household refrigerators. Systems containing a zeolite coated with $KMnO_4$ are now available and are meant to be used in consumer refrigerators. Two examples are Mrs. Green's Extra Life cartridges from Dennis Green (USA) and Fridge Friend (TM) sachets from Ethylene Control (US patent 5278112) (Vermeiren et al., 1999).

Activated carbon has also been used for the adsorption and breakdown of ethylene. Charcoal containing PdCl as a metal catalyst was effective at 20 °C in reducing the rate of softening in minimal processed kiwi fruits and bananas and in reducing chlorophyll loss in spinach leaves but not in broccoli by scavenging ethylene (Abe and Watada, 1991). Films can also be used as the carriers for the C_2H_4 absorbing materials like finely dispersed zeolite, clay, and Japanese oya, but these materials alter the permeability of the packaging films. The CO_2 and C_2H_4 will tend to diffuse more rapidly and the permeability with respect to O_2 also increases to a great extent in these films as compared to the pure polyethylene (PE) films.

Flavour and odor releasers / absorbers

The odor and flavour absorbers can be used to remove the undesirable flavor substances that are produced due to various oxidative and biochemical processes occurring in the food during the deterioration of a product. These odors and flavors get accumulated in the headspace of the packages and on being exposed, they are detected by organoleptic senses of the consumers. In order to remove these undesirable odors, scavengers can be used for scalping but sometimes scalping can be detrimental to food quality. Odor/flavor is one of the important sensory attribute that is used by the consumers for judging the quality of food product, there are certain risks associated with its removal. By the removal of such undesirable flavor components, food is likely to be consumed even if it is spoiled. The flavor and odor scavengers have been utilized for debittering the grapefruit juice. The causes of bitter taste are naringin (aglycosidicflavanone) and limonin (triterpenoid lactone). Limonin is formed as a result of heat treatment of the citrus juice during processing and chemical action in the acidic medium. To remove this, an active thin cellulose acetate layer was applied inside the packaging material. This layer consists of fungal derived enzyme naringinase, consisting of α-rhamnosidase and β-glycosidase, which hydrolyses naringin to naringenin and pruning, both non-bitter compounds (Vermeiren et al., 2000).

Preservative releasers

To control the growth of undesirable microbes in foods, anti-microbial agents can be incorporated in or coated onto the food packaging material. Antimicrobial packaging material helps to extend the lag phase of microbes and reduce the growth rate of microbes to prolong the shelf-life and maintain food safety.

4.3.2 Selection of the scavengers / emitters

The factors that need to be considered while selecting the absorbers and/or releasers (Day and Potter, 2011) for a food package are:

- The size and weight of the product,
- The temperature and relative humidity to which the food will be exposed,
- Permeability (water vapor permeability and gas transmission rate) of the packaging material,
- The desired shelf life of the product,
- Water activity of the product,
- Sensitivity of the scrubber and product to moisture, and
- Absorption capacity and kinetics of the scrubber.

4.3.3 Active packaging of banana

The project was taken up to study the effect of active packaging components (ethylene absorbent, moisture adsorbent and CO_2 absorbent) on shelf-life and quality of fresh banana (Cv. Singapuri), a local popular cultivar of West Bengal, India. At 15 °C and 85% relative humidity, the storage life of all active packaged bananas was three times higher than that of the control samples (without any packaging). This implies that the active packaging technology has lower down the rate of metabolic activities resulting in an extended ripening process (Figure 5).

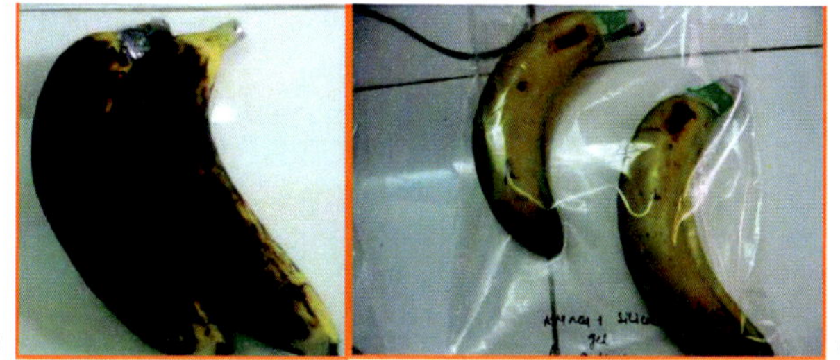

(a) (b)

Fig. 5: (a) Control samples after seven days and (b) packaged samples with 5g ethylene scrubber after 21 days of storage (See colour version on page 336)

4.3.4 Active packaging of tomatoes

The shelf life of tomatoes with AP has been successfully extended to 55 days which is around four times the shelf life of control tomatoes (without scrubbers)as shown in Figure 6. The firmness of the tomatoes after 55 days was found to be 1327 ± 42.52 g which was comparable to the firmness of fresh tomatoes 1365 ± 24.66 g.

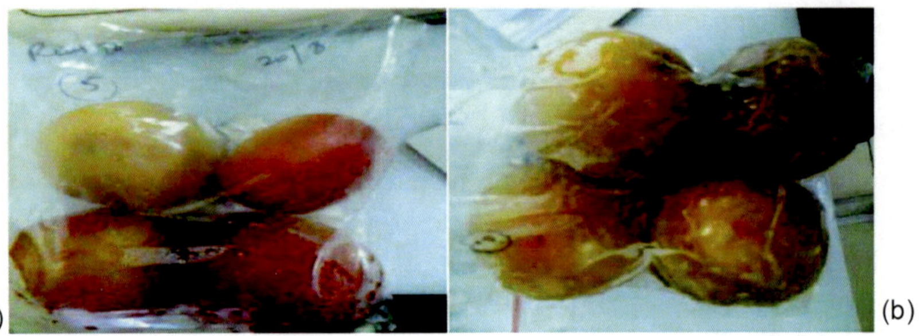

Fig. 6: (a) Actively packaged tomatoes after 55 days and (b) control tomatoes after 15 days (See colour version on page 336)

4.3.5 Active packaging of guava

Fully mature (firmness 1900 ± 27 g), free from any injury, medium sized guavas were used for the study. About 250 g of the fruit was heat sealed in polypropylene (PP) pouches of 25 μ thickness and was stored at 10 °C and 90 % RH in a thermostatically controlled environment test chamber. Samples were drawn at 7 days interval and analyzed for important physico-chemical parameters till a firmness of 750 ± 23 g, pH 4.87 ± 0.01 was observed. Three types of scavengers were used viz. silica gel as moisture adsorber, potassium permanganate as ethylene absorber and calcium hydroxide as CO_2 absorber. The concentration of scavengers was varied at 5-levels ranging from 0-4 g for silica gel, 0-1 g for ethylene scavenger and 0-4 g for carbon dioxide scavenger using central composite rotatable design (CCRD).

The experiment having 4 g silica gel, 0.15 g ethylene scrubber and 4 g carbon dioxide scrubber per 250 g of sample implicated 30 days shelf life (Figure 7). Table 3 shows important physico-chemical properties of freshly harvested guava and active packaged guava after 30 days of storage.

Table 3: Physico-chemical properties of fresh guava and active packaged guava after 30 days of storage at 10 °C and 90% RH

Sample	Firmness, g	Colour (L*,a*,b*)	pH	TSS °Bx
Freshly harvested mature guava	1900 ± 13.87	52.37, -6.84, 27.98	3.57 ± 0.01	3.6 ± 0.21
Active packaged guava after 30 days storage	750 ± 5.59	63.63, 6.74, 41.49	4.17 ± 0.01	5.25 ± 0.11

(a)   (b)

Fig. 7: (a) Actively packaged guava at 0 day, and (b) after 30 days of storage at 10 °C and 90% RH (See colour version on page 337)

4.4 Edible coating

Edible coatings can generally be defined as thin layers of protective edible materials applied on foods, or between food components, by wrapping, brushing, immersing or spraying. These coatings have the ability to reduce respiration rate, control mass transfer, colour changes, provide mechanical protection or add sensory appeal to food products, retain volatile flavors and reduce microbial growth (Gennadios et al., 1993). Coatings involve formation of films directly on the surface of the object they are intended to protect or enhance in some manner. In this sense, coatings become part of the product and remain on the product throughout use and consumption. Edible coatings formulated from protein, polysaccharide, and lipid substances often incorporated with plasticizers, surfactants and emulsifiers have been suggested as a means of food protection and preservation. A few examples have already found commercial use: meat casings from collagen, waxes for fruits and vegetables, and corn zein-based coatings for nutmeats and candies.

4.4.1 Selection criteria of edible films and coatings

Biopolymers, including proteins, polysaccharides and lipids or their combination, have been extensively studied to prepare edible coatings or films. Their functional properties are affected by the polymer structure, plasticizer concentration, solvent and other factors related to film dissolution, permeability and diffusion properties (Monaco et al., 2008). A proper understanding of the role of constituents on network formation is required for designing edible coating or film with the desired functional properties. The success of edible coating mainly depends upon the careful selection of parameters that give desirable internal composition of gaseous atmosphere for the controlled movement of O_2, CO_2 and water vapors suitable for specific products.

For edible coatings to be used in foods, there are several requirements to be considered, such as appropriate gas and water barrier properties; good mechanical strength and adhesion; reasonable microbial, biochemical and physico-chemical stability; effective carrier for antioxidant, flavour, color, nutritional or antimicrobial additives; safe for human consumption (free of pathogenic microorganisms and hazardous compounds); acceptable sensorial characteristics; low cost of raw materials; and simple technology for production. The optimization of edible coating composition is in one of the most important steps of the research in this field, since they must be formulated according to the properties of the fruits and vegetables to which they have to be applied (Rojas-Grau et al., 2009). Thus, it is very important to characterize and test different coating solutions on fresh and minimally processed food, since each one of them has different quality attributes to be maintained and enhanced during the storage time (Figure 8).

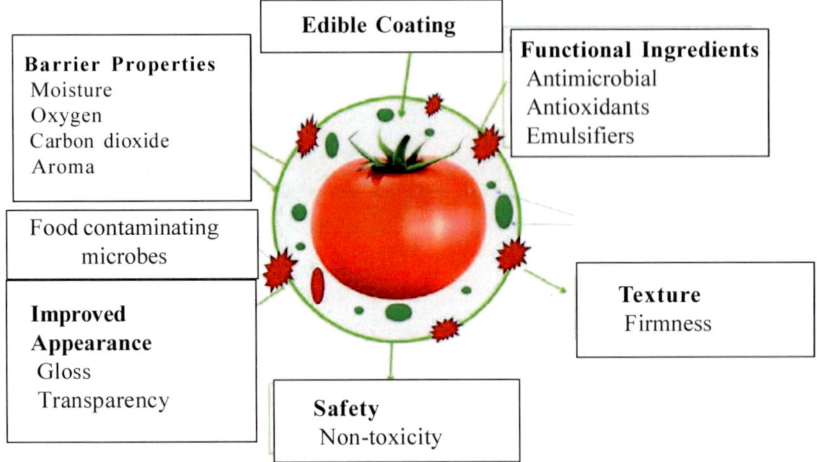

Fig. 8: Quality attributes provided by an edible coating to food products

The coating concentration or thickness of the film have to be carefully controlled, if it is too thick. It may adversely affect the quality of the food product because internal oxygen is below a desirable or beneficial level and consequently there is an increase in CO_2 concentration which is above critical tolerable level. These conditions lead to the anaerobic atmosphere which causes anaerobic fermentation. This situation can be controlled by taking some of points into consideration such as:

(i) Developing several edible coatings,

(ii) Controlling wettability of edible coatings, and

(iii) Measuring gas permeation properties of selected coatings,

(iv) Measuring diffusion properties of skin and flesh of selected foods

(v) Predicting internal gas compositions for fruits coated with edible films, and

(vi) Observing coating effects on quality changes of fruits.

4.4.2 Edible coating of tomato and mushroom

Coating of tomatoes

The coating solutions were optimized in two steps, firstly a CCRD design with two independent variables, i.e. concentration of sodium alginate (0 to 1.5 % w/v) and glycerol (0 to 2 % v/v) was used and experiments were conducted to finalize the concentration of sodium alginate and glycerol. In the second study, a factorial design (3-levels) was used to optimize the coating concentration which consisted of sunflower oil (1-3%, v/v), clove oil (0-0.5%, v/v), sodium alginate (0-1.5%, w/v) and glycerol (0-2%, v/v). The concentration of components in coating formulation affected the physico-chemical parameters of the coated tomatoes (Figure 9).

(a) (b)

Fig. 9: (a) Coated tomatoes after 48 days storage and (b) Uncoated tomatoes after 15 days storage (10 °C and 85% RH) (See colour version on page 337)

Coating of mushrooms

Coating of mushrooms was done so as to prevent it from excessive moisture loss. Mushrooms were applied with coating solution prepared with the same procedure as discussed for the preparation of the sodium alginate based coating. Coating of mushrooms with carboxy methylcellulose in composite form extended shelf-life its from 2 days to 7 days at refrigerated conditions i.e. 10°C and 85 % RH (Figure 10).

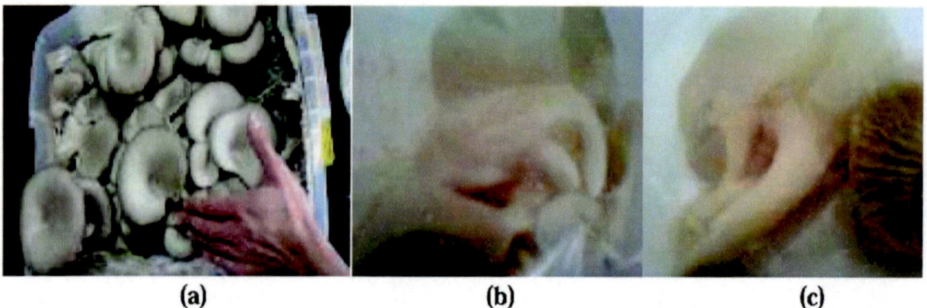

(a) (b) (c)

Fig. 10: (a) Fresh mushrooms (b) Control sample (c) Mushrooms coated with composite coating

4.4 Combination technology

The controlled atmosphere storage, active packaging, edible coating and modified atmosphere packaging have proven to be promising technologies for the extension of shelf life of fresh fruits and vegetables. The efficacy of these technologies can be studied for their potential synergistic effect when used in combination of one or more technologies. The synergistic effect of edible coating (EC) and CA storage was studied for tomatoes at IIT Kharagpur, wherein an optimized coating mixture was used to coat the tomatoes that were kept in the CA storage unit at 10 °C and 85% RH with varying gas concentrations ranging from 3-9% O_2 and 0-5% CO_2. The synergistic effect of EC and CA storage showed an increase in shelf life of coated tomatoes from 48 days to 65 days (Figure 11).

(a) (b)

Fig. 11: (a) Uncoated tomatoes in CA storage unit after 21 days, and (b) coated tomatoes in the CA storage unit after 65 days of storage

5. Potential for Commercial Application

Because postharvest produce is living and highly perishable, its packaging presents unique challenges compared with many other food products. As a consequence, matching the properties of the package to the needs of the produce has often been difficult to achieve and has involved compromises. The package must not only act as an inert barrier to the external environment, but also take into consideration respiration issues. The demands on package performance continue to increase, inûuenced in particular by market and social changes, and there is a need for packaging innovations such as active packaging that can further enhance the performance of the package so that the properties of the package more closely match the needs of the produce. For instance, the growing demand by consumers for packaged fresh-cuts introduces further challenges, such as the need to deal with elevated microbial loads, as well as exacerbating condensation and respiration issues.

The demand for longer shelf-lives in order to access distant markets and the increasing demand by major retailers for produce in shelf-ready packaging are other examples of future pressures on packaging. It is likely that the trend of integrating active packaging technologies into the packaging material itself will continue, eventually culminating in the development of a truly interactive package that responds directly to the needs of the produce. The feasibility of such interactive packaging is approaching rapidly with the recent advances in the development of flixible electronic components, such as batteries and sensors. Incorporation of active packaging technologies into edible coatings is another relatively unexplored area that has the potential to grow in the future. Although a number of sachet-based active packaging technologies are used commercially for the preservation of fruits and vegetables, further uptake of active packaging technologies (especially plastic-based materials) will be dependent on the demonstration of a clear beneût to the horticultural industry through increased shelf-life and quality to justify any cost associated with their implementation. This is a major challenge in an industry where there are typically low proût margins.

6. Summary

The CA/MA storage has always played a very important role in the preservation of food by extending their shelf life. But active packaging is an emerging and exciting area of food technology that may confer wide range of applications in terms of preservation of various food products. The objective of the technology includes extension in the shelf life of food products while maintaining the sensory and nutritional quality, also ensuring the microbial safety. The future of any innovation in packaging depends upon the extent to which it can satisfy the requirements of the product packaged. Commercial development, therefore, will be driven by needs as perceived in the food industry or in other industries with related problems. The use of oxygen scavengers is the major sub category which is the most effective and plays a crucial role in extending the shelf life. It is predicted that the demand of use of these scavengers in forms other than sachets like scavengers in packaging films, bottles and bottle caps etc. will stimulate the market in near future. Edible coatings are also one such area which has great potential to give further protection against oxidation, respiration and microbial safety and can be used for delivering the desired active materials to the food surface. For designing the active packaging system, the characteristics and properties of food to be packaged should be fully understood. The collaboration of edible coating with active packaging may advance the possible benefits of both the technologies synergistically.

References

Abe, K. and Watada, A.E. 1991. Ethylene absorbent to maintain quality of lightly processed fruits and vegetables. *Journal of Food Science,* 56: 1589-1592.

Aharoni, N, Rodov, B., Fallik, E., Afek, Uzi, Chalupowicz, D., Aharon, Z., Maurer, D, Orenstein, J. 2007. Modified Atmosphere Packaging for Vegetable Crops using High water Vapor Permeable Films. In: Charles L. Wilson, editor. Intelligent and active packaging for fruits and vegetables. CRC Press: Taylor & Francis Group. 73-112.

Anon, 1995.Ensuring product freshness with new desiccant product. *Food Marketing and Technology*, 10: 75-77.

Day, B.P.F. and Potter, L. 2011. Active packaging. In: Food and beverage packaging technology. 2nd edition.John Wiley & Sons. 257-258.

Fonseca, S.C., Oliviera, F.A.R., Frias, J.M., Brecht, J.K., Chau, K.V. 2002. Modeling Respiration Rate of Shredded Galega Kale for Development of Modified Atmosphere Packaging. *Journal of Food Engineering*, 54: 299-307.

Gennadios, A., Weller, C., Testin, R.E. 1993. Modification of Physical and Barrier Properties of Edible Wheat Gluten Based Films. *Biological Systems Engineering – Cereal Chemistry*, 70(4):426-29.

Hewage, S.K., Wainwright, H., Wijerathnam, S.W., Swinburne, T. 1995. The Modified atmosphere storage of bananas as affected by different temperatures. In: Ait-Oubahou A and El-Otmani M (Eds.). Postharvest Physiology and Technologies for Horticultural Commodities: Recent Advances. InstituteAgronomique and Vetérinaire Hassan II, Agadir. Morocco. p. 172 - 176.

Kader, A.A. 1986. Biochemical and physiological basis for effects of controlled and modified atmospheres on fruits and vegetables. *Food Technology,* (USA).

Lee, D.S., Hagger, P.E., Lee, J., Yam, K.L. 1991. Model for fresh produce respiration in modified atmosphere based on principles of enzyme kinetics. *Journal of Food Science*, 56(6): 1580–1585.

Mangaraj, S., Goswami, T.K., Mahajan, P.V. 2009. Applications of plastic films for modified atmosphere packaging of fruits and vegetables: a review. *Food Engineering Reviews,* 1(2): 133-158.

McLaughlin, C.P., O'Beirne, D. 1999. Respiration rate of a dry coleslaw mix as affected by storage temperature and respiratory gas concentrations. *Journal of Food Science ,* 64: 116 - 119.

Monaco, R.D., Giancone, T., Cavella, S., Masi, P., 2008. Predicting texture attributes from microstructural, rheological and thermal properties of hazelnut spreads. You have full text access to this content. *Journal of Texture Studies*., 39 (5): 460-480.

Nguyen, C., Charlin F., 1994. The microbiology of minimally processed fresh fruits and vegetables. *Critical Reviews Food Science Nutrition*, 34(4): 371-401.

Nguyen, T.B.T., Ketsa, S., Van Doorn, W.G. 2004. Relationship between browning and the activities of polyphenol oxidase and phenylalanine ammonia lyase in banana peel during low temperature storage. *Postharvest Biology and Technology,* 30: 187 - 193.

Nicolai, B.M., Hertog, M.L.A.T.M., Ho, Q.T., Verlinden, B.E., Verboven, P. 2009. Gas Exchange Modeling.In Yahia, E.M. (Ed). Modified and Controlled Atmospheres for the Storage, Transportation, and Packaging of Horticultural Commodities. CRC Press, Taylor & Francis Group. 93-110.

Rojas-Grau, M.A., Soliva-Fortuny, R., Martin-Belleso. 2009. Edible Coatings to Incorporate Active Ingredients to Fresh Cut Fruits – a review. *Trends in Food Science & Technology,* 20: 438-47.

Sen, C., Mishra, H.N., Srivastav PP. 2012. Modified atmosphere packaging and active packaging of banana (Musa spp.): a review on control of ripening and extension of shelf life. *J. Stored Prod. Journal of Stored Products and Postharvest Research,* 3(9): 122-132.

Sen, C., Srivastav, P.P., Mishra, H.N. 2014. Active Packaging- An Approach to Minimize Post Harvest Loss of Banana. In: Pomogaibog, O (Ed). Lambert Academic Publishing, Saarbrucken, Germany. 44-46.

Trujillo, J.P.F., Nock, J.F., Watkins, C.B. 2001. Superficial Scald, Carbon Dioxide Injury, and Changes of Fermentation Products and Organic Acids in 'Cortland' and 'Law Rome' Apples after High Carbon Dioxide Stress Treatment. *Journal of the American Society for Horticultural Science*, 126 (2): 235-41.

Vermeiren, L., Devlieghere, F., BeestMv, KruijfNd, Debevere. 1999. Developments in the Active Packaging of Foods. *Trends in Food Science & Technology,* 10: 77-86.

Vermeiren, R., A. De Clippele, and D. Deboutte 2000. "A descriptive survey of Flemish delinquent adolescents." *Journal of Adolescence* 23(3): 277-285.

Watkins, B.C., Silsby, K.J. and Goffinet, M.C. 1997.Controlled atmosphere and antioxidant effects on external CO2 Injury of Empire apples.*Hort Science*, 32(7).1242-46.

Yahia, E.M. 2007. Needs for active packaging in developing countries. In: Charles L. Wilson, editor. Intelligent and active packaging for fruits and vegetables. CRC Press: Taylor & Francis Group. 263-88.

9

Product Formulation and Shelf Life Evaluation Models

Danie Shajie A, R Upadhyay, S Chakraborty, H N Mishra

1. Introduction

Product development is a basic activity in the food industry. Each company has a product mix, often including hundreds of products, which are constantly evolving - old products dying, products reaching maturity, products contributing to rapid growth and new products being introduced.

In the food industry, just as any other industry, product and process development is considered a vital part – indeed the lifeblood – of smart business strategy. Failure to develop new and improved products relegates firms to competing solely on price which favours the players with access to the lowest cost inputs (land, labour etc). Product and process development (commonly referred to as product development) is systematic, commercially oriented research to develop products and processes satisfying a known or suspected consumer need. Product development is a method of industrial research in its own right. There are essentially four basic stages in the model product development process.

- Product strategy development,
- Product design and development,
- Product commercialization, and
- Product launch and post-launch.

Each stage has activities which produce outcomes (information) upon which management decisions are made (Figure 1). In practice, some of the activities performed in the product development process can be truncated, or some stages can be omitted or avoided based on a company's accumulated knowledge and experience.

The quality of the food product developed is the overall degree of acceptability of the food and it is influenced by a group of variable properties with individual units. Being a physico-chemically and biologically active system, the quality of any food is always prone to deteriorate except few cases like aging and maturity. There exists a finite length of time after production to retain the organoleptic and microbiological safety of the food. This time duration is generally termed as shelf life of the food product.

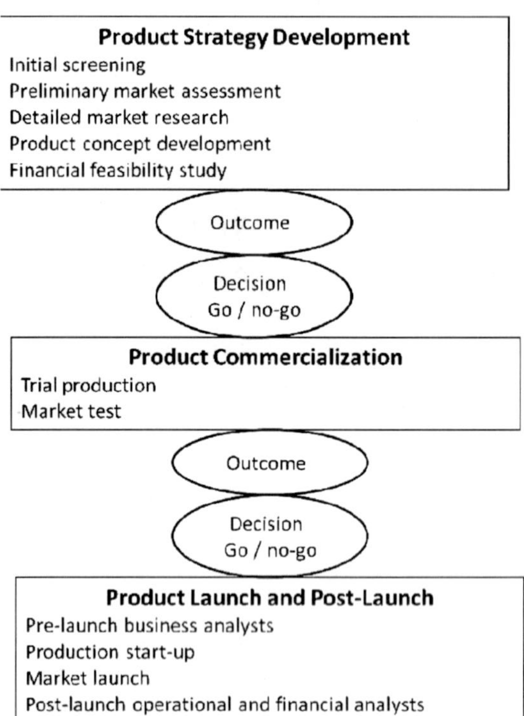

Fig. 1: Different stages of product developmemt process
Source : Siriwongwilaichat, 2001; adapted from Earle and Earle, 2000.

The emergence of rapid analytical methods over the past one decade leads to rise in the application of accelerated shelf life testing methods. The oxidation kinetic modeling and multivariate statistical tools are increasingly exploited to shorten the time duration of shelf life tests. The complicating factor in the understanding of oxidative stability of oils and fats is the wide array of means by which oxidation and stability are assessed and defined. The various methods and assays in these determinations can differ greatly not only in methodology, but also in the compounds and reactions being assessed (Mancebo Campos et al., 2008). Thus, it is highly possible the results and conclusions of a study may be just as thoroughly affected by method-selection as by the true stabilities of the assessed samples. It is, therefore, important to choose methods and techniques that are compatible with both the product and the objectives of the study. This chapter is an attempt to provide recent insight into the methodologies for product development, kinetics in shelf life modeling and predictive modeling based shelf life estimation.

2. Mathematical Modeling

Decision makers (such as consumers, producers, and policy makers) are often concerned with how to do things "best." Analytical solutions for simple constrained optimization problems can sometimes be obtained via the classical methods of algebra and calculus. However, optimal solutions for more complex problems typically require the use of numerical algorithms. A model represents the essential features of an object, system, or problem without unimportant details. The models in this supplement have the important aspects represented in mathematical form using variables, parameters, and functions. Analyzing and manipulating the model gives insight into how the real system behaves under various conditions. From this the best system design or action is determined. Mathematical models are cheaper, faster, and safer than constructing and manipulating real systems.

2.1 Benefits of mathematical models

The main benefit of optimization models is the ability to evaluate possible solutions in a quick, safe, and inexpensive way without actually constructing and experimenting with them. Other benefits are as follows.

2.1.1 Structures the thought process

Constructing an optimization model of a problem forces a decision maker to think through the problem in a concise, organized fashion. The decision maker determines what factors he or she controls; that is, what the decision variables are. The decision maker specifies how the solution will be evaluated (the objective function). Finally, the decision maker describes the decision environment (the constraints). Modeling acts as a way of organizing and clarifying the problem.

2.1.2 Increases objectivity

Mathematical models are more objective since all assumptions and criteria are clearly specified. Although models reflect the experiences and biases of those who construct them, these biases can be identified by outside observers. By using a model as a point of reference, the parties can focus their discussion and disagreements on its assumptions and components. Once the model is agreed on, people tend to live by the results.

2.1.3 Makes complex problems more tractable

Many problems in managing an organization are large and complex and deal with subtle, but significant, inter-relationships among organizational units. For example, in determining the optimal amounts of various products to ship from geographically dispersed warehouses to geographically dispersed customers and the routes that should be taken; the human mind cannot make the billions of simultaneous tradeoffs that are necessary. In these cases, the decision maker often uses simple rules of thumb, which can result in less than optimal solutions. Optimization models make it easier to solve complex organization-wide problems.

2.1.4 Make problems amenable to mathematical and computer solution

By representing a real problem as a mathematical model, one can use mathematical solution and analysis techniques and computers in a way that is not otherwise possible.

2.2 Mathematical programming for new product development

Mathematical programming (MP) relates to the use of mathematical models for solving optimization problems. A typical MP model involves the selection of values for a limited number of decision variables (often called activities), focusing attention on a single objective function to be maximized (or minimized, depending on the context of the problem), subject to a finite set of constraints that limit the selection of the decision variables. Linear programming (LP) is the simplest and most widely used form of MP in which the objective and constraints are linear functions of the decision variables.

Linear programming (LP) is a branch of mathematics which deals with modeling a decision problem and subsequently solving it by mathematical techniques. The problem is presented in the form of a linear function which is to be optimized (i.e maximized or minimized) subject to a set of linear constraints. All LP problems have four properties in common.

(i) LP problems seek to maximize or minimize some quantity (usually profit or cost). This property is referred as the objective function of an LP problem. The major objective of a typical firm is to maximize dollar profits in the long run. In the case of a trucking or airline distribution system, the objective might be to minimize shipping costs.

(ii) The presence of restrictions, or constraints, limits the degree to which one can pursue objectives. For example, deciding how many units of each product in a firm's product line to manufacture is restricted by available labor and machinery. It is, therefore, desired to maximize or minimize a quantity (the objective function) subject to limited resources (the constraints).

(iii) There must be alternative courses of action to choose from. For example, if a company produces three different products, management may use LP to decide how to allocate among them its limited production resources (of labor, machinery, and so on). If there were no alternatives to select from, LP would not be useful.

(iv) The objective and constraints in linear programming problems must be expressed in terms of linear equations or inequalities.

2.3 Assumptions and requirements of linear programming

Linear programs make the following implicit assumptions.

Proportionality with linear programs

It is assumed that the contribution of individual variables in the objective function and constraints is proportional to their value. That is, if the value of a variable is doubled, the contribution of that variable to the objective function and each constraint in which the variable appears is doubled. The contribution per unit of the variable is constant. For example, suppose the variable xj is the number of units of product j produced and cj is the cost per unit to produce product j. If doubling the amount of product j produced doubles its cost, the per unit cost is constant and the proportionality assumption is satisfied.

(a) Additivity

Additivity means that the total value of the objective function and each constraint function is obtained by adding up the individual contributions from each variable.

(b) Divisibility

The decision variables are allowed to take on any real numerical values within some range specified by the constraints. That is, the variables are not restricted to integer values. When fractional values do not make a sensible solution, such as the number of flights an airline should have each day between two cities, the problem should be formulated and solved as an integer program.

(c) Certainty

It is assumed that the parameter values in the model are known with certainty or are at least treated that way. The optimal solution obtained is optimal for the specific problem formulated. If the parameter values are wrong, then the resulting solution is of little value.

Linear programs have objective functions that are to be maximized or minimized, linear constraints that can be of three types (less than or equal to, equal to, and greater than or equal, and non-negative decision variables (or activities). A constraint whose left-hand side (LHS) is less than or equal to (greater than or equal to) the constraint constant on the RHS is termed maximum (or minimum) constraint. A constraint whose LHS is equal to its RHS is called an equality constraint. A mathematical formulation of a standard form of a maximum LP problem with maximum constraints can be given as described in the following paragraph with example of formulation of a food product with specified characteristics using multiple ingredients.

The objective function as defined in the linear programming is of the form of Equation 1.

$$F = a\,X_1 + b\,X_2 + c\,X_3 + d\,X_4 + e\,X_5 \qquad (1)$$

Where, a, b, c, d and e are cost of the ingredients X_1 (oilseeds), X_2 (milk powder), X_3 (cereal and/ or legume flour), X_4 (sugar/glucose) and X_5 (refined vegetable oil) respectively.

The individual protein, lipids, and moisture content of each ingredient that contributes to the total protein, lipids and moisture content of the formulation are given in equations 2 to 4, respectively.

$$X_6 \text{ (total protein in grams)} = f\,X_1 + g\,X_2 + h\,X_3 \qquad (2)$$

Where, f, g, h are the protein content in X_1, X_2 and X_3 respectively.

$$X_7 \text{ (total lipids in grams)} = j\,X_1 + k\,X_2 + l\,X_3 + m\,X_5 \qquad (3)$$

Where, j, k, l, m are the lipids content in X_1, X_2, X_3 and X_5, respectively.

$$X_8 \text{ (total energy in kcal)} - n\,X_1 \mid o\,X_2 + p\,X_3 + q\,X_4 + r\,X_5 \qquad (4)$$

Where, n, o, p, q, r are the energy content content in X_1, X_2, X_3, X_4 and X_5, respectively.

The constraints were defined based on the range of protein, lipids and energy content to be present in the product as per the guidelines requirement of one process or product. The constraints defined accordingly are given in Eq(s) 5 to 7.

$$s \leq X_6 \leq t \qquad (5)$$

$$u \leq X_7 \leq v \qquad (6)$$

$$w \leq X_8 \leq z \qquad (7)$$

Where, s & t, u & w, and w & z are the numerical values for one respecting constraint. The total weight of the ingredient equals to 100 and the values of X_i are greater than or equal to zero for all values of i (where i = 1,2,3,4...) as defined in Eq(s) 8 & 9, respectively.

$$X_1 + X_2 + X_3 + X_4 + X_5 = 100 \qquad (8)$$

$$X_1, X_2, X_3, X_4, X_5 \geq 0 \quad (9)$$

Linear programming in MatLab was done using the "linprog" solver tool available in the optimization tool box. The data is defined in the matrix and the solver is called. The results are obtained once the optimization is terminated.

3. Application of Reaction Kinetics in Shelf life Modeling

Once is the product is formulated and the process parameters are optimized, in order to maintain the shelf life of any food, the meaning and criteria of organoleptic quality and acceptability limits have to be accurately defined. The study of the chemical and biological reactions and physical changes that occur in the food during and after processing allows identifying the most important one related to its safety, integrity and overall quality. A systematic analysis and interpretation of the quality changes lead to a more meaningful way of assessing the shelf life. In addition, application of kinetics in food quality loss is essential for efficiently designing appropriate tests and analyzing the obtained results. The overwhelming majority of the food reactions that have been studied have been characterized as pseudo-zero or pseudo-first order. The general forms of the quality loss reactions and their half-lives are presented in Table 1.

Table 1: Forms of biochemical reactions and their half-lives

Apparent order of reaction	Function	Half life
0	$C_t = C_o - k_o t$	$C_o / 2k_o$
1	$C_t = C_o e^{-k_1 t}$	In $2 / k_1$
2	$\frac{1}{C_t} - \frac{1}{C_0} = k_2 t$	$1/(k_2 C_0)$
n (n ¹ 1)	$\frac{1}{n-1}\left[C_t^{1-n} - C_0^{1-n}\right] = k_n t$	$C_0^{1-n}\left[\frac{2^{n-1}-1}{k_n(n-1)}\right]$

C_t and C_0 are the concentration of the compounds at treatment time of 't' day and '0' day, respectively; k_j is the rate constant (j = 0, 1, 2 and n) and t is the storage time in day.

The overall quality of frozen foods and non-enzymatic browning reactions are considered to follow zero-order kinetics. On the other hand, vitamin loss, microbial death / growth, oxidative color loss and texture loss during thermal processing are reported to follow the first order reaction kinetics. However, it is necessary to confirm the reaction order as the failing to calculate order may lead to over- or under- estimation of shelf-life.

3.1 Effect of environmental factors on the reaction rate

The environmental factors strongly affect the deteriorating reactions in the food system during storage are temperature, relative humidity, total and partial pressure of different gases, light, and mechanical stresses. The storage temperature is the important factor considered. It not only affects the reaction rates but also modifies the quality directly unlike other factors which are generally controlled by the food packaging.

3.1.1 Temperature

The most prevalent and widely used model to describe the temperature dependency of the reaction rate is the Arrhenius relation (Eq. 10), derived from thermodynamic laws as well as statistical mechanics principles.

$$\ln k_i = \ln k_{ref} + \frac{E_a}{R}\left(\frac{1}{T_{ref}} - \frac{1}{T}\right) \tag{10}$$

Where k_{ref} is the rate constant at reference temperature; E_a is the activation energy ($kJ.mol^{-1}$); T_{ref} and R represent reference temperature (323 K) and universal gas constant ($8.314\ kJ.mol^{-1}.K^{-1}$), respectively.

The experimental error associated with the calculated values of k can lead an amplification of a substantial error in E_a from only two points. Therefore, it is recommended to have at least three points in ln k vs (1/T) plot, so that from the slope, activation energy value can be calculated. The temperature dependence has been traditionally expressed in the literature as Q_{10}, the ratio of the reaction rate constants at temperatures differing by 10 °C or the change of shelf-life of the food stored at a temperature higher by 10°C (Eq. 11).

$$\text{In } Q_{10} = \frac{E_a}{RT}\left(\frac{10}{T+10}\right) \tag{11}$$

The majority of the earlier food literature reports endpoint data rather than complete kinetic modeling of quality loss. The shelf-life plots are true straight lines only for narrow temperature ranges of 10 to 20 °C. In most of the quality reactions, the activation energy ranges between 50 and 150 $kJ.mol^{-1}$; whereas, the Q_{10} fluctuates at lot with storage temperature as presented in Table 2. For such a narrow interval, data from an Arrhenius plot will give a relatively straight line in a shelf-life plot.

Table 2: Dependence of Q_{10} of various reactions with E_a and storage temperature (Taoukis et al., 1997)

Type of reactions	Q_{10} at different temperatures			E_a (kJ·mol^{-1})
	4 °C	21 °C	35 °C	
Enzymatic, hydrolytic	2.13	1.96	1.85	50
Nutrient loss, lipid oxidation	4.54	3.84	3.41	100
Nonenzymatic browning	9.66	7.52	6.30	150

In case of thermal processing, another term is used to characterize the temperature dependence of rate of food quality loss is the z-value. It is the temperature change that causes a 10-fold change in the reaction rate constant as given by Eq. 12.

$$z = \frac{(\ln 10)RT^2}{E_a} \tag{12}$$

3.1.2 Other factors

The moisture transfer models are useful for predicting the shelf life of moisture sensitive packaged foods as they include the water activity (a_w) as a critical parameter. It describes the degree or extent of bound water available towards the chemical reactions. Hence, it has a critical role in determining the rate of spoilage reactions that occur in the food system (Figure 2). It is clear from Figure 2 that, each every type of biochemical reactions needs a minimum water activity level to initiates and subsequently reaches a maximum. Textural quality is greatly affected by the moisture content of the food thus indirectly related to a_w. Besides, the pH of the system has a key role in determining the rate of microbial, enzymatic and biochemical reactions as these types of reactions happens at an optimum pH level. The functionality and solubility of proteins strongly depend on pH and temperature of the medium.

The gaseous environment or composition inside a package is one of the important factors controlling the rate of spoilage reactions in the food. The oxidation reactions are greatly affected by the oxygen concentrations in the surrounding gaseous medium. Further, the presence and relative amount of other gases, especially carbon dioxide strongly affects the biological and microbial reactions in fresh meat, fruit and vegetables.

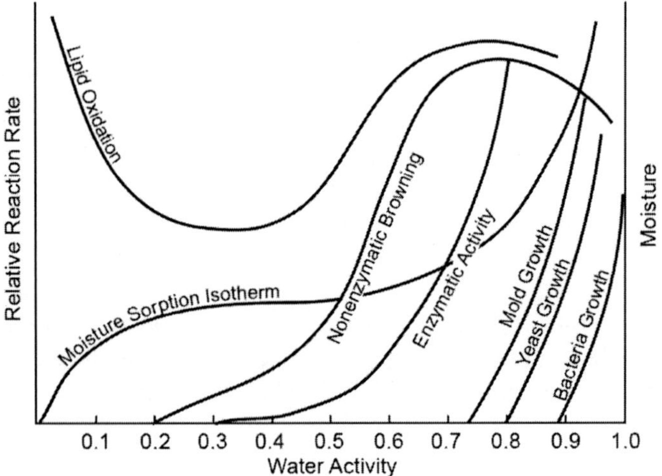

Fig. 2: Global food stability plot as a function of water activity (Labuza and others, 1969)

3.2 Prediction of shelf life

Considering all these factors, the overall quality (OQ) of a can be food written as a function of a number of different factors like temperature, pH, a_w, gaseous compositions and so on (Eq 13).

$$OQ = f\left(\sum Q_i\right) = f\left(\sum k_i t\right) \tag{13}$$

Where, Q_i is the i th quality indicator; k_i is the deterioration rate of Q_i; t is the storage time. Again, the rate of deterioration (k_i) is a function of environmental and medium factors (Eq.14).

$$k_i = f\left(T, a_w, pH, p_{O2}, p_{CO2}, \ldots\right) \tag{14}$$

The change in all of the factors is also a function of storage time (t). Therefore, OQ can be written as (Eq. 15):

$$OQ = \int_0^t k_i dt \tag{15}$$

The remaining fraction of OQ, denoted as $\hat{O}_r$ can be set as a tolerance limit and the remaining shelf life (u_r) can be calculated using Eq. 16:

$$\theta_r = Min\left[\frac{\Phi_r}{k_i}\right] \tag{16}$$

Where, the rate constants (k_i) are calculated for a quality index assuming remaining conditions kept constant.

This method of shelf life prediction is applicable for a complex system under variable conditions. The critical thing before applying this technique is to identify the major deterioration modes, determination of the corresponding quality functions and estimating the effects of different factors on the rate constant.

3.3 Accelerated shelf life testing (ASLT) method

The accelerated shelf life testing (ASLT) method has been used to predict shelf life at normal storage conditions based on data collected at high temperature and high humidity conditions which accelerate the spoilage reactions. It helps in extrapolating the data of regular storage conditions through Arrhenius model, thereby, reduce the experimental duration substantially. Designing a shelf-life test is a systematic approach (Figure 3) correlating all aspects of food like engineering, analytical & physical chemistry, microbiology and their regulatory aspects.

The level of test and control storage temperatures depends on the type of food being stored. An outline of the general criteria for selecting the storage temperature has been presented in Table 3.

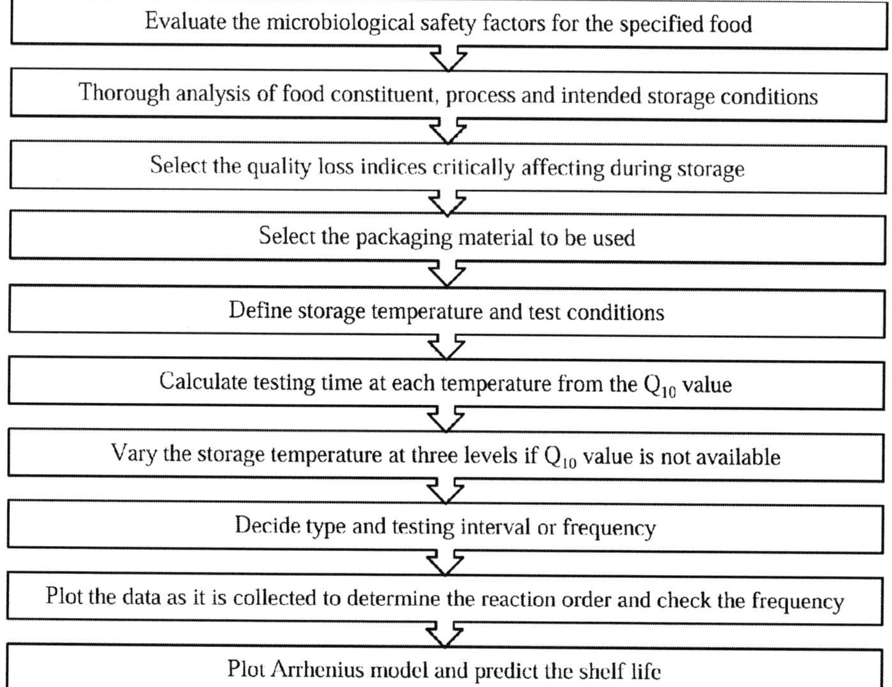

Fig. 3: Systematic approach in designing an experiment to determine the shelf life of food

Table 3: Guideline for selecting the storage temperature for different types of food products (Taoukis et al., 1997)

Product type	Storage temperature (°C)	Control temperature (°C)
Canned	25, 30, 35, 40	4
Dehydrated	25, 30, 35, 40, 45	-18
Chilled	5, 10, 15, 20	0
Frozen	-5, -10, -15	<-40

3.4 Case study

3.4.1 Shelf life of sweetened condensed milk

Patel et al., (1996) performed a storage study of commercial sweetened condensed milk at 7, 15, 30, 45 and 55 °C; the shelf life was predicted according to the kinetics of non-enzymatic browning and color changes. The formation of hydroxyl methyl furfural (HMF) followed first order kinetics, whereas, the color change in the powder showed a zero order behavior. The range of Q_{10} and activation energy values for each quality loss reactions have been summarized in Table 4. The rate of color change during storage ranged between 0.0012 to 2.8 d^{-1} within 7-55 °C, whereas, the rate of HMF formation varies from 0.0013 to 0.21 d^{-1}. The Q_{10} values for these variables ranged from 1.21 to 1.84 for temperatures up to 30 °C and were nearly 2-3 times higher above 30 °C. The sensory color score was negatively correlated with absorbance of color change values. The Arrhenius relationship was applied for predicting the product's shelf-life. When the lower limit of color change was set at 10, the predicted shelf life was 147 days.

Table 4: Summary of Q_{10} and E_a of sweetened condensed milk at different storage temperature

Reactions	Q_{10} values at			E_a ($kJ \cdot mol^{-1}$) at		
	7 °C	30 °C	55 °C	7 °C	30 °C	55 °C
Color change	1.84	1.84	5.01	45	45	140
HMF formation	1.21	1.21	3.23	14	14	101

3.4.2 Storage study of pineapple puree

Chakraborty et al., (2015) evaluated the storage stability of four different pineapple puree samples (S1, untreated; S2, treated at 600 MPa/50 °C/13 min; S3, treated at 600 MPa/70 °C/20 min; and S4, treated at 0.1 MPa/95 °C/12 min) packed in EVOH pouch and stored at 5, 15 and 25 °C up to 120 days. The total color change (ΔE*) and browning index (BI) of the sample increased with storage duration and temperature following zero-order kinetics. The degradation of ascorbic acid (AA), total phenolic content (TPC) and total antioxidant capacity (TAC) during storage followed first-order decay. The rate constant for AA degradation (k × 10^{-2}, d^{-1}) varied from 0.31 to 2.46, 0.33 to 2.68 and 0.47 to 3.35 d^{-1} for S2, S3 and S4, respectively; the highest value of k (3.35×10^{-2} d^{-1}) was obtained for S4 at 25 °C. In S3 and S4, the count for aerobic mesophiles (AM) and yeast & mold (YM) were < 6 Log·cfu·g^{-1} up to 120 days at 5 °C; whereas, S1 spoiled after 7, 5 and 2 days at 5, 15 and 25 °C, respectively. The sensory quality and color stability (OA > 5 and ΔE* < 12) of sample S2 were maintained up to 110, 50 and 20 days at 5, 15 and 25 °C, respectively. The activation energy values for the degradation of ascorbic acid and ΔE* are summarized in Table 5. Based on ascorbic acid content, the predicted shelf life of S2 and S3 at 5 °C was 143 and 132, respectively.

Table 5: Estimated shelf-life of pineapple puree processed under different conditions based on ascorbic acid content and color change values

Sample	Shelf life (d) for retaining 200 mg.kg^{-1} AA				Shelf life (d) up to ΔE* ≤ 12.0			
	5 °C	15 °C	25 °C	E_a (kJ·mol^{-1})	5 °C	15 °C	25 °C	E_a (kJ·mol^{-1})
S2	143	78	34	68	146	80	41	60
S3	132	46	27	69	136	53	29	64
S4	26	6	4	65	117	33	18	63

AA, ascorbic acid; ΔE*, total color change.

4. Shelf Life Prediction of Edible Oils

In the absence of reliable means to accurately predict the shelf life according to composition, the most dependable techniques to ascertain oil stability continue to involve the monitoring of its oxidative state (Farhoosh & Hoseini-Yazdi, 2013). The most historically classical methods to monitor current oxidative state (peroxide value, conjugated dienes and trienes, *p*-anisidine, and 2-thiobarbituric acid reactive substances) have been in use for more than six decades. These assays are not comprehensive measures of oxidation, but are considered still to be of merit for their simplicity, minimal equipment requirements, and documented associations with sensory scores. However, the advances in machinery and methodologies have given rise to rapid and accelerated methods to determine oxidation and oxidative stability (Farhoosh & Hoseini-Yazdi, 2013).

The aim of this section is to provide an update on the methods that are of considerable current use for the monitoring the shelf life of oils. The emergence of novel approaches, design of accelerated storage tests and rapid measures of stability are also reviewed to discuss their modern presence in scientific literature, and their established benefits and concerns.

4.1 Oxidative stability

Oxidative stability of oils can be considered as the sum total of the relevant intrinsic and extrinsic effects occurring throughout its life-cycle (Upadhyay & Mishra, 2015). Thus, the evaluation of the oxidative stability of oils requires the precise measurement of lipid degradation over the time. There are different approaches to determine lipid oxidation within a time frame which are mentioned in Table 6.

The first approach, the oil sample is placed in a highly controlled environment and lipid oxidation is monitored over time with a variety of conventional physico-chemical methods (Farhoosh & Hoseini-Yazdi, 2013; Mancebo Campos et al., 2008). However, due to the high costs associated with the bulk storage of oils and time required to perform such tests (often 2 year), the second approach of accelerated storage tests are increasingly employed. The accelerated tests involve exposing the oil samples to extreme conditions for short periods of time and monitoring their degradation (*e.g.* Rancimat test or oxygen bomb method) (Farhoosh & Hoseini-Yazdi, 2013). Although, rapid test provides quick assessment of oxidative status of oils, there are certain disadvantages associated with shelf life estimation which often leads to under/over prediction (Farhoosh & Hoseini-Yazdi, 2013). Therefore, the third approach which involves an integrated approach utilizing the multivariate Statistical analysis are rapidly popularizing for their ability to provide reliable estimation of shelf life (Upadhyay & Mishra, 2015; Farhoosh & Hoseini-Yazdi, 2013). These tests will be discussed in more detail later in this review. The next section will discuss about the modeling of oxidation kinetics for shelf life estimation.

4.1.1 Modeling of oxidation kinetics

In this section, the selective application of univariate and multivariate analysis has been described to model the kinetics of lipid oxidation in oils.

Univariate analysis

A kinetic approach to oxidative stability assessment of extra-virgin olive oil was examined to estimate the shelf life (Mancebo Campos et al., 2008). The kinetic behavior of different oxidation indices *viz.,* peroxide value (PV), UV oxidation indices at 232 and 270 nm (K_{232} and K_{270}) and unsaturated fatty acids during storage of oil samples in darkness at different temperatures (25–60 °C) was successfully performed. The PV and K_{232} followed apparent pseudo zero-order kinetics, whereas the K_{270} followed pseudo first-order kinetics. Besides, the temperature-dependent changes in oxidation indices followed Arrhenius kinetics between 25 and 60 °C. The time required to reach the upper limits for

Table 6: A comparative overview of different storage stability tests

Type of method	Experimental conditions	Quality indicators	Advantages	Disadvantages	Selected references
Conventional	Ambient to moderately high temperature (25–60°C)	Peroxide value, para-anisidine value, iodine value, free fatty acid, thiobarbituric acid value, conjugated diene and triene value, carbonyl value, total polar matter	Easy to perform, popular and acceptable	Time and resource consuming, Requires different chemicals and reagents, different cut-off limits for quality parameters leads to ambiguous results	Mancebo Campos et al., (2008); Upadhyay & Mishra, (2015); Zanoni et al., (2005)
Accelerated	High temperature (≥ 100 °C) under saturated air conditions (15–20 L/h)	Induction period, oil stability index	Rapid, automated, reproducible to obtain shelf life data	Often leads to under and/or over prediction of shelf life	Upadhyay & Mishra, (2015)
Combined	Based on combination of conventional and accelerated methods	Use of chemometrics to develop shelf life prediction models	Combines conventional and accelerated method which gives better predictability of shelf life, can be effectively used as initial check of oxidative stability	Use of complicated mathematical analysis	Farhoosh,& Hoseini-Yazdi, (2013), Upadhyay & Mishra, (2015)

PV, K_{232} and K_{270} (EU norms) were successfully used to model the predictive shelf life equation under normal storage temperature.

Another interesting research on shelf life modeling of olive oil dealt with mass transport phenomenon taking place in the oil-package system (Kanavouras & Coutelieris, 2006). A mathematical model was developed using the basic mass transport equations to investigate the oxidative deterioration of olive oil. The evolution of hexanal over time was considered as indicator of oxidative deterioration in oil samples packaged in glass and plastic containers for one year under light and dark conditions. They successfully introduced the concept of probability for the packaged olive oil not to reach the end of its shelf life. This model undoubtedly exhibited wider application and can be adopted while considering the oil – packaging material interactions.

Multivariate analysis

The screening of stability/instability indices was carried out using multivariate statistical analyses on the 63 chemical and 18 sensory parameters measured for virgin olive oil (Zanoni et al., 2005). The data processing indicated that only a few stability indices (peroxide value and UV oxidation indices) were related to lipid degradation in oil samples. Furthermore, a mathematical relationship was derived between the content of oleic acid and other fatty acids, which further reduced the influential parameters to acidity, oleic acid content and bitter taste of oil samples. This phenomenological model was a successful attempt to predict the stability of extra virgin olive oil at commercial scale based on combined stability/instability indices.

In another investigation, the predictive models for Tuscan virgin olive oil stability during commercial activities were set up (Pagliarini et al., 2000). The oxidative stability was studied for five lots of a batch of oil sample to simulate different commercial activities. The chemical, physical, and sensory analyses were routinely carried out and the experimental data were processed by multivariate analyses to select significant parameters and by regression analyses to set up kinetic models. A few parameters were found to be significant: hydroxytyrosol and tyrosol contents, carotenoid absorbance at 475 and 448 nm, α-tocopherol content, Rancimat induction time, and K_{232}. It was also shown that the stability of this oil samples was not significantly influenced by different uncontrolled bottling, transport, and storage conditions in supermarkets. Empirical models were set up to predict the time to reach a reference value for K_{232}.

4.2 Modern approaches towards shelf life testing of edible oils

Recently, Farhoosh and Hoseini-Yazdi (2013) developed an interesting shelf life prediction model by combining the oxidative stability measures at different temperatures. According to their findings, the conjugated diene values can be used as an index of oxidative stability at low temperatures. The mathematical approach described by them was rather effective in accurately estimating the shelf life of oil samples. In a similar study, a unified model was developed which estimated the shelf life of sunflower oil blends stabilized with natural antioxidants with an error of less than 10% (Upadhyay & Mishra, 2015). The chemometric approach was used to develop the shelf life prediction models at different temperatures and tested for their individual performance.

References

Chakraborty, S., Rao, P.S., Mishra, H.N. 2016. Changes in quality attributes during storage of high pressure and thermally processed pineapple puree. *Food Bioprocess Technology,* 9(5)768-791.

Earle, M.D., and Earle, R.L. 2000. Building the future on new products. Leatherhead Publishing, England. 112 pages.

Farhoosh, R., & Hoseini-Yazdi, S. Z. 2013. Shelf-life prediction of olive oils using empirical models developed at low and high temperatures. *Food Chemistry*, 141(1):557-565.

Kanavouras, A., & Coutelieris, F. A. (2006). Shelf-life predictions for packaged olive oil based on simulations. *Food Chemistry*, 96(1): 48-55.

Labuza, T.P., Tannenbaum, S.R., Kavel, M. 1969. Water content and stability of low-moisture and intermediate-moisture foods. *Food Technology,*24, 35–39.

Mancebo, Campos, V., Fregapane, G., & Desamparados Salvador, M. 2008. Kinetic study for the development of an accelerated oxidative stability test to estimate virgin olive oil potential shelf life. *European Journal of Lipid Science and Technology*, 110(10):969-976.

Pagliarini, E., Zanoni, B., & Giovanelli, G. 2000. Predictive study on Tuscan extra virgin olive oil stability under several commercial conditions. *Journal of Agricultural and Food Chemistry,* 48(4):1345-1351.

Patel, A.A., Gandhi, H., Singh, S., Patil, G.R. 1996. Shelf life modeling of sweetened condensed milk based on kinetics of maillard browning. *Journal of Food Processing Preservation,* 20(6): 431-51.

Siriwongwilaichat, P. 2001. Technical information capture for food product innovation in Thailand. PhD Thesis. Massey University, New Zealand.

Taoukis, P.S., Labuza, T.P., Saguy, I.S. (1997). Kinetics of food deterioration and shelf-life prediction. *In: Handbook of Food Engineering Practice*, 361-403.

Upadhyay, R., & Mishra, H. N. 2015. Predictive modeling for shelf life estimation of sunflower oil blended with oleoresin rosemary (*Rosmarinus officinalis* L.) and ascorbyl palmitate at low and high temperatures. *LWT-Food Science and Technology*, 60(1):42-49.

Zanoni, B., Bertuccioli, M., Rovellini, P., Marotta, F., & Mattei, A. 2005. A preliminary approach to predictive modelling of extra virgin olive oil stability. *Journal of the Science of Food and Agriculture*, 85(9):1492-1498.

10

Rapid Methods for Food Quality Analysis

S Tripathi, V R Sinija, S K Bag, VK Shibby, R Pande, A Deswal, HN Mishra

1. Introduction

Foods are complex biologically and chemically active systems that require close monitoring of their manufacturing, distribution and storage conditions in order to maintain their nutritive value, safety and sensory qualities by suitable chemical. The determination of trace impurities in food system presents considerable difficulties owing to complexity as it contains thousands of major and minor compounds (Nilufer and Boyacio, 2002). The choice of the method of analysis depends on the sample, the analyte to be assayed, accuracy, limit of detection, cost and time to complete the analysis (Aboul- Eneim et al., 2000). The chemical methods employed for food analysis and quality control are generally time consuming and require tedious or destructive sample preparation. For development of the new detection method, emphasis should be on development of simplified, cost-effective and efficient procedures that complies with the legislative requirements (Stroka and Anklem, 2002; Enker, 2003).

The molecular spectroscopy is a fast, accurate, easy and non-destructive technique that can be used as a replacement of time-consuming methods. Near infrared (NIR) spectroscopy has proved to be a powerful analytical tool used in the agricultural, nutritional, petrochemical, textile and pharmaceutical industries (McGlone and others, 2002; Esteban-Diez and others, 2004). Infra red spectroscopy is based on the absorption of infra red light by the substance to be measured. This absorption excites molecular vibrations and rotations, which have frequencies that are the same as those in the infrared range of the electromagnetic spectrum (Pande and Mishra, 2015). This application uses a fiber optic coupled FT-NIR system operating in the 10,000 – 4,000 cm^{-1} region in reflectance mode. It involves the analysis and quantification of molecular responses to introduced signals of known energy or frequency. All molecules (e.g. water, glucose, protein) have a definite amount of energy. When alternate energy (e.g. infrared radiation) is introduced, an energy exchange occurs between the introduced energy and the energy contained within the molecule. This is expressed as absorbance (energy is absorbed resulting in a loss of introduced energy), attenuated (energy is scattered resulting in a loss of introduced energy) or emitted (energy is released resulting in a gain on the introduced energy). As individual molecules and molecular groups (e.g. alcohols, nucleic acids, proteins, sugars and fats) have definite energy, by applying external energy of a known amount/type, one can structurally identify, quantify and even determine the natural state of these molecules within complex samples and mixtures.

NIR spectroscopy has been used in past for determining various biochemical parameters in foods/ beverages and also in the area of food quality control. Since the 1990s, attempts have been made to simultaneously predict water, alkaloids and phenolic substance content in tea leaves using NIR spectroscopy (Hall and others, 1988; Schulz and others, 1999). Studies on the application of NIR spectroscopy to quantitative analysis of total antioxidant capacity in green tea were also reported recently (Lupaert and others, 2003). Near-infrared reflectance spectroscopy (NIRS) is also being used to predict D-chloro-inositol, vitexin, and isovitexin contents in Mung bean using (Yang, Xu-zhen, & Gui-xing, 2011). Near-infrared calibration models were developed by partial least squares (PLS) and test-set validation on 364 sorghum samples to predict crude protein and moisture content on whole-grain and milled flour samples (Brauteseth, 2009). Few studies have reported applications of NIR spectroscopy to analyse and quantify antinutritional factors (phytic acid) in pulses (Pande and Mishra, 2015).

Various studies have been done on application of NIR spectroscopy in the area of mycotoxin detection. A rapid PLS regression model for prediction of mycotoxins deoxynivalenol in wheat kernels by Infratec and NIR spectroscopy at levels just above the permissible levels proposed by European Union in wheat flour has been developed (Petterson and Aberg, 2003). The possibility of reflactance and transmittance NIR spectroscopy to detect fumonisin in single corn kernel in the range of 10-100 ppm for online detection and sorting was explored (Dowell et al., 2002). NIR spectroscopy was also used to predict scab, vomitoxin, and ergosterol in single wheat kernels (Dowell et al., 1999). FT-NIR spectroscopy coupled with multivariate analysis was used for identification of bacteria (*E. coli*, *Pseudomonas aeruginosa*, *Bacillus subtilis*, *B. cereus* and *B. thuringiensis*) in apple juice (Saona et al., 2004).

Sensory analysis is also an important area of research in food science and technology. The disadvantage of the human sensory system is that no two brains are alike, and the same brain may react differently from one day to the next, depending on an individual's health, mood or environment, making the data subjective. To overcome this problem, techniques like headspace or dynamic headspace gas chromatography coupled with mass spectrometry (GC–MS) (Miller et al., 2008), and solid-phase micro-extraction sampling followed by GC–MS (Sanches-Silva et al., 2008) have been utilized for objective evaluation of sensory profiling of foods (Torri et al., 2010; Jaffres et al., 2011). In this regard, electronic nose system might bring on a number of convenience as they require less technical skills, lower capital cost of the operating system and use the automated artificial intelligence software to utilize a vast bank of time series spoilage qualitative pattern recognition and evaluation techniques (Schaller et al., 1998; Falasconi et al., 2014).

Thus, NIR spectroscopy based rapid methods or other economical methods (like electronic nose) are gaining importance and applications in industries for the quality control due to its advantages over analytical techniques, mainly in easy or no sample preparation and prediction of chemical and physical sample parameters from a single spectrum enabling several components to be determined simultaneously based on the use of multivariate calibrations (Moros et al., 2006; Kim et al., 2007). The low molar absorptivity of near-infrared (NIR) bands permits the measurement of solid or powdered samples with little or no sample preparation, thus avoiding human experimental errors.

2. Principles & Practices Involved

2.1 Fourier transform near infrared (FT-NIR) spectroscopy

The technique of FT-NIR spectroscopy involves electromagnetic spectroscopy and measures the electromagnetic spectra absorbed or released by given atoms, molecules and molecular groups. These spectra are waves of energy of defined frequencies (measured in Hertz), and subsequently wavelength (measured in metres), which are characteristic for individual molecular groups, molecules and atoms. The infrared region of the electromagnetic spectrum may be divided both instrumentally and functionally into near, middle and far infrared spectra. A near-infrared spectrum (800nm-2500nm) represents combination and overtone bands that are harmonics of absorption frequencies in the mid-infrared region. These combination and overtone bands correspond to the frequencies of vibrations between the bonds of the atoms making up the material. Because each different material is a unique combination of atoms, no two compounds produce the exact same near-infrared spectrum. Therefore, near-infrared spectroscopy can result in a positive identification (qualitative analysis) of each different material. In addition, the size of the peaks in the spectrum is a direct indication of the amount of material present. Fourier transformed instruments are based on interferometers that are widely used in modern spectrometers (Osborne et al., 1993). The Fourier transform technique is based on the use of an interferometer (mostly of the Michelson -type) that is able to detect intensities of several spectral frequencies in a composite signal.

The Fourier transform of the recorded interferogram is the infrared spectrum. The purpose of an interferometer (Smith, 1996 and Bertrand, 2002) is to split a beam of light into two beams and to introduce a difference in their respective traveling distances. The plot of the intensity versus the optical path difference is called an interferogram. When the source is polychromatic (Smith, 1996), radiation of different wavelengths undergoes destructive and constructive interference at different optical path differences. Each wavelength of light leads to an interferogram with a specific path difference, resulting in intensity typical of their frequency that can be measured by the detector. The signal passing through the sample is the sum of each specific interferogram and therefore contains intensity information about all the wavelengths contained in the band passing the sample. The interferogram (intensity versus time function) is then Fourier transformed to obtain the final NIR spectrum, which is intensity versus frequency function.

Due to the dramatic decrease in band intensity of the higher overtones, the NIR spectrum is usually dominated by overlapping overtone and combination bands of the structurally lighter groups (e.g. CH, NH and OH). Overtones are the result of two or more molecular absorptions of energy quantum. As a quantum of energy is applied to a given molecule, the molecule absorbs that energy, and vibration corresponding to a simple "near" harmonic motion is produced (e.g. wave). The overtone band of this simple harmonic motion has approximately twice the frequency of the principle vibrational frequency. The intensity of the overtone band is dictated by the level of "anharmonicity" of the principle vibrational band. Chemical bonds that vibrate with large amplitude (e.g. hydrogen bonds) have a high degree of anharmonicity. These bonds are very common and as a consequence, can dominate the observable NIR bands.

As such, molecular overtones and combinational vibrations characteristic of NIR produce very broad, complex, overlapping spectral outputs that can conceal *specific* information on chemical assignments. This is overcome through the use of multivariate calibration, chemometric techniques and multivariate methods such as principle component analysis and partial least squares (Koljonen et al., 2008; Cozzolino et al., 2009; Gishen et al., 2005). The aim of the multivariate calibration development is to develop a model that correlates the NIR spectra to the values obtained by the standard independent reference methods (Beebe and Kowalaski, 1987). Calibration is, therefore, a relationship between the log (1/R) or absorbance values for the amount of the component obtained from the set of the samples determined by the NIR instrument and the values for the amount of component determined by reference method and is expressed as an approximation involving some form of regression equation (Hruschka, 2001). The calibration model developed then can be used for a new set of same sample (within the range) to determine the amount of component (for which model was developed) in that sample. There are mainly two groups of multivariate analysis method qualitative and quantitative analysis. The present study was focused on quantitative analysis hence it is only discussed in the following text. Partial least squares (PLS) are now dominating the practice of multivariate calibration, because of the quality of the calibration models produced and the ease of their implementation (Lavine, 1998). This algorithm was developed in the beginning of the 80´s (Tenenhaus, 1998; Eriksson et al., 2000). There are several other important chemometrics methods, such as principal component regression, multiple linear regression which are another widely used multivariate calibration methods other then PLS (Blanco et al., 1998).

When calibration is finished the predictive abilities of the developed PLS based calibration model is tested by various validation procedures. Two important validation methods include independent validation (or test validation) where independent data sets (not used in developing model) are used to verify the model and cross validation (which uses calibration data set only). Predictive residual sum of squares (PRESS) or root mean square error of cross validation (RMSECV) gives an estimation of the predictive ability of the model. The number of significant PLS components is usually calculated to be the minimum number for which the PRESS and RMSECV value is not significantly different from the lowest PRESS and RMSECV value (Haaland and Thomas, 1988).

2.2 Electronic nose

The electronic nose has been defined as an instrument which comprises an array of electronic chemical sensors with partial specificity and an appropriate pattern recognition system, capable of recognizing simple or complex odours (Gardner and Bartlett, 1994). The electronic nose system parallels the human olfactory system as it assesses the mixture of volatiles released from a sample, while other instrumental methods usually separate the aroma into its individual components. The sensors of electronic nose are mostly based on conductive polymers, metal oxides, or surface acoustic waves utilizing the piezoelectric effect for detection of aroma profile of injected sample (Miettinen et al., 2002).

The arrangement of electronic nose is shown in Figure 1. The electronic nose consists of auto sampler, injection system, temperature controlled sensor chamber with sensor array, mass flow controller and a computer major components. The carrier gas is pure air at a pressure of 5 psi.

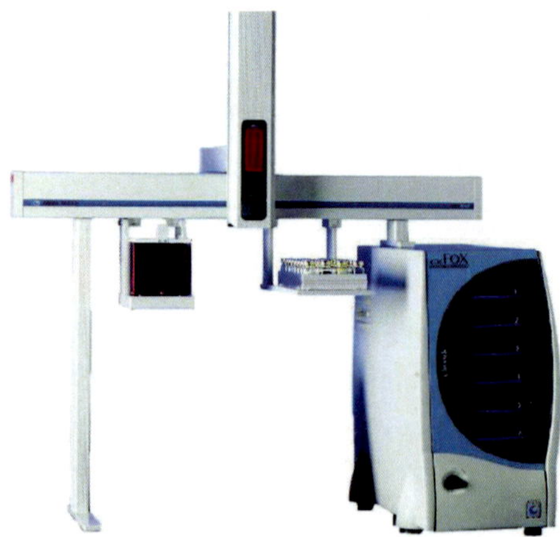

Fig. 1: Alpha-MOS electronic nose

3. Materials & Methodologies

Various studies on application of FT-NIR spectroscopy for detection of several biochemical components in foods/beverages and/or quality parameters were carried out in Food Chemistry and Technology Laboratory of Agricultural & Food Engineering Department, IIT Kharagpur. FT-NIR spectroscopy coupled with multivariate analysis was used to determine moisture, caffeine and total polyphenol content in green tea granules and instant green tea powder; moisture content in bael pulp; moisture content in dahi powder; detection of adulteration of starch in milk and detection of aflatoxin B_1 (AFB_1) spiked red chilli powder and phytic acid content in green gram seeds. Electronic nose technique as an alternative to classical sensory analysis, microbial and chemical analysis for determining microbiological and sensory life of oat milk at room and refrigerated temperature was also developed.

3.1 Sample preparations for FT-NIR analysis

FT- NIR spectra were recorded on MPA™- Multipurpose analyzer (Bruker optics, Germany; Figure 2) equipped with a quartz beam splitter; an integrated michelson interferometer; highly sensitive PbS 12.800 - 3.600 cm^{-1} detector, multiple NIR measurement accessories for different sampling techniques combined with Qpus / quant 5.5 software. This spectrometer utilized the Fourier-transform and had distinct advantages compared to dispersive spectrometers because of presence of cube-corner mirrors in a ROCKSOLID™-interferometer. The spectrometer with an integrated Michelson interferometer utilized the Fourier -transform had distinct advantages compared to dispersive spectrometers. The spectra generated over a range of wave numbers from 12,000 to 4,000 cm^{-1} were interpreted based on the overtones of different functional groups in the product.

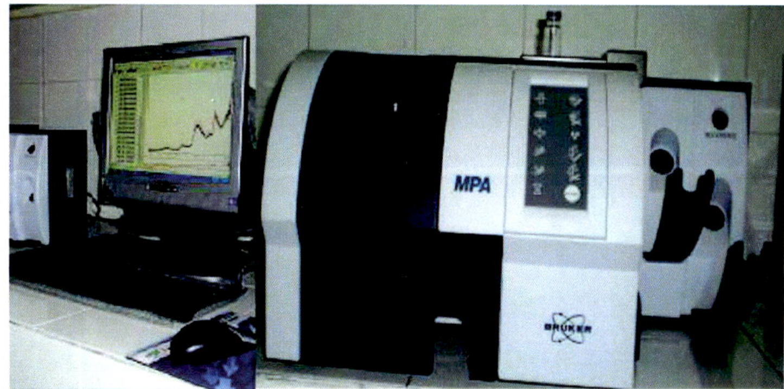

Fig. 2: FT-NIR MPATM Spectrometer

3.1.1 Green tea quality parameters

Method for moisture determination

Instant green tea and green tea granules with different moisture (3–40%) contents were used for the model development. Total 30 samples were used for the model development and cross validation was followed for validating the developed model (Sinija and Mishra, 2008b).

Method for caffeine estimation

A one step extraction process with chloroform was followed for the sample preparation of caffeine estimation. Pure caffeine powder was dissolved in chloroform to give samples of different caffeine concentration in the range between 0 and 90 mg/100 mL. Out of 40 samples of different concentrations, 34 samples were used to develop the calibration model, and the rest were used for test validation. Each sample was analyzed in triplicate (Sinija and Mishra, 2009).

Method for total polyphenols determination

Instant green tea and green tea granules prepared under different processing conditions were used for the model development. Total 30 samples were used for the model development and cross validation was followed. Total polyphenols of individual samples used for calibration were determined by standard procedure using FC Reagent (Sadasivam and Manickam, 1996). To avoid over fitting of the model, the rule of thumb, which states that the number of PLS factors should not be more than one sixth the number of independent specimens in the calibration set (Kemsley, 1998) was adopted in all the above models.

Moisture content in green tea granules and instant green tea powder

The instant green tea and green tea granules samples were prepared and the spectra of 30 samples were measured in diffused reflectance mode. For each sample (8-10 g in sample bottles), three spectra were recorded at three different points by rotating the sample bottle by 120°. The average of these three spectra was taken for analysis.

3.1.2 Moisture content in bael pulp

The spectra were measured by keeping 8-10 gm of sample in small sample bottle using diffused reflectance mode. For each sample, three spectra were recorded at three different points by rotating the sample bottle by 120°and the average was taken. Total 30 *bael pulp* samples with different moisture (75-95%) contents were used for the model development and cross validation was followed for validating the developed model.

3.1.3 Moisture content in dahi powder

The FT-NIR spectra for absorbance values of plain *dahi* and *dahi* pre-concentrated by cloth bag method and centrifugal separation, and dahi powder samples were compared for a range of wave numbers (12000-4000 cm^{-1}) in reflectance mode. The objective of this study was to demonstrate that FT-NIR can be used to determine the moisture content in dry cultured milk products like *dahi* powder. The FT-NIR spectra of samples were generated in integrating sphere module. Total 20 powder samples, with known reference values stored in individual 22mm glass vials were used.

3.1.4 Adulteration in milk

Sinija et al. (2008a) studied the feasibility of measuring adulteration in milk using FT NIR technique. Milk samples adulterated with different concentrations of water and starch were scanned using FT NIR spectroscopy. A partial least square model was developed with vector normalization method in near infra red region (4000- 12000 cm^{-1}). Suitable spectral wave number regions by principal least square (PLS) regression models development. Maximum correlation coefficient (R^2) values were obtained for the validation models developed for adulterants, water and starch, respectively.

3.1.5 Aflatoxin B_1 in red chilli powder

Red chillies (*Capsicum annum* L var. Sannam) of good quality was procured from local market and powdered in laboratory mixer. The powdered chilli samples were checked for presence of any microbial contamination initially by selective plating. AFB_1 standard was procured from Hi-media Laboratories, India. For the current study spectra's were collected in diffuse reflectance mode with sphere macrosample integrating sphere measurement channel. The normal chilli powder was spiked with the different concentrations of standard AFB_1 prepared in methanol. The complete sample size was (n =65) and the concentration of the AFB_1 in the prepared chilli powder samples ranged from 500 μg/kg to 15.75 μg/kg. The calibration set contained 40 samples and 25 samples were kept for cross validation. Fifty grams of spiked chilli powder was densely packed into sample cup (Duroplan, Germany) and placed in sample holder in rotating sample wheel accessory for high reproducibility and to remove any in-homogeneity in sample. The scanner speed was 10 Khz and each spectra was the average spectrum of 32 scans. Each sample was collected three times after rotating the sample cup by 360° and the mean of three spectra of the same sample was used in further spectral analysis steps.

3.1.6 Phytic acid content in green gram seeds

Green gram (Cv. Malviya12) seeds were procured from the experimental farm of Banaras Hindu University, Varanasi, India. Green gram samples were cleaned and any foreign matter was removed manually before its use in the experiment (Pande and Mishra, 2015). Pure phytic acid powder purchased from Sigma Aldrich was dissolved in 3% trichloro acetic acid (TCA) to give samples of different phytic acid concentration in the range between 100 and 1500 mg/100 g. Total volume of each sample prepared was 10 mL. To avoid over fitting of the model, the rule of thumb, which states that the number of PLS factors should not be more than one sixth the number of independent specimens in the calibration set (Kemsley, 1998; Sinija & Mishra, 2009; Pande and Mishra, 2015) was adopted. Hence approximately out of 35 samples of different concentrations, 30 samples were used to develop the calibration model, and the rest were used for validation. Each sample was analysed in triplicate (Sinija and Mishra 2009; Pande and Mishra, 2015). The scanner speed was 10 kHz and each spectrum was the average spectrum of 32 scans. A background spectrum was recorded with the help of empty beaker and background was active for 24 h. This rapid and non-destructive method of food analysis was adopted as a novel technique to check the level of phytic acid in green gram seeds. The spectra were measured by keeping 10 mL of sample in small beaker measurement with MPAFibre 2 solid probe. For each sample, three spectra were recorded by instrument and the average was taken (Pande and Mishra, 2015).

3.2 Chemometrics: multivariate analysis

The spectral data were analyzed using PLS regression with various preprocessing techniques using *OPUS 5.5* statistical software. Multivariate analysis by partial least square algorithm was used for quantitative and qualitative analysis. These methods with original and some preprocessed spectra were used to develop calibration models. The performance of the final PLS model was evaluated in terms of root mean square error of cross-validation (RMSECV) for cross validation and root mean square error of prediction (RMSEP) during test validation, the root mean square error of estimation (RMSEE) and the correlation coefficient (r^2). For RMSECV, a leave-one-sample-out cross-validation is performed. The spectrum of one sample of the training set is deleted and a PLS model is built with the remaining spectra of the training set. The left-out sample is predicted with this model and the procedure is repeated with leaving out each of the samples of the training set. In test validation a set of samples are identified as test data and with the remaining data set calibration model will develop first and then the test spectra will be used for the validation of developed model.

The number of PLS vectors used is defined in the OPUS software by the size of the 'rank'. The optimum PLS rank can be calculated only if the number of calibration spectra is sufficiently high (e.g. one component and 20 calibration spectra). The PLS regression has the advantage that the PLS factors are arranged in correct sequence, according to their relevance to predict the component values.

The residual (Res) is the difference between the true and fitted value. Thus the sum of squared errors (SEE) is the quadratic summation of these values, (Eq. 1).

$$SSE = \sum [\mathrm{Re}\, s_i]^2 \tag{1}$$

The root mean square error of estimation (RMSEE) is calculated from this sum, with '*n*' being the number of samples and '*r*' the rank, Eq. 2

$$RMSEE = \sqrt{\frac{1}{n-r-1} \times SEE} \tag{2}$$

The determination coefficient, r^2 in Eq. 3, gives the percentage of variance present in the true component values, which is reproduced in the regression. r^2 the approaches 100% as the fitted concentration values approach the true values.

$$r^2 = \left(1 - \frac{SSE}{\sum (y_i - y_m)^2}\right) \times 100 \tag{3}$$

Where, y_m is the mean of the reference results for all samples. The r^2can be negative (in some cases) for low ranks, when the residual are larger than the variance in the true values (y_i). In case of cross validation the RMSECV is calculated using Eq. 4

$$RMSECV = \sqrt{\frac{\sum_{i=1}^{n} (\bar{y}_i - y_i)^2}{n}} \tag{4}$$

Where, n is the number of samples in the training set, y_i is the reference measurement for the sample i, and $y_i^-$is the estimated result for sample i, When the model is constructed with the sample i removed. The number of PLS factors included in the model is chosen according to the lowest RMSECV. For the test set, the root mean square error of prediction (RMSEP) is calculated as follows, Eq. 5.

$$RMSEP = \sqrt{\frac{\sum_{i=1}^{n} (y_i - \hat{y}_i)^2}{n}} \tag{5}$$

Where, n is the number of samples in the test set, y_i is the reference measurement result for test set sample i and $y_i^\wedge$ is the estimated result of the model for test sample i.

3.3 Electronic nose technology for oat milk quality parameters

3.3.1 Sample preparation and measurement

The enzymatically developed oat milk was filled in sterile clear glass bottles under aseptic conditions and stored at room temperature (24 °C ± 2 °C) and refrigerated conditions (4 ± 0.5 °C). Measurements were performed every 4th hour for sample stored at room temperature and every third day for samples stored at refrigerated conditions up to 15 days. For every measurement, two replicate samples were withdrawn for microbiological (Total plate count), electronic nose response, sensory and chemical analysis (pH).

Oat milk sample (1 mL) was filled in glass vials with the help of a micropipette. Glass vials were then hermetically closed with aluminium caps and samples were kept in e-nose tray for analysis. The auto sampler injected a specified amount of sample into the instrument through the injection port and the sample was analyzed. Gaseous compounds (samples) pass through the chambers that contain 6 sensors each. The carrier gas pushes the headspace injected and every sensor reacts with the chemical compounds in the air flux. Sensors resistivity was monitored for 120 s. Other analytical conditions of electronic nose selected for the study are shown in Table 1.

Table 1: Analytical condition of electronic nose

Parameter	Measure
Quantity of sample in the vial, mL	1
Headspace generation times	300
Headspace generation temperature, °C	50
Syringe volume, mL	2.5
Injected volume, mL	1
Injected speed, mL	1
Acquisition times	120
Headspace generation times	300

3.3.2 Multivariate analysis for electronic nose data

Multivariate statistical techniques like principal component analysis (PCA), discriminant factorial analysis (DFA) and partial least square (PLS) were applied to the electronic nose sensors output. PCA was used to evaluate discrimination performance; DFA was used to identify unknown samples in one of the training groups and PLS was used to correlate the sensors data with quantitative characteristics.

PCA is a linear method that has been shown to be effective for discriminating the response of an electronic nose to simple and complex odour (Gardner, 1991). By using PCA, data was expressed and presented in such a as to highlight the similarities and differences in analyzed samples (Dutta, et al. 2003). In the vector space, PCA identified the major directions, and the corresponding strengths, of variation in the data. PCA achieved this by computing the *eigen* vectors and *eigen* values of the covariance matrix of the dataset. Keeping only a few eigen vectors corresponding to the largest eigen values, PCA can be also used as a tool to reduce the dimensions of the dataset while retaining the major variation of the data (Abdi and Williams, 2010; Tripathi et al., 2012). During this process, the data was transformed into two-dimensional or three dimensional coordinates with uncorrelated projection vectors that retain the most important information from the original data (Feng et al., 2011).

DFA was used to develop predictive models for classification of samples based on grouping variables. The DFA method is based on a search for directions along which the groups are as far apart as possible and the samples of the same group are as close together as possible (AlphaMOS release October, 2002).

Partial least square method (PLS) was used to correlate the sensors data with quantitative characteristics (Di Natale et al., 2003; Lozano et al., 2007). For this purpose, pre-processing of the raw data was done to obtain the best linear correlation between sensor response (S) and compound concentration (C). The pre-processing was a "log conversion" which transforms the relation between (S) and (C) as seen from Eq. 6.

$$S= k (C)^n \Rightarrow \text{Log } S = \text{Log } k + n \text{ Log } (C) \tag{6}$$

4. Quality Issues and Challenges

The methodology of these rapid technologies discussed above is promising however still there are few technical challenges. Due to the low absorption of the overtone and combination bands in the NIR range, it is not necessary to dilute the samples in contrast to MIR measurements. Advantages and disadvantags of one rapid technologies are briefly summarised as follows:

Advantages	Disadvantages
• Fast (5-10 sec. Typical)	• Not a "primary" method
• No sample preparation, simple to operate	• Perceived as complex
• No waste, no pollution	• To make the method robust and replicable extremely large number of samples are required for calibration model
• Simultaneous determination of multiple components per measurement	• Samples used for testing need to have same physical and chemical properties as were used in model/method development; any change in the sample property will not be identified by the method
• Highly precise and accurate	• The developed method can only measure in the range as used in calibration models; the instrument is not precisely sensitive for over and under range
• Measure through packaging	• The equipment requires trained personnel's for working
• Transferable methods	• Materials, solvents, preparation time and waste disposal are factors which can be expensive and add to the cost of spectroscopic analyses
• Real time monitoring for process	

5. Potential for Commercial Application

NIR spectroscopy can be used for many qualitative and quantitative applications in various industries dealing with chemicals, pharmaceuticals, cosmetics, polymers, rubber, textiles, food stuffs, or feedstock. It is the ideal technique for quality assurance (QA) or quality control (QC). The rapid methods developed in laboratory covered wide range of products and parameters. Thus it finds applications in food analysis covering different industrial sectors like dairy products, beverages (tea, *bael* pulp, and oat milk), pulses (antinutritional factor), and spices (AFB_1 in spiked red chilli powder). Dedicated sampling accessories are available for any type of liquid, powdered, semi-solid or solid sample.The methods developed generally included few components or parameters of the product. For further application of the methodology the calibration models can be improved by simultaneous inclusion of other quality parameters. Also, the sample size needs to be increased further for more robustness and reproducibility. The methods developed included the preparation of different concentrations of the analyte (biochemical parameter under investigation) by use of commercial standards (e.g. spiking of AFB_1). However, the methodology finds promising application if the variations of concentrations in the samples used for calibration and validation are found naturally in the product. This makes the developed method more suitable for quality control studies and real time monitoring.

6. Developed Rapid Methods of Food Quality Analysis

6.1 FT-NIR methods for determination of green tea quality

6.1.1 Determination of moisture content in green tea granules and instant green tea powder

Figure 3 shows the characteristic absorption spectra of the calibration data set. Spectra are similar to those reported by (Chen et al., 2007). From Figure 3 it is seen that the water absorption bands were around 5155 and 7000 cm^{-1} corresponding to O-H stretching and O-H deformation. The most intensive band in the spectrum belong to the vibration of the second overtone of the carbonyl group (5352 cm^{-1}), followed by the C-H stretch and C-H deformation vibration (7212 cm^{-1}), the $-CH_2$ (5472 cm^{-1}) and the $–CH_3$ overtone (5808 cm^{-1}). The vibration of the carbonyl group, -CH and $-CH_2$ vibrations are caused by the constituents such as polyphenols, alkaloids, protein, volatile and nonvolatile acids and some aroma compounds.

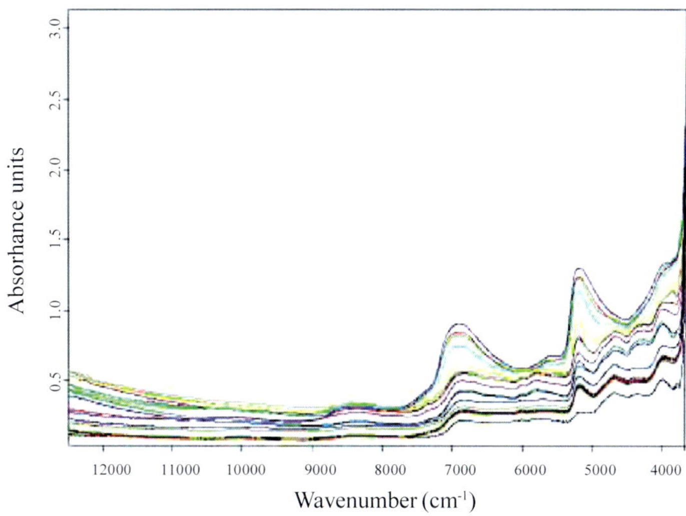

Fig. 3: FT-NIR absorption spectra of green tea samples used for calibration

Figure 4 show values of RMSECV corresponding to each PLS factor for determining moisture content with different spectral pre-processing methods. From Figure 4, vector normalization preprocessing method with highest value for r^2 and lowest value for RMSECV, is superior to other methods. The RMSECV value decreases sharply with initial factors, however, it gradually decreases as PLS factor increases. This model needs five PLS factors.

The cross-validated calibration model used the entire spectral region (4000 to 12000 cm^{-1}). PLS- regression method gave r^2 values of 0.9975 for calibration data set and RMSECV value of 0.83 for cross validation (Sinija and Mishra, 2008b). The results of this study clearly demonstrated the capability of FT-NIR for this application. This result was validated with fresh tea samples analyzed by standard gravimetric method and by moisture analyzer (AND MX- 50 model). Fresh samples were measured using the above 3 methods in triplicate and the values obtained were analyzed statistically (Table 2).

Table 2: Average moisture content (% wb) of green tea samples by different methods

Method	Instant green tea	Green tea granules
Gravimetric method	4.31	5.57
FT-NIR method	4.53	5.78
Moisture Analyzer	4.63	5.84

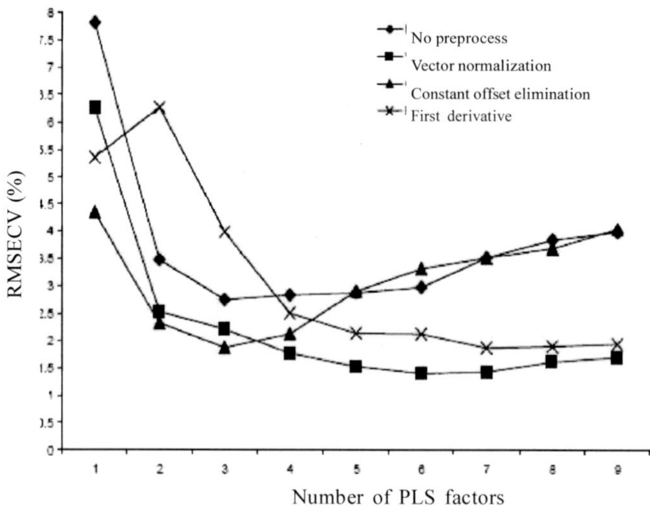

Fig. 4: RMSECV and PLS factors for various pre- pprocessing methods

6.1.2 Determination of caffeine content in green tea samples

In this study the spectra of caffeine was loaded with pure chloroform as the background (blank) for measurement. Figure 5a shows the FT-NIR spectra of pure caffeine which has major peaks at absorbance bands (wave numbers) of 4201.4, 5285.1, 7137.4, and 8712.6 cm^{-1}. These true peaks were selected after smoothing the spectrum to avoid interference due to noise. Major peaks at absorbance bands or wave numbers of 5285 and 4201 cm^{-1} may be due to the stretching vibrations of >C=O, C=C and C=N bonds of caffeine. Peaks at 7137 and 8712 cm^{-1} may be due to -CH bonds of methyl ($-CH_3$) groups. Caffeine molecule has 3 methyl groups on a cyclic structure; hence the peak may play an important role in the estimation of caffeine. Some minor peaks observed in the pure caffeine spectrum may be due to unknown bond vibrations. Figure 5b shows the spectrum of tea samples extract in the chloroform which indicates the characteristic absorption spectra of the calibration data set. Spectra are similar to those reported by Chen et al. (2007). The most intensive band in the spectrum belong to the vibration of the second overtone of the carbonyl group (5285 cm^{-1}), followed by the -CH (7137 cm^{-1}), the $-CH_2$ (5472 cm^{-1}) and the $-CH_3$ overtone (5808 cm^{-1}). The vibration of the C=O, -CH and $-CH_2$ are caused by ingredients such as polyphenols, alkaloids, protein, volatile as well as non-volatile acids and by some aroma compounds (Paradkar and Irudayaraj, 2002).

Table 3 shows the RMSEP and r^2 values for caffeine content determination with different spectral preprocessing methods. From table it is clear that the highest value of r^2 and lowest value of RMSEP was obtained for 1st derivative plus straight line subtraction method. The test-validated calibration model used the entire spectral region (4000 to 12000 cm^{-1}). PLS- regression method gave r^2 values of 0.9903 for calibration and 0.9809 for validation set data and RMSEP value of 1.95 for validation and RMSEE 1.86 for calibration set. The results of this study clearly demonstrated the efficacy of the developed FT-NIR method for caffeine determination in green tea (Sinija and Mishra, 2009).

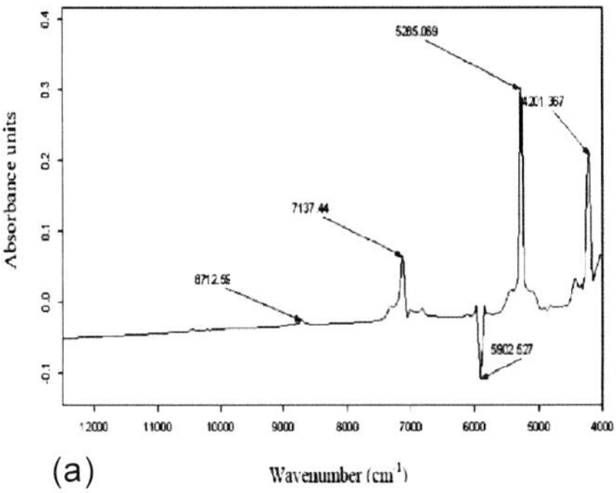

Fig. 5a: FT-NIR Spectra of pure caffeine

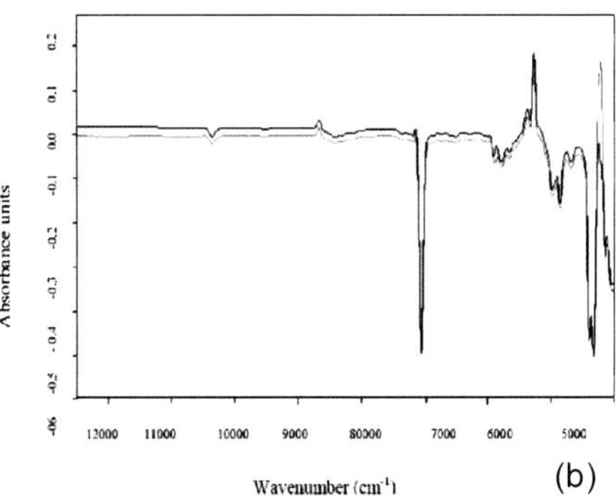

Fig. 5b: Spectrum of tea caffeine extracts in chloroform

Table 3: RMSEP and r^2 values for different spectral pre-processing techniques in caffeine content determination

Pre-processing technique	r^2 (Calibration)	RMSEE	r^2 (Validation)	RMSEP	PLS factor
No preprocessing	96.58	2.36	95.23	4.01	7
1st derivative+ Straight line subtraction	99.03	1.86	98.09	1.95	6
1st derivative + vector normalization	91.66	9.03	84.38	10.7	4
Straight line subtraction	95.89	6.62	90.30	8.44	5
2nd derivative	99.78	1.81	98.36	3.08	8

Results obtained from FT-NIR spectroscopy were compared with that of the conventional UV spectroscopic method (Table 4). The FT-NIR method was further validated through recovery studies by determining the caffeine content in artificially spiked tea samples. FT-NIR prediction in the recovery study was found to be successful and comparable to the result obtained by the UV spectroscopic method (Table 5).

Table 4: Comparison of results obtained by FT-NIR and UV spectroscopic methods for determination of caffeine in green tea samples

Products	Caffeine content by UV spectroscopic method (mg/100 mL)		Caffeine content by FT-NIR method (mg/100 mL)	
	Mean	± SD	Mean	± SD
Instant green tea powder	40.46	1.34	43.28	2.86
Green tea granules	46.36	2.43	49.51	3.14

Table 5: Results of caffeine recovery estimates from UV and FT-NIR spectroscopy

Product	Added caffeine (mg)	Caffeine recovery by UV spectroscopy method (%)	Caffeine recovered by FT-NIR method (%)
Instant green tea powder	306090	103.9101.8101.2	98.399.199.8
Green tea granules	306090	104.2102.4101.8	97.999.6 100.1

6.1.3 Determination of polyphenols in green tea samples

Figure 6 shows the FT-NIR spectra of tea samples used for developing the calibration model. These true peaks were selected after smoothing the spectrum to avoid interference due to noise. The most intensive band in the spectrum belong to the vibration of the second overtone of the carbonyl group (5285 cm^{-1}), followed by the -CH (7137 cm^{-1}), the $-CH_2$ (5472 cm^{-1}) and the $-CH_3$ overtone (5808 cm^{-1}). The vibration of the C=O, -CH and $-CH_2$ are caused by ingredients such as polyphenols, alkaloids, protein, volatile as well as non-volatile acids and by some aroma compounds (Paradkar and Irudayaraj, 2002).

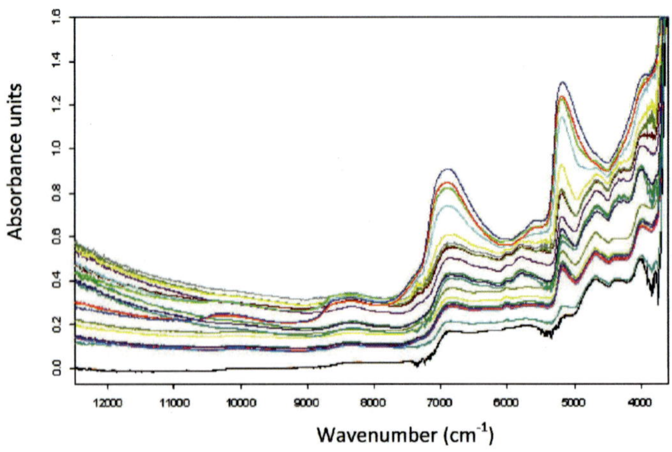

Fig. 6: FT-NIR spectra of calibration data set for green tea samples

Cross validation method as described above was followed in this method. The lowest RMSECV value equals to 1.45 was obtained after the vector normalization spectral pre-processing and also maximum value for coefficient of determination was obtained for the same preprocessing method. PLS- regression method gave r^2 values of 0.978 and RMSECV value of 1.45 for cross validation. The results of this study clearly demonstrated the capability of FT-NIR for this application. The predictive abilities of the calibration models were evaluated by the residual predictive deviation (RPD) also, which was defined as the ratio of the standard deviation of the reference data to the standard error of predicted data for the population tested (Pink and others, 1998). RPD value for the present model is 4.5, which is greater than three and considered to be desirable for prediction purposes as per the previous references (Pink et al., 1998; Nieuwoudt et al., 2006).

The result obtained by the developed method was validated with fresh tea sample and cross checked by standard spectrophotometric method. Total polyphenols content of freshly prepared tea samples were measured in triplicate by the above two methods and the values obtained were analyzed statistically (Table 6).

Table 6: Total polyphenols content (mg/g) of green tea samples

Method	Instant green tea	Green tea granules
FT-NIR method	246.34	217.93
FC Reagent method	242.15	224.83

6.1.4 Versatility of developed methods

All methods developed were tested with fresh tea leaves and black tea samples and found that these can be used for the determination of moisture, caffeine and total polyphenols in any form of tea as the matrix of the tea samples remain the same, only difference will be in the concentration of various constituents. The spectra of different tea samples are shown in Figure 7 which shows that all the spectra looks similar except the difference in heights of peaks.

These methods were used to determine the amount of various constituents in instant green tea powder and green tea granules during storage study. Two market samples (one black tea and one green tea) were also tested by the developed methods for moisture content, caffeine and total polyphenols and the results are given in Table 7. From the table it is clear that there is no appreciable variation in the values determined by chemical analysis and the developed methods. The overall results demonstrate that NIR spectroscopy with multivariate calibration could be successfully applied as a rapid tool to determine simultaneously the various constituents in tea.

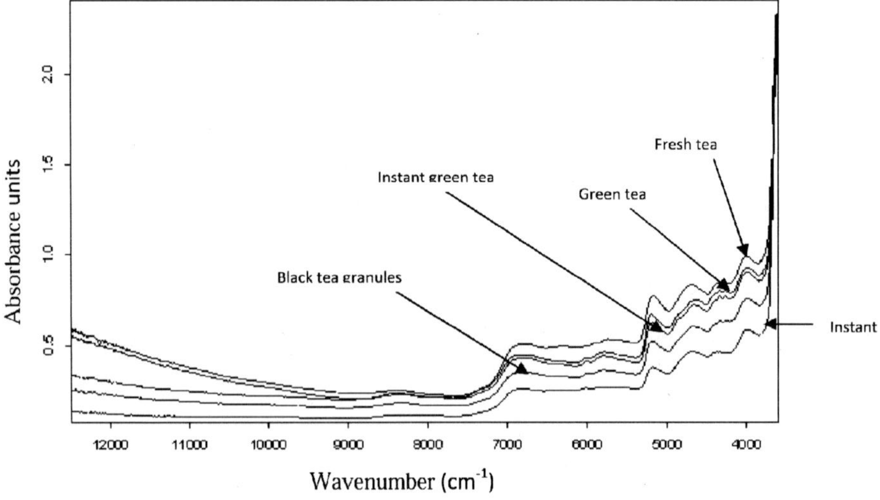

Fig. 7: FT-NIR spectra for various forms of tea

Table 7: Results of market sample analysis by FT-NIR method and chemical method

	Moisture (% wb)		Caffeine (mg/100 mL)		Total polyphenols (mg/g)	
	Gravimetric method	FT-NIR method	Spectro photometric method	FT-NIR method	Spectro photometric method	FT-NIR method
Black tea	4.31	4.56	54.35	55.70	235.12	239.35
Green tea	4.50	4.71	56.16	57.51	256.32	259.36

6.2 Determination of moisture content in *bael* pulp and pulp powder

Figure 8 shows the characteristic absorption spectra of the distribution data set (n=30) of bael (stone apple) pulp samples with moisture ranging from (70 to 95%). The spectra which illustrate the lowest molecular absorptivities are in the short wavelength region (12500–8700 cm^{-1}) with higher values in the first overtone region around 7100 cm^{-1} and still higher absorbance levels in the combination region (6000–4250 cm^{-1}). It is readily apparent that all the spectra were quite homogeneous. A small baseline offsets and bias were present due to the light scattering or concentration variation.

6.2.1 Spectral properties

The absorbance spectra for bael pulp samples are dominated by water absorption bands. The peaks at 10282 cm^{-1} and 7100 cm^{-1} correspond to respective second and first overtone stretch vibration modes of O–H bond. The absence of such bands in the spectra of dried bael pulp powder (Figure 8) where moisture has been reduced to 2.52% by foam mat drying, following the reported procedure (Lijuan et al., 2009) confirms these bands to be overtones of water molecules. Earlier report suggested absorption band at around 5,155 and 7,000 cm^{-1} corresponding to O–H stretching and O–H deformation (Sinija and Mishra, 2009).

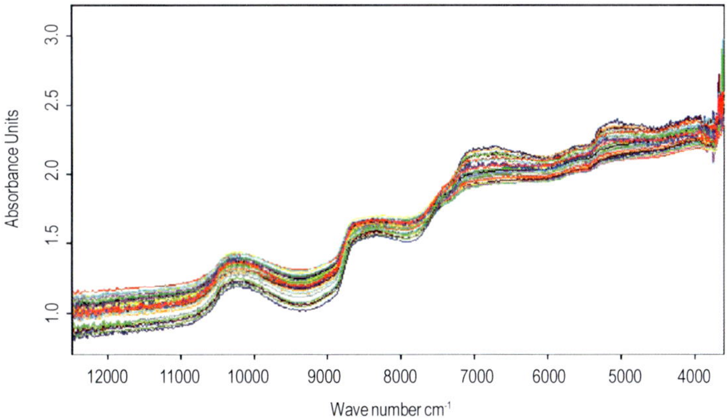

Fig. 8: NIR absorption spectra used for the calibration data set for *bael* pulp powder
(See colour version on page 338)

The absorption at 6670 cm^{-1} in case of powder sample corresponds to C–H first overtone stretch vibration modes in CH_3 and CH_2 groups whereas that at 8600 cm^{-1} (in pulp) and 8283 cm^{-1} (in powder) ascertain as the second overtones of C-H stretch vibration modes (Jukka et al., 2000). The characteristics sugar C–H combination bands absorbs in 4350-4550 cm^{-1} region (Inon et al., 2005; Leon et al., 2005), which is predominantly observed in the spectra of the powder. The band at 5234 cm^{-1} in the pulp spectrum belongs to the vibration of the second overtone of the C=O stretch, which could be observed at 5175 cm^{-1} in case of powder. The vibration of the carbonyl group, –CH and –CH_2 vibrations, are caused by the constituents such as wave number cm^{-1} absorbance units polyphenols, alkaloids, protein, volatile and nonvolatile acids, and some aroma compounds (Sinija and Mishra, 2009).

6.2.2 Spectral pre-processing and PLS model building

In this study, vector normalization, min-max normalization and multiplicative scatter correction methods were tried. In vector normalization, the spectrums are normalized by first calculating the average intensity value and subsequent subtraction of this value from the spectrum. Then, the sum of the squared intensities is calculated, and the spectrum is divided by the square root of this sum. In min-max normalization method spectrum first subtracts a linear offset and then sets the Y-maximum to a value of 2 by multiplication with a constant. Then it is used similar to the vector normalization. Multiplicative scatter correction performs a linear transformation of each spectrum for it to best match the average spectrum of the whole set. This method is often used for spectra measured in diffuse reflection.

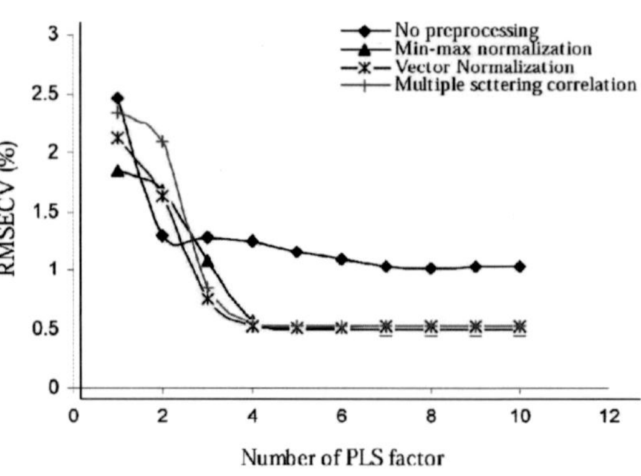

Fig. 9: Effect of number of PLS factors on RMSECV for various pre processing methods for determining moisture content in *bael* samples

Figures 9 and 10 show values of RMSECV and R^2 values corresponding to each PLS factor for determining moisture content in bael samples with different spectral pre-processing.

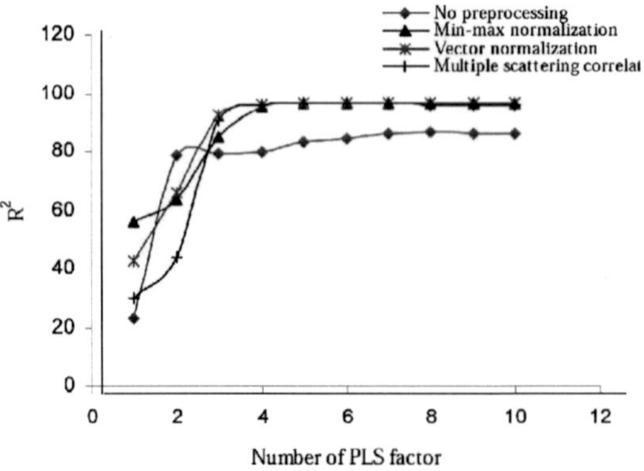

Fig. 10: Effect of number of PLS factors on R^2 for various pre processing methods for determining moisture content in *bael* samples

From the Figure 9 and 10 it is evident that the pre-processing greatly effects in improving the quality of model and min-max normalization pre-processing method is superior to others which has the highest value for R^2 and minimum value for RMSECV. The RMSECV value decreases sharply with initial factors. Compared with others, the lowest RMSECV is equal to 0.485 obtained after the min-max normalization spectral pre-processing. This model needs five PLS factors. In this application, min-max normalization seems to perform better than other pre-processes.

The linear regression plot of the calibration and validation data sets for the best model (MMN pre-processing) was developed. The equation of the straight line for this cross validation plots of the calibration data sets is represented as $y = 1.0021x - 0.0716$ and has reasonably high R^2 value (93) and low RMSEE (0.266) showing good performance by this model for predicting moisture content of the bael pulp samples in the range of 70–95% (wb). The test validation of also suggested that the method was very good in predicting moisture content to accuracy with high value of correlation coefficient (R^2= 97) and low values of

RMSECV. The offset and the slope for the equation of the regression line for this set was 1.4274 and 0.9835 respectively. The predictive abilities of the calibration models were evaluated by the residual predictive deviation (RPD) and the RPD value for the present model is more than 3 and considered to be desirable for prediction purposes as per the previous references (Pink et al., 1998; Nieuwoudt et al., 2006). The results of this study clearly demonstrated the usability of FT-NIR for this application.

6.3 Determination of moisture content in dahi powder

Unique spectra of plain and pre concentrated *dahi* samples was generated. Study also aimed at comparison of the spectra of reconstituted samples in the near infrared region using FT-NIR spectroscopy (Shiby, 2008).The FT-NIR spectra for absorbance values of plain *dahi* and *dahi* preconcentrated by cloth bag method and centrifugal separation and dahi powder samples were compared for a range of wavenumbers (12000-4000 cm^{-1}) (Figure 11). The peak shift in the range 5000 to 5250 cm^{-1} represents the changes in moisture due to pre concentration and drying. Similar spectra have been obtained for other dairy products such as cheese and butter (Ru and Glats, 2000). The near infra red absorption bands for protein –peptides due to –NH deformation are in the range of 2080-2220 nm and 1560-1670 nm. The lipid absorption bands are seen in the range of 2300-2350 nm due to methylene –CH stretch and in the range 1680-1760 due to $-CH_2$ and $-CH_3$ stretch. The carbohydrate absorption bands due to C-O, O-H stretching combination is in the wavelength range of 2060-2150 (Nielson, 1998). The changes in the FT-NIR spectra of pre-concentrated *dahi* samples are mainly due to the loss of moisture and the resulting increased concentration of other constituents. Lactose, being water soluble, is mainly lost in the whey during pre-concentration.

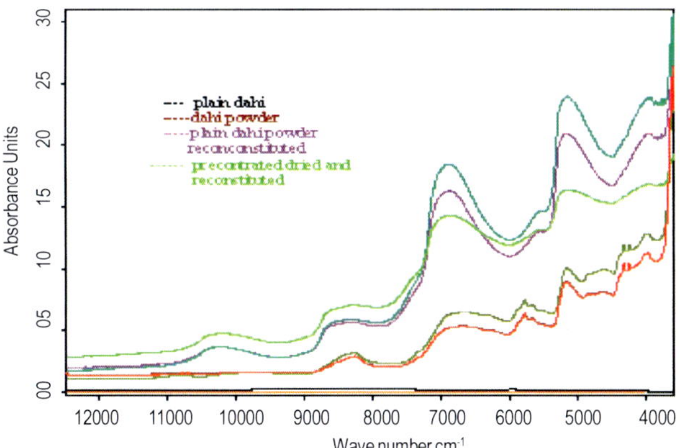

Fig. 11: FT-NIR spectra of (a) plain dahi, (b) dahi powder and (c) reconstituted dahi (plain) and (d) reconstituted dahi (pre-concentrated and dried) (See colour version on page 338)

The spectra obtained for dahi powder were similar to milk powder spectra recoded by previous authors (Ru and Glats, 2000). NIR is a well-established method for the analysis of milk powder for moisture, fat and protein content, online as well as in the laboratory. Samples with moisture in range 0.02-0.36% were analyzed. Spectra of two of the dahi powder samples are shown in Figure 12. The spectral differences at around 8500 cm^{-1} and 5100 cm^{-1} can be directly correlated to the different moisture contents within the two samples. The results of this study clearly demonstrated the capability of FT-NIR for this application with cross validation accuracies (RMSECV) of 1350 ppm for moisture where a prediction accuracy of 0.135% is achieved for moisture. The models were having high value of coefficient of determination 0.9926. Hence our results proved that FT-NIR spectroscopy is a rapid and efficient tool for online quality control of dahi powder on the basis of its moisture content. FT-NIR spectroscopy can be successfully used for determining stability of dahi powder during the drying process.

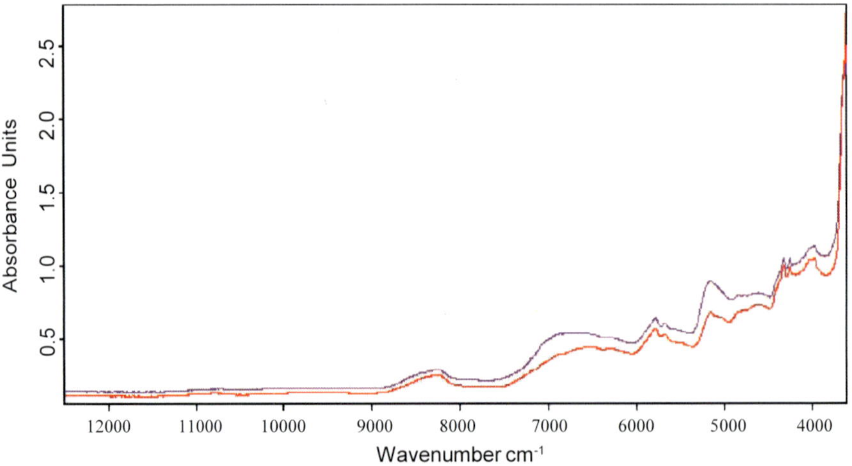

Fig. 12: FT-NIR absorption spectra (12000-4000 cm^{-1}) of two dahi powder samples with different moisture contents

6.4 Determination of AFB_1 in red chilli powder

The study the feasibility of FT-NIR spectroscopy in AFB_1 detection the AFB_1spiked red chilli powder was used. The water absorption bands at 5155 cm^{-1} and 7000 cm^{-1} were excluded selectively as they were interfering with other components of interest in the sample. The wavelength regions of 6900 - 4998 cm^{-1} and 4902 - 3999 cm^{-1} were selected for better interpretation of the relationship between the model and chemical structure of the compound in interest (Figure 13). The best preprocessing method used for developing calibration model with high predictive ability having optimum number of factors and frequency ranges (wavenumber cm^{-1}) is selected on the basis of highest value of R^2 while lowest values for RMSEE. Similarly, the optimal spectral range was selected for validation model on the basis of minimum values for RMSECV, PLS factors and relatively higher R^2. To test the performance of the optimized calibration model; RMSECV was plotted as function of PLS factors (Figure 14) for searching best spectral preprocessing method with minimum error for predicting AFB_1 content. From the Figure 14, it is clearly seen that preprocessing greatly affects in improving the quality of model. The unprocessed spectra covering the entire scan range of 4000 – 12000 cm^{-1} had higher RMSECV values from 7.56 to 1.88% and correspondingly lower R^2 (72%) compared to other preprocessed models. From the Figure 14, it can be inferred that SLS preprocessing method is superior and simpler than other methods; as SLS based model required comparatively fewer (five) PLS factors compared to nine factors in MMN and seven in COE models. The SLS preprocessed model also had lowest RMSECV value (0.65).

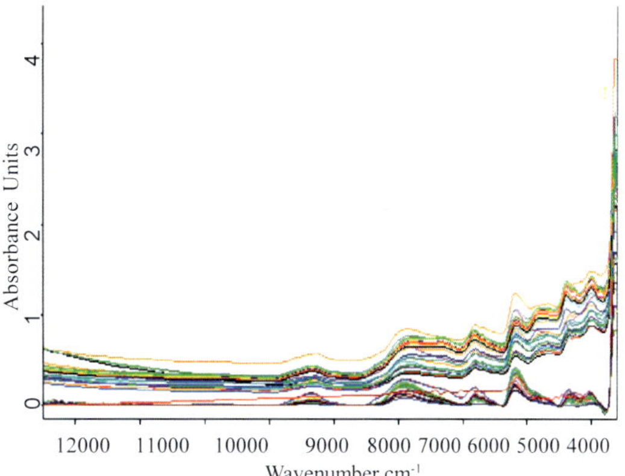

Fig. 13: FT-NIR Absorption spectra of the chilli powder samples for the whole data set (See colour version on page 338)

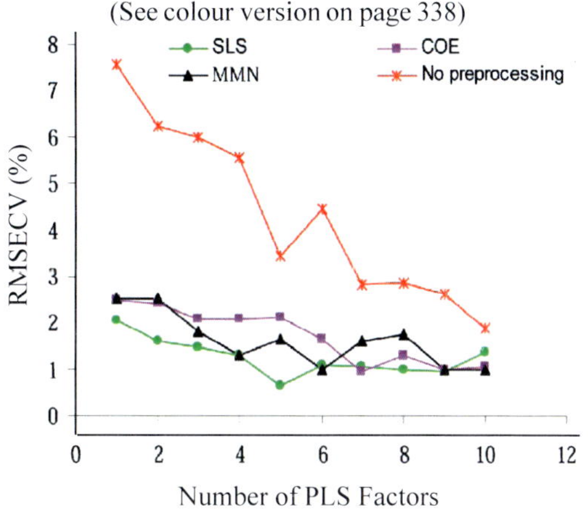

Fig. 14: Effects of number of PLS factors on RMSECV for the calibration model using different preprocessing techniques (SLS, COE, MMN, and Unprocessed)

The plot of residuals for the calibration and validation data sets is shown in Figure 15 and 16 respectively also show the normal distribution of the residuals around zero. This implies that the error or difference in predicting the response (aflatoxin concentration) by the model and as actually determined is relatively low. In building the best calibration models when very low concentrations of AFB_1 (<15 μg/kg) were used, error was seen and model was unable to predict the concentrations appropriately (results not shown) so the final model was developed for the concentration ranges above 15 μg/kg. These results are in agreement with the previous reports on predicting mycotoxin deoxynivalenol in wheat kernels just above the accepted levels (Petterson and Aberg, 2003; Dowell et al., 1999).

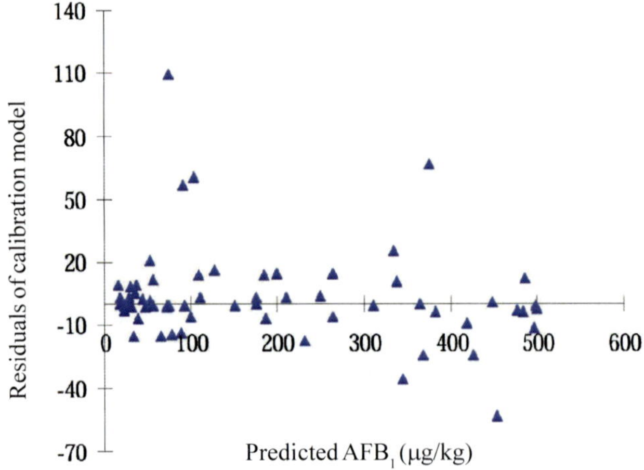

Fig. 15: Plot of residuals for the calibration model data set for AFB_1 content in spiked red chilli powder

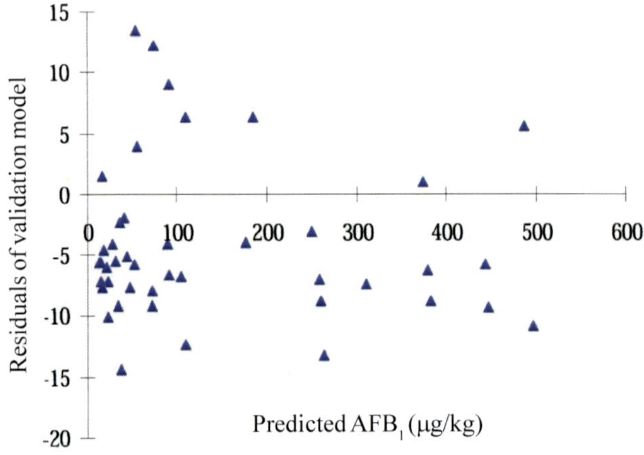

Fig. 16: Plot of residuals for the validation data set for AFB_1 content in spiked red chilli powder

The SLS preprocessing method was found superior and simpler than other preprocessing methods for predicting AFB_1 content with most accuracy. The PLS factors required in SLS based model was five compared to nine factors in MMN and seven in COE models. The SLS preprocessing enhanced the characteristic absorption bands in the spectra compared to unprocessed spectra. The equation of the straight line for the linear regression plot of the calibration data sets for the best model (SLS preprocessing) is represented as y = 0.9798x + 3.4119 with R^2 = 98%. The RMSEE calculated for this calibration model was 0.27. The RMSECV and R^2 values for the best validation model (SLS preprocessing) were 0.65 and 96.7% respectively.

The correlation plots (linear regression) between the data found by FT-NIR and reference procedures showed good comparability between the methodologies assayed. The FT-NIR method was developed by artificial spiking the AFB_1 standard (known concentration) in the red chilli powder. Similar type of model calibrations for FT-NIR giving good results have been practiced in past (Saona et al., 2004) so this technique was followed. The FT-NIR model for detecting AFB_1 in red chilli powder was optimized by different spectral preprocessing methods and validated by cross validation procedure. The quantification of AFB_1 in red chilli powder was possible in the range of 15 to 500 µg/kg. AFB_1 detection was possible without any sample preparation and use of chemicals in just 10 sec. Though the limit of detection is slightly more than as approved by European Union but the developed method can be used for bulk sorting of the chilli samples from the infected ones before going for any other chemical procedures. The results are more significant in Indian context as spices are exported widely to Europe and USA from India and aflatoxin contamination is a major constraint in increasing exports. The presented results offer wide opportunities in future for further improving the model for simultaneous detection of other aflatoxins and using the naturally contaminated sample for model building.

6.5 Determination of phytic acid content in green gram

Figure 17 shows the FT-NIR spectra of pure phytic acid which has major peaks at absorbance bands (wavenumbers) of 4201.4, 4500.543, 5385.069, and 6105.43 cm^{-1} (Pande and Mishra, 2015). These true peaks were selected after smoothing the spectrum to avoid interference due to noise. The strong signals (peaks) are due to overtones and combination bands of functional group vibrations in the near infrared region. The peak may play an important role in the quantification of phytic acid (Pande and Mishra, 2015). Figure 17 shows the characteristic absorption spectra of the calibration data set. The three combination bands were present for hydroxyl, water and carboxyl group. 4201.400 cm^{-1} corresponding to O–H stretching, water absorption bands were around 4500.543 cm^{-1} corresponding to O–H stretching and O–H deformation and 5385.069 cm^{-1}. The most intensive band in the spectrum belongs to the vibration of the first overtone of the hydroxyl group 6200 cm^{-1} Pande and Mishra, 2015). The above mentioned region was also defined by Sundaram, Kandala et al., (2009). Cross validation was carried out using thirty-five samples; five samples were used for test validation. The best number of factor was determined by the lowest RMSEP value and highest value for R^2. The RMSEP and R^2 values for phytic acid content determination with different spectral preprocessing methods are presented in Table 8 (Pande and Mishra, 2015).

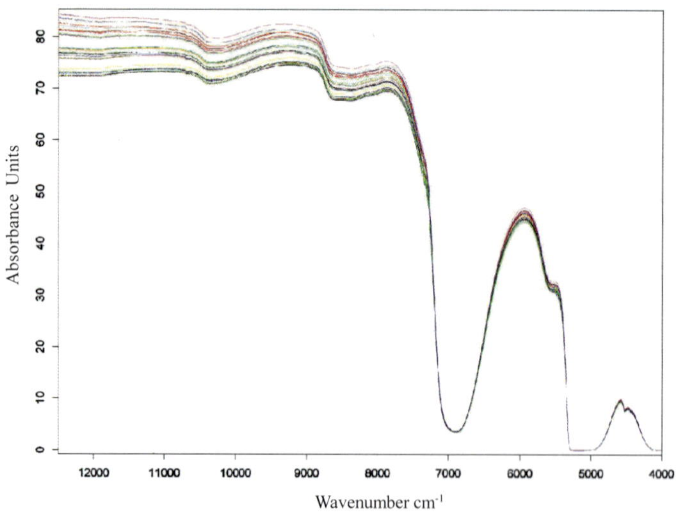

Fig. 17: FT-NIR absorption spectra of standard phytic acid and green gram seed sample containing phytic acid used for calibration (See colour version on page 339)

From Table 10 it is observed that the highest value of R^2 (97) and lowest value of RMSECV (0.727) for calibration data set and highest value of R^2 (96) and lowest value of RMSEP (1.715) for validation data set were obtained for first derivative plus vector normalization method. The minimum error and maximum R^2 was obtained for preprocessing method first derivative plus vector normalization suggesting its superiority over other. Thus, this method gave the best model for the prediction of phytic acid content in the green gram seed (Pande and Mishra, 2015).The test validated model used the entire spectral region (4000–12,000 cm^{-1}).The linear regression analysis of cross validation and test validation showed the R^2 values of 97% and 96%, respectively. Also the lower value of RMSECV and RMSEP for cross and test validation showed very low value of 0.727 and 1.715, respectively (Pande and Mishra, 2015). The developed method was compared with chemical analysis method. The results of this study clearly suggested the efficacy of the developed FT-NIR method for phytic acid determination in green gram sample. Sample of green gram seed used in experiments dissolved in 3% trichloroacetic acid was analysed by the developed method to quantify phytic acid content. The FT-NIR method showed the phytic content 591 mg/100 g and standard spectroscopic method showed 593 mg/100 g. The results were statistically compared by using SPSS 17.0 with paired sample t-test; no statistically significant differences in the means of FTNIR method and standard method were observed. The samples

Table 8: RMSEP and R^2values for different spectral pre-processing techniques for phytic acid content determination

Wavenumber region (cm^{-1})	Preprocessing method	PLS factors	RMSECV (%)	R^2(%)	RMSEP (%)	R^2 (%)
4601.4-4246.6	First derivative plus vector normalization	2	0.727	97.0	1.715	96
4601.4-4246.6	Second derivative	3	0.803	95.0	1.768	93
12000 - 4000	No spectral data preprocessing	5	0.972	94.6	1.796	92
7501.9-5450	MinMax normalization	4	1.01	92.0	1.826	89

were found to be highly correlated ($R^2 = 0.96$). Thus it can be inferred that precision of measuring phytic acid content by the paired two methods do not differ significantly. The comparable results show the efficiency of developed method over laborious chemical methods (Pande and Mishra, 2015).

6.6 Electronic nose methods

The aim of this work was to investigate the application of commercial electronic nose based on metal-oxide sensors in monitoring the change in the volatile compounds of oat milk during storage at different temperatures.

6.6.1 Microbial flora evaluation

The initial total microbial count of freshly prepared oat milk was 1.2 log10 cfu/g. This limit came for samples stored at room temperature after 20 hours of storage when total microbial count was in the range of 7.04 to 7.4 log10 cfu/g. For oat milk stored at refrigerated conditions, microbial load on Day 11 of storage varied from 5.2 to 5.41 log10 cfu/g. On 13th day of evaluation, a high microbial count corresponding to a value of 7.3 – 7.43 log10 cfu/g was reported. As the bacterial population increases, pH decreases in both the cases. This can be attributed to the uncontrolled fermentation resulting from action of bacteria on nutrients present in oats, converting carbohydrates to acid. The mean value of the pH at refrigerated temperature, on the day of collection (1st reading) was found to be 6.65, which decreased to 4.4 and 4.7 for milk stored at room temperature and refrigerated temperature, respectively when microbial load reached the upper limit for safe storage in both cases.

6.6.2 Electronic nose sensor response analysis

In Figure 18, the response of all the 18 sensors of the electronic nose to the volatiles generated is shown. The response intensity on the ordinate axis

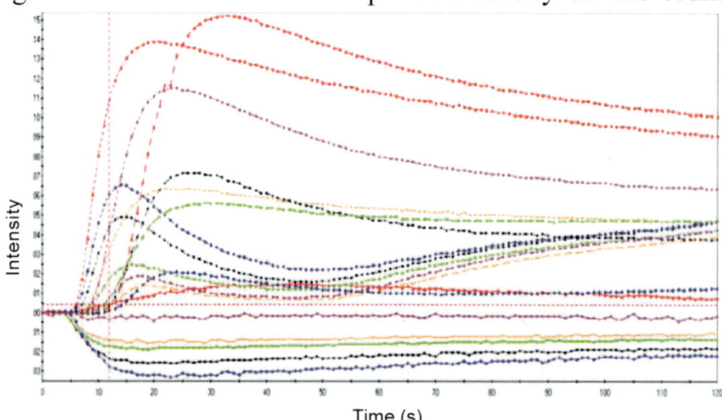

Fig. 18: Variation of response of 18 sensors with time of analysis

(See colour version on page 339)

corresponds to the rate of change of the relative resistance of different sensors $[(R_0-R)/R_0]$, where R_0 is the resistance value at t = 0, and R is the resistance value with changes in time (Labreche et al., 2005).

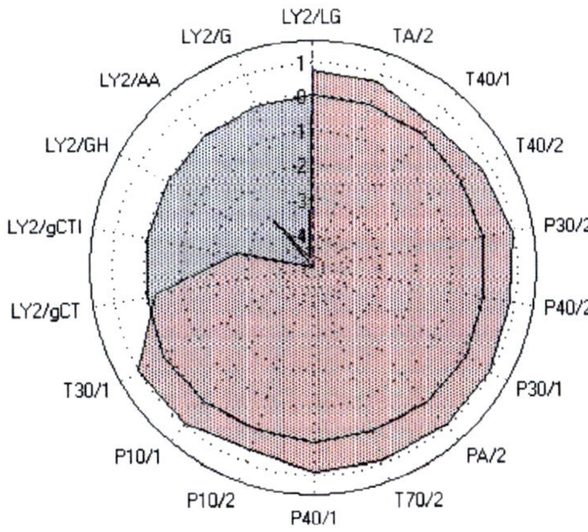

Fig. 19: Variation of sensor responses for control and spoiled oat milk samples

It can be seen from Figure 18 that all the 18 sensors responded differently to the volatiles generated by the sample. It is this sensor response which is used as a fingerprint for a particular food sample under a standard condition. These response data are further analyzed using different statistical tools to understand the results better. In addition, the radar graphs for the analyzed samples show a clear difference in the sensor responses for fresh and spoiled samples. Figure 19 shows the superimposed images of sensor responses for control and spoiled oat milk sample. It can be seen from the radar graph that there is a clear demarcation between the sensor responses to the volatiles generated in fresh and spoiled oat milk samples.

6.6.3 Classification of samples using PCA analysis

As a first step, in order to evaluate the ability of the electronic nose to discriminate among different oat milk samples, the sensor responses were elaborated by principal component analysis (PCA). PCA was performed on covariance matrix of sensor responses. For the oat milk samples stored at room temperature, the two first principal components, PC1 and PC2, accounted for 99.16 % of the total variance in sensor responses, 81.91 % and 17.25 %, respectively, as seen from Figure 20 (a). Similarly, Figure 20 (b) shows that for oat milk stored at refrigerated conditions, the two first principal components, PC1 and PC2, accounted for 97.72 % of the total variance, 95.42 and 2.3 %, respectively. A scatter plot in Figure 20 shows the sample separation according to the storage time. Considering the map of the PCA performed on the data sets obtained from electronic nose, fresh samples of oat milk mostly presented negative score values according to PC1 and as the storage period increased at both the storage conditions, samples showed positive scores according to PC1. It is clear from the PCA plots that electronic nose is able to differentiate between the samples with different degree of spoilage and sensory parameters as assessed earlier.

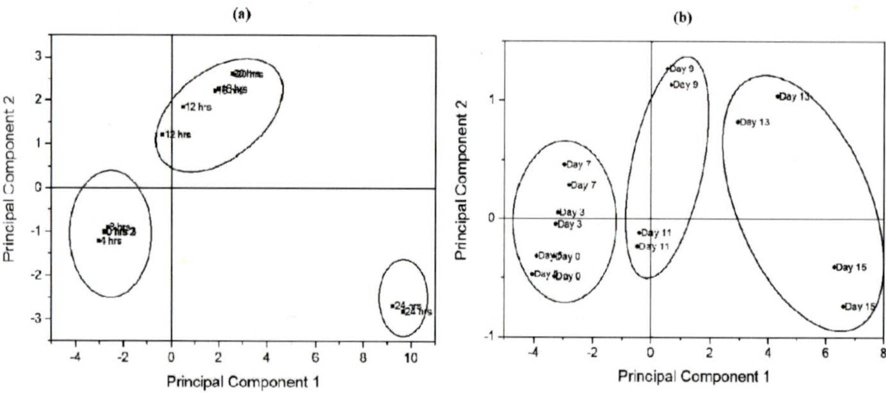

Fig. 20: PCA plots for oat milk samples (a) room temperature, and (b) refrigerated temperature

6.6.4 Classification of samples using DFA analysis

To classify the samples into different groups, the data obtained by electronic nose was analyzed using DFA as done by other researchers (Gardner et al., 1992; Young et al., 1999; Canhoto and Magan, 2003; Lebrun et al., 2008). The electronic nose was able to classify the odor changes in oat milk samples based on microbial counts. Figure 21 shows results from the DFA using the discriminant functions for oat milk samples at different storage conditions. Microbial counts and sensory data were used as grouping variables, and 18 electronic nose sensor outputs as independent variables. In the conducted study, the DFA model built classified oat milk samples based on the total microbial population into "unspoiled" (microbial counts < 6 log10 cfu/g) and "spoiled" (microbial counts ≥ 6 log10 cfu/g). As shown in Figure 21 (a), samples stored for more than 12 hours constituted the spoiled group and other samples constituted an unspoiled group. Similarly, for refrigerated oat milk, data for day 13 and 15 formed a spoiled

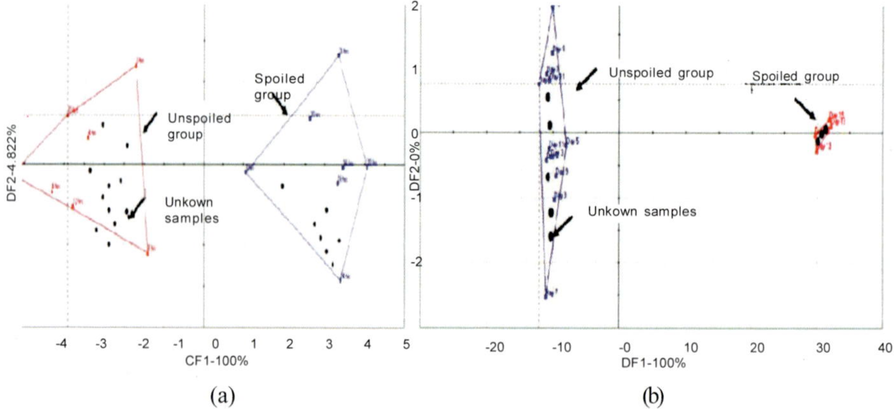

Fig. 21: DFA of odours of oat milk samples based on microbial counts and electronic nose readings (a) room temperature, and (b) refrigerated temperature

group and other samples constituted an unspoiled group as shown in Figure 21 (b). Further to validate the accuracy of the built model, some samples with known microbial count were projected on the DFA and classified correctly in the spoiled and unspoiled groups with accuracies of more than 90 % for both the storage temperatures.

DFA separated the data into different groups based on sensory scores received by oat milk samples with correct classification rates of 100 % for both the storage conditions. The scatter plots obtained with the discriminant functions at each temperature are given in Figure 22 (a), (b). Three groups were identified on the basis of sensory score for the samples stored at room temperature. During 4 hours of storage, the sensory properties of oat milk were same as fresh milk (Group 1). However, after 8 hours of storage, the oat milk started to show significant variation in sensory properties (Group 2) and at the end of 20 hours, oat milk became unacceptable (Group 3).

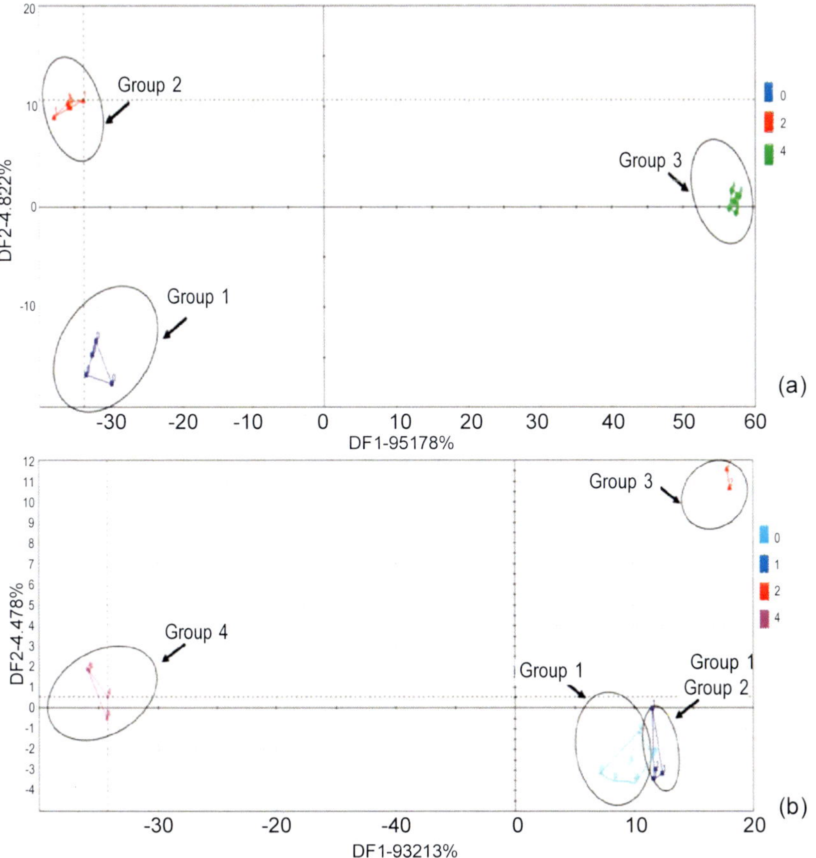

Fig. 22: DFA of odours of oat milk samples based on sensory score and electronic nose readings (a) room temperature, and (b) refrigerated temperature

For the samples stored under refrigerated conditions, sensory properties were similar to fresh samples till 5 days of storage (Group 1). Then, a slight difference in sensory properties was observed. For day 7 and day 9, samples received sensory score value of 1 (Group 2) and day 11 sample received sensory score of 2 (Group 3). And at the end of 13th day, the sensory properties of oat milk were found to be significantly different from fresh control samples (Group 4). Similar type of differentiation between samples on the basis of microbial data and sensory score has been established for dairy milk (Korel and Balaban, 2002).

6.7.5 Correlation of electronic nose with sensory and microbial data using PLS

Using the algorithm given by Eq. 6, high value of correlation coefficients, viz., 0.99 and 0.92 were obtained between the electronic nose signals and bacteriological data for samples stored at room and refrigerated temperature, respectively. This indicates that the electronic nose system can become a simple and rapid technique for estimating the microbial quality of oat milk. This algorithm makes it possible to enter 0 or a negative value as a sensory score. The obtained data was found to correlate well with sensory data with a correlation coefficient value of 0.95 for oat milk. Similar findings were observed by other researchers as well (Bleibaum et al., 2002; Tikk et al., 2008).

6.7.6 Shelf-life prediction model using electronic nose

In this method, by calculating distances from the reference (for example a fresh sample), an odor unit was determined. The higher distance means a more important difference between reference and aged sample in terms of odour, taste and chemical composition. In the study, separate shelf-life models were built for both the storage conditions. It can be seen from Figure 23 (a) that the odour unit for most spoiled sample for oat milk stored at room temperature is 3.57 ± 0.03 whereas for oat milk stored at refrigerated conditions, the odour unit for most spoiled sample of Day 15 is 1.61 ± 0.02 as depicted in Figure 23 (b). This difference in value of odour unit signifies that deterioration of sensory properties of oat milk was much more at room temperature than at refrigerated conditions for the conducted study. Similar conclusion was reached by sensory panelists.

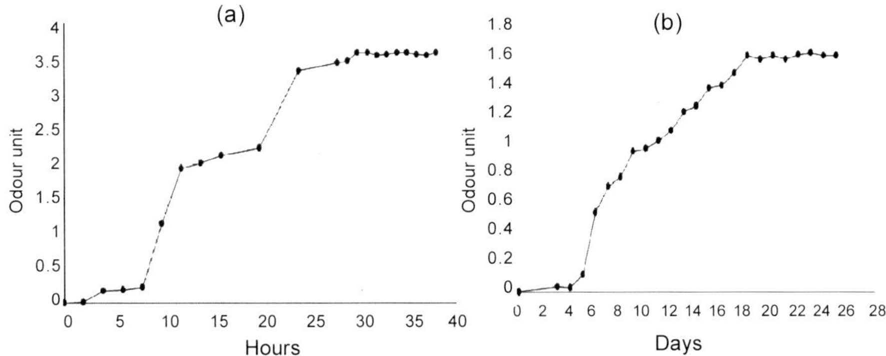

Fig. 23: Shelf-life model for oat milk samples (a) room temperature, and (b) refrigerated temperature

7. Summary

Molecular spectroscopic techniques have been proved to be a non-destructive, non-evasive, with faster response time and 'easy-to-handle' assay, and hence, emerging out as the preferred mode. Out of the several molecular spectroscopic techniques, FT-IR and NIR convincingly addresses the identification of the functional groups of various, molecules present as ingredients in food by exploiting molecular vibrations upon absorption of energy in the IR and/ or NIR region of the spectrum. Fabrication and implementation of such a highly efficient technique can lead to rapid assay for the analysis of various food ingredients and thus, potentially be widely used in futuristic industrial applications.

The studies discussed in the chapter clearly indicates that FT-NIR spectroscopy can be used successfully to estimate certain quality parameters in tea powder and granules (moisture, polyphenol content, caffeine); quality of dairy products like curd and milk (moisture content and starch addition); quality of processed products (moisture content in bael pulp and powder); microbial safety of red chili powder (presence of AFB_1) and certain antinutritional factors (phytic acid content in green gram seeds).The model was developed using the spectral region 4000–12,000 cm^{-1}. The developed method provides a fast, specific, simple, and easily automatable method for quantitative detection. Detection was possible either with no sample preparation or just one step solvent extraction procedure.

The limitation of the technique is difficult in its calibration for particular component of interest which requires initial calibration or standardization of the instrument with the classical biochemical techniques used for detecting that component. Since the calibration model which is developed for estimation of component of interest relies on the standard biochemical procedures, the sensitivity of the model developed for FT-NIR analysis thus is not better than the biochemical procedure used to calibrate the instrument. Also large sample size is required to make the model more robust.

The developed FT-NIR method was successfully validated by classical standard biochemical procedures. Since this is a new and emerging field of study in food quality assurance and has major advantage of fast and non-destructive estimation of organic compounds. Still it has advantage of non destructive detection possible without the use of any chemical.

Being highly optical based technique, the sample for Fourier transformed IR and NIR absorption spectroscopic analysis has to be contamination free. Further, interfering stray light has to be avoided to get highly resolved peak of vibrational absorption with appropriate signal-to-noise ratio. There are further scopes of improvement in this methodology by increasing number of samples in developing calibration model, more recovery studies, use of more sensitive biochemical procedures (like HPLC rather than UV spectroscopy or artificial spiking) to develop the calibration models, presence of natural variation in the samples used for calibration and validation models, improving the range of estimation, simultaneous multi-component analysis.

Similarly, electronic nose can also be successfully used as an alternative rapid method for determination of microbiological and sensory life of oat milk at room and refrigerated temperature. The study demonstrated that in terms of sensory properties, oat milk was same as fresh sample upto 4 hours and 9 days at room temperature and refrigerated conditions, respectively as judged by a sensory panel and electronic nose. Same technique can be used for other type of food products also to develop rapid method for determination of their shelf-life using electronic nose. The technique is more sensitive and precise than olfactory system in humans.

References

Abdi, H. and L. J. Williams, 2010. Principal component analysis. Wiley Interdisciplinary Reviews: *Computational Statistics,* 2(4): 433-459.

Aboul-Eneim, H. Y., Stefan, R.I. and Baiulescu, G.E. 2000. *Quality and Realibility in Analytical Chemistry,* New York: CRC Press.

Beebe, K.R., and Kowalaski, B. R. 1987. An introduction to multivariate calibration and analysis. *Analytical Chemistry,* 59: 1007-1015

Bertrand, D. 2002. La spectroscopie proche infrarouge et ses applications dans les industries de l´alimentation animale. *INRA Productions Animales* 15(3): 209-219.

Blanco, M. Coello, J., Iturriaga, H., Maspoch, S. and Pezuela, A.C. 1998. Near infrared spectroscopy in the pharmaceutical industry. *Analyst,* 123(8): 135R-150R.

Bleibaum, R.N., H. Stone, et al. 2002. Comparison of sensory and consumer results with electronic nose and tongue sensors for apple juices. *Food Quality and Preference,* 13(6): 409-422.

Brauteseth, E.M. 2009.The mutagenesis of Sorghum bicolour (L.) Moenchtowords improved nutrition and agronomic performance. *PhD Thesis*, School of Agricultural Sciences and Agribusiness, University of KwaZulu-Natal.

Canhoto, O.F. and Magan, N. 2003. Potential for detection of microorganisms and heavy metals in potable water using electronic nose technology. *Biosensors and Bioelectronics,* 18(5): 751-754.

Chen, Q., Zhao J., Fang, C.H., Wang, D. 2007.Feasibility study on identification of green, black and Oolong teas using near-infrared reflectance spectroscopy based on support vector machine (SVM). *Spectrochimica Acta Part A: Molecular and Biomolecular Spectroscopy,* 66: 568-574.

Cozzolino, D., Cynkar, W., Shah, N., Dambergs, R., Smith, P. 2009. A brief introduction to multivariate methods in grape and wine analysis. *International Journal of Wine Research,* 1: 123–130.

Di Natale, C., Macagnano, A., Martinelli, E., Paolesse, R., D'Arcangelo, G., Finazzi-Agrò, A., D'Amico, A. 2001. The evaluation of quality of post-harvest oranges and apples by means of an electronic nose. *Sensors and Actuators B: Chemical,* 78 (1–3): 26-31.

Di Natale, C., Macagnano, A., Martinelli, E., Paolesse, R., D'Arcangelo, G., Finazzi-Agrò, A., D'Amico, A. 2003. Lung cancer identification by the analysis of breath by means of an array of non-selective gas sensors. *Biosensors and Bioelectronics,* 18(10): 1209-1218.

Dowell, F. E., Pearson, T. C., Elizabeth, B., Maghirang, F. X. and Wicklow, D. T. 2002. Reflectance and Transmittance Spectroscopy Applied to Detecting Fumonisin in Single Corn Kernels Infected with *Fusarium verticillioides*. *Cereal Chemists,* 79(2): 222–226.

Dowell, F.E., Ram, M.S. and Seitz, L.M. 1999. Predicting Scab, Vomitoxin, and Ergosterol in Single Wheat Kernels Using Near-Infrared Spectroscopy. *Cereal Chemists,* 76 (4):573–576

Dutta, R., Hines, E.L., Gardner, J.W. 2003. Tea quality prediction using a tin oxide-based electronic nose: an artificial intelligence approach. Sensors and Actuators B: *Chemical,* 94(2): 228-237.

Enker, C.G. 2003. *The Art and Science of Chemical Analysis,* New York: John Wiley and Sons.

Eriksson, L., Trygg, J., Johansson, E., Bro, R. and Wold, S. 2000. Orthogonal signal correction, wavelet analysis, and multivariate calibration of complicated process fluorescence data. *Analytica Chimica Acta,* 420 (2): 181-195.

Esteban-Diez, I., Gonzalez-saiz, J.M., Pizarro, C. 2004.An evaluation of orthogonal signal correction methods for the characterization of Arabica and robusta coffee varieties by NIRS. *Analytica Chimica Acta,* 514:57–67.

Falasconi, M. E. Comini, et al., 2014. Electronic Nose and Its Application to Microbiological Food Spoilage Screening. Sensing Technology: Current Status and Future Trends II. A. Mason, S. C. Mukhopadhyay, K. P. Jayasundera and N. Bhattacharyya, Springer International Publishing. 8: 119-140.

Feng, T., Zhuang, H., Ye, R., Xie, Z. 2011. Analysis of volatile compounds of Mesona Blumes gum/rice extrudates via GC–MS and electronic nose. *Sensors and Actuators* B: *Chemical,* 160 (1): 964-973.

Gardner, J.W., Shurmer, H.V., Tan V et al. 1992. Application of an electronic nose to the discrimination of coffees. *Sensors and Actuators B: Chemical,* 6(1): 71-75.

Gardner, J.W. and Bartlett, P.N. 1994. A brief history of electronic noses. Sensors and Actuators B: *Chemical,* 18(1): 210-211.

Gardner, J. W. 1991. Detection of vapours and odours from a multisensor array using pattern recognition Part 1. Principal component and cluster analysis. *Sensors and Actuators B: Chemical,* 4(1–2): 109-115.

Gishen, M., Dambergs, R., Cozzolino, D. 2005. Grape and wine analysis – enhancing the power of spectroscopy with chemometrics.A review of some applications in the Australian wine industry. *Australian Journal of Grape and Wine Research,* 11: 296-305.

Haaland, D.M. and Thomas, E.V. 1988. Partial least square methods for spectral analysis. 1. Relation to other quantitative calibration methods and the extraction of qualitative information. *Analytical Chemistry*, 60 (11): 1193-1202.

Hall, M.N., Robertson, A., Scotter, C.N.J. 1988. Near-infrared reflectance prediction of quality, theaflavin content and moisture content of black tea. *Food Chemistry*, 27: 61–75

Hruschka, W.E. 2001. Data nalysis: Wavelength selection methods. In: Near infrared technology in the agricultural and food industries, 2nd ed., 39-58, St.Paul USA: *American Association of Cereal Chemists.*

Inon, F.A., Llario, R., Garrigues, S., De la Guardia, M. 2005. Development of a PLS based method for determination of the quality of beers by use of NIR: Spectral ranges and sample-introduction considerations. *Analytical and Bioanalytical Chemistry*, 382(7): 1549–1561.

Jaffres, E., Lalanne, V., Mace, S., Cornet, J., Cardinal, M., Serot, T., Dousset, X., Joffraud, J.J. 2011. Sensory characteristics of spoilage and volatile compounds associated with bacteria isolated from cooked and peeled tropical shrimps using SPME–GC–MS analysis. *International Journal of Food Microbiology*147(3): 195-202.

Jukka, R., Eetu, R., Jussi, T., Markku, K., Jukka-Pekka, M., JoukoYliruusi. 2000. In-line moisture measurement during granulation with a four-wavelength near infrared sensor: an evaluation of particle size and binder effects *European Journal of Pharmaceutics and Biopharmaceutics* 50 (2): 271 – 276.

Kemsley, E.K. 1998. Case studies. In E.K. Kemsley (Ed.), Discriminant analysis and class modeling of spectroscopic data (pp. 110–164).Chichester, UK: John Wiley and Sons Ltd.

Kim, Y., Singh M. and Kays, E.S. 2007. Near-infrared spectroscopic analysis of macronutrients and energy in homogenized meals. *Food Chemistry,* 105: 1248-1255.

Koljonen, J., Nordling, T., Alander, J. 2008. A review of genetic algorithms in near infrared spectroscopy and chemometrics: past and future. *Journal of Infrared Spectroscopy,* 16 (3): 189-197.

Korel, F. and Balaban, M. 2002. Microbial and sensory assessment of milk with an electronic nose. *Journal of Food Science,* 67(2): 758-764.

Labreche, S., Bazzo, S.S., Cade, S., Chanie, E. 2005. Shelf life determination by electronic nose: application to milk. Sensors and Actuators B: *Chemical,* 106(1): 199-206.

Lavine, B.K. 1998. Chemometrics. *Analytical Chemists* 70(12): 209R.228R.

Lebrun, M., Plotto, A; Goodner, K; Ducamp, MN; Baldwin, E. 2008. Discrimination of mango fruit maturity by volatiles using the electronic nose and gas chromatography. *Postharvest Biology and Technology*, 48: 122–131.

Leon, L., Kelly, J.D., Downey, G. 2005. Detection of apple juice adulteration using near-infrared transflectance spectroscopy. *Applied Spectroscopy*, 59(5): 593–599.

LijuanXie, Xingqian Ye, Donghong Liu, Yibin Ying. 2009. Quantification of glucose, fructose and sucrose in bayberry juice by NIR and PLS. *Food Chemistry*, 114: 1135–1140.

Lozano, J., Santos, J.P., Gutiérrez, Horrilloet, M.C. 2007. Correlating e-nose responses to wine sensorial descriptors and gas chromatography–mass spectrometry profiles using partial least squares regression analysis." *Sensors and Actuators B*: *Chemical,* 127(1): 267-276.

Lupaert, J., Zhang, M.H., Massart, D.L. 2003. Feasibility study for the using near infrared spectroscopy in the qualitative and quantitative of green tea, *Camellia sinensis* (L.). *Analytica Chimica Acta,* 487: 303 – 312

McGlone, V.A., Jordan, R.B., Seelye, R., Martinsen, P.J. 2002. Comparing density and NIR methods for measurement of Kiwifruit dry matter and soluble solids content.*Post Harvest Biology and Technology,* 26:191 – 198

Miettinen, S.M., Piironen, V., Tuorila, H., Hyvönen, L. 2002. Electronic and human nose in the detection of aroma differences between strawberry ice cream of varying fat content. *Journal of Food Science* 67(1): 425-430.

Miller, E.I., F.M. Wylie, et al. 2008. Simultaneous Detection and Quantification of Amphetamines, Diazepam and its Metabolites, Cocaine and its Metabolites, and Opiates in Hair by LC-ESI-MS-MS Using a Single Extraction Method. *Journal of Analytical Toxicology,* 32(7): 457-469.

Moros, J., Armenta, S., Garrigues S and Guardia, L.D.M. 2006. Univariate near infrared methods for determination of pesticides in agrochemicals. *Analytica Chimica Acta,* 579: 17-24.

Nieuwoudt, H.H., Pretorius, I.S., Bauer, F.F., Nel DG, Prior, B.A. 2006. Rapid screening of the fermentation profiles of wine yeasts by Fourier transform infrared spectroscopy. *Journal of Microbiological Methods,* 67: 248 – 256.

Nilufer, D. and Boyacio, D. 2002. Comparative Study of Three Different Methods for the Determination of Aflatoxins in Tahini. *Journal of Agricultural and Food Chemistry,* 50: 3375-3379.

Osborne, B.G., Fearn, T. and Hindle, P.H. 1993. Practical NIR spectroscopy with applications in food and beverage analysis, 2nd edition, Longman, Harlow.

Pande, R., Misha, H.N. 2015. Fourier transform near-infrared spectroscopy for rapid and simple determination of phytic acid content in green gram seeds (*Vignaradiata). Food Chemistry,* 172:880-884.

Paradkar, M.M., Irudayaraj, J. 2002. A rapid FTIR spectroscopic method for estimation of caffeine in soft drinks and total methylxanthines in tea and coffee. *Journal of Food Science,* 67(7): 2507-2511.

Petterson, H. and Aberg, L. 2003. Near infrared spectroscopy for determination of mycotoxins in cereals. *Food Control,* 1(4): 229–232.

Pink, J., Naczk, M., Pink, D. 1998. Evaluation of the quality of frozen minced red hake: use of Fourier transform infrared spectroscopy. *Journal of Agricultural and Food Chemistry*, 46: 3667-3672.

Ru, Y.J. and Glats, P.C. 2000. Application of near Infrared Spectroscopy (NIR) for monitoring quality of milk, meat and fish-Review. *Asian Australian Journal of Animal Science,* 13 (7): 1017-1025.

Sadasivam, S., Manickam ,A. 1996. Biochemical Methods. New Age International (P) Ltd. Publishers, New Delhi.

Sanches-Silva, A., S. Pastorelli, et al. 2008. Development of an Analytical Method for the Determination of Photoinitiators Used for Food Packaging Materials with Potential to Migrate into Milk. *Journal of Dairy Science,* 9 1(3): 900-909.

Saona, L.E.R., Khambaty, F.M, Fry, F.S., Dubois, J. & Calvey, E.M. 2004. Detection and Identification of bacteria in a juice matrix with Fourier Transform-Near Infrared spectroscopy and Multivariate Analysis. *Journal of Food Protection,* 67(11): 2555-2559.

Schaller, E., Bosset, J.O., et al. 1998. Electronic Noses' and Their Application to Food. *LWT - Food Science and Technology,* 31(4): 305-316

Schulz, H., Engelhardt, U.H., Wengent, A. 1999. Application of NIRS to the simultaneous prediction alkaloids and phenolic substance in green tea leaves. *Journal of Agricultural and Food Chemistry,* 475: 5064–5067.

Shiby, V.K. 2008. Dahi powder: Process Technology,Storage and Utilisation.Ph.D thesis submitted to Indian Institute of Technology Kharagpur India.

Sinija, V.R., Shiby, V.K. and Mishra H.N. 2008a. Rapid method for detection of adulteration in milk using FTNIR spectroscopy *by Beverage & Food World Journal*, 35 (3):43-45.

Sinija, V.R., Mishra, H.N. 2008b. Moisture Sorption Isotherms and Heat of Sorption of Instant (Soluble) Green Tea Powder and Green Tea Granules. *Journal of Food Engineering*, 86 (4): 494–500.

Sinija, V.R., Mishra, H.N. 2009. FTNIR spectroscopy for determination of caffeine in green instant tea powder and tea granules. *LWT- Food Science and Technology,* 42: 998- 1002.

Sinija, V.R. 2009. Green Tea Powder and Granules: Process technology, storage and quality evaluation. *Ph.D Thesis*, Agricultural and Food Engineering Department, Indian Institute of Technology Kharagpur, India.

Smith, B.C. 1996. Fundamentals of Fourier Transform infrared spectroscopy. CRC Press, Florida.

Stroka, J. and Anklem, E. 2002. New Strategies for the Screening and Determination of Aflatoxins and Detection of Aflatoxin-producing Moulds in Food and Feed. *Trend in Analytical Chemistry,* 21: 90–95.

Sundaram, J., Kandala, C., Butts, C. 2009. Application of near infrared spectroscopy to peanut grading and quality analysis: overview. *SensInstrum Food Quality Safety,* 3(3):156–164.

Tenenhaus, M. 1998. La regression PLS, *Theorie et pratique*. Editions Technip, Paris.

Tikk, K., J.-E. Haugen, et al. 2008. "Monitoring of warmed-over flavour in pork using the electronic nose–correlation to sensory attributes and secondary lipid oxidation products." *Meat science,* 80(4): 1254-1263.

Torri, L., Sinelli, N. et al. 2010. Shelf life evaluation of fresh-cut pineapple by using an electronic nose. *Postharvest Biology and Technology,* 56(3): 239-245.

Tripathi, A., A. Mohanty, Mohanty, M.N. 2012. Electronic Nose For Black Tea Quality Evaluation Using Kernel Based Clustering Approach. *International Journal of Image Processing,* 6(2): 86.

Yang, Y., Xu-zhen, C., Gui-xing, R.E.N. 2011. Application of near-infrared reflectance spectroscopy to the evaluation of D-chiro-Inositol, vitexin, and isovitexin contents in mung bean. *Agricultural sciences in China*, 10(12):1986–1991.

Young, H., Rossiter, K., Wang, M., Miller, M. 1999. Characterization of Royal Gala apple aroma using electronic nose technology-potential maturity indicator. *Journal of Agricultural and Food Chemistry*, 47: 5173–5177.

11

Enzymatic Detoxification of Aflatoxin B_1 in Foods

C Das Mukhopadhyay, S Tripathy, H N Mishra

1. The Aflatoxin Molecule

Aflatoxins are chemically difuro-coumarin derivatives. Presently, 18 different types of aflatoxins have been identified with aflatoxin B_1, B_2, G_1, G_2, M_1 and M_2 being the most common. Of these Aflatoxin B_1 (AFB_1) and G_1 occur most frequently, with AFB_1 being the most potent. Physicochemical and biochemical characteristics of the AFB_1 molecule reveal two important sites for toxicological activity (Heathcote and Hibbert, 1978). The first site is the double bond in position 8,9 of the furofuran ring (Figure 1). The interactions of aflatoxin, DNA and proteins, which occur at this site alter the normal biochemical functions of these macromolecules and leads to deleterious effects at the cellular level. Monteiro et al. (1966) studied the *in vitro* interaction of groundnut proteins namely arachin and conarachins with AFB_1 molecule. The second reactive group is the lactone ring in the coumarin moiety (Lee et al., 1981). The lactone ring is easily hydrolyzed and is therefore vulnerable to degradation.

1.1 Types, detection and biosynthesis of aflatoxin

So far 18 different types of aflatoxins have been identified. Aflatoxin B_1 (AFB_1) is the most potent amongst them. To exert any biological activity AFB_1 must be converted to its reactive epoxide in the tissues particularly in the liver. This epoxide is highly reactive and can form derivatives with several cellular macromolecules, including DNA, RNA and protein. Guanine alkylation by AFB_1 produces exo-8,9-epoxide. Methods for detection of aflatoxin are broad and complex which are categorized into two main groups, i.e, physico-chemical and biological. Former one includes chromatography, fluorodensitometry, and minicolumn detection. These methods were thoroughly reviewed by Elllis et al. (1991). Latest detection methods also involve bioassay, preparation of monoclonal antibody against the toxin, RIA and ELISA (Chu, 1984). Pon's method and Romer's method are officially accepted for extraction and purification of aflatoxin (AOAC 1995).

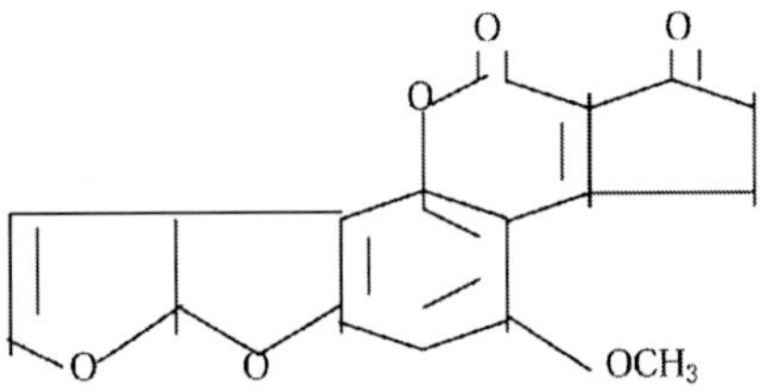

Fig. 1: Structure of aflatoxin B_1 molecule

Aflatoxins are secondary metabolites derived through polyketide pathway. Secondary metabolism occurs in small number of species optimally after a phase of balanced growth. The biosynthesis is initiated by condensation of acetyl Co A to two molecules of malonyl Co A the first stable precursor of aflatoxin biosynthesis is anthrone. This is formed by cyclisation and aromatization of acetyl and malonyl Co A, which then is oxidized to norsolorinic acid. The later undergoes several metabolic conversions involving decaketide intermediates to form aflatoxin B_1. Intermediates are mostly enzyme bound and transient. Figure 2 is a schematic representation of the pathway of aflatoxin biosynthesis.

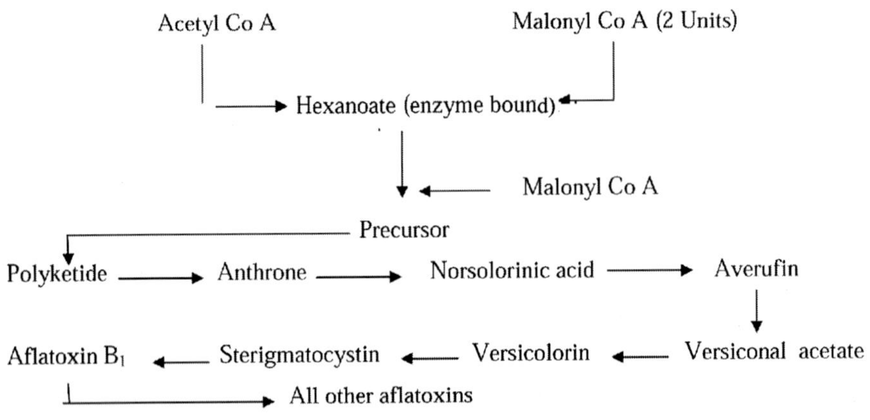

Fig. 2: A schematic diagram of aflatoxin biosynthetic pathway

1.2 Mechanism of action and clinico-pathological impacts of aflatoxin

Aflatoxins may be considered as biosynthetic inhibitors both *in-vivo* and *in vitro*, with large doses causing total inhibition of biochemical systems and lower doses affecting different metabolic systems. They inhibit O_2 uptake in whole tissues by acting on electron transport chain. Aflatoxin also reduces hepatic glycogen level probably by inhibiting glycogenesis or depression of glucose transport to liver cells or acceleration of glycogenosis. At the molecular level aflatoxins impart their effect either interfering with DNA replication or transcription of messenger RNA into protein. Biological effects as already stated are broadly categorized as carcinogenicity, hepatotoxicity, mutagenicity and teratogenicity (Ellis et al., 1991). The effects are influenced by species variation, sex, age, nutritional status and effect of other chemicals. In addition, the dose level and period of exposure of the organism to the toxin are very important. The toxicological effects of AFB_1 occur after the metabolic activation of the molecule by the microsomal mixed function oxidase system. These enzymatic reactions involve metabolism and detoxification. The metabolic phase leads to the formation of reactive AFB_1 epoxide and the next phase involves degradation of the molecule.

Mutation is caused due to binding of the aflatoxin molecule to DNA and the subsequent erroneous protein synthesis. Since aflatoxins are potent protein synthesis inhibitors, they impair differentiation in sensitive primordial cells (Moss and Smith, 1985) and lead to teratogenic effect on certain animals. Acute and chronic effects of aflatoxins in farm and laboratory animals as well as human beings are well documented. Dietary intake of aflatoxin containing food or feed leads to the disease of livestock named aflatoxicosis. The risk posed by aflatoxin depends on the level and type of aflatoxin in diet, the strain of animal and its nutritional status. Repeated ingestion of aflatoxin causes bile duct proliferation, hepatic necrosis, osteosclerosis of bone, childhood cirrhosis, immune – suppression and hepatic veno-occlusive lesions. The reported outbreaks of aflatoxicosis in men were due to consumption of staple foods such as maize and not due to the consumption of groundnut. Bhat (1989) reported that groundnut meal contaminated with aflatoxin caused Indian childhood cirrohsis and liver cancer. Aflatoxins poses serious human health hazards by modulating the immune system (Williams et al., 2004, Jiang et al., 2005), stunted growth in children (Gong et al., 2002). Immunologic suppression, impaired growth and nutritional interference have also been established by other researchers, *viz*, Patten (1981), Cullen and Newberne (1994), Fung and Clark (2004) and Williams et al. (2004).

1.3 Factors affecting growth of fungi and production of aflatoxins in food

Aflatoxins are a group of closely related heterocyclic compounds produced predominantly by two filamentous fungi, *Aspergillus flavus* and *Aspergillus parasiticus*. These molds are frequent contaminants of food and agricultural commodities. *Aspergillus* species are capable of growing on a variety of substrates and under a variety of environmental conditions. Therefore, most foods are susceptible to aflatoxigenic fungi at some stage of production, processing, transportation and storage. The capacity of fungi to produce toxic metabolites has been known since 1913 when scientists of United States Department of Agriculture conjectured that the products of mould growth might be involved in diseases. Eruption of Turkey X disease in England in 1960 resulted in a drastic reappraisal of the problem. The relative toxicity and mutagenicity of aflatoxins were reviewed by Samarajeeva et al. (1990). Wantabe et al (1996) studied biosynthetic pathway for aflatoxin and its precursor. Investigations revealed that the precursor undergoes several cyclization, aromatization and other metabolic conversions through formation of unstable intermediates.

A large portion of early mycotoxin research was carried out by Japanese and Russian scientists. In liquid culture these fungi normally have a period of active synthesis and accumulation of the aflatoxin in the medium (Hamid and Smith, 1987). They also produce toxin using several food grains as solid substrates (Sharma et al., 1994). A study was conducted wherein the growth of fungus and the production of aflatoxin in seven varieties of spices were studied. These spices include cumin, cinnamon, clove, black pepper, cardamom, ginger and coriander. It was concluded that clove was the most resistant to the development of aflatoxins and cumin was the most susceptible spice among all the seven that were studied. Aflatoxin content in red chilli powder was reviewed by (Aydin et al., 2007). Researches propose that the factors held responsible for the contamination of food by aflatoxin include the genotype of the crop planted, the soil type, the climate of the region, the daily minimum and maximum temperatures of the region and the daily net evaporation of the region (Wilson and Payne, 1994; Brown and Chen 2001; Bankole and Mabekoje, 2004; Fandohan and Gaonlonfin, 2005). It has also been seen that aflatoxin contamination is also promoted by insect activity, poor timing of harvest, inadequate drying of the crop plant before storage , heavy rainfall both during harvest and post harvesting times (Ono et al., 2002; Hawkins and Windham, 2005; Turner and Sylla, 2005). The humidity and temperature of the storage are equally important determinant of the aflatoxin formation. Lack of aeration during drying and storage also leads to aflatoxin production and subsequent contamination. Effect of different strains of toxigenic mold, nutrient composition of the growth media and all other factors controlling aflatoxin production has been briefed in Table 1.

Table 1: Factors affecting aflatoxin biosynthesis

Factors	Specific effects
Biological strain variability	*A. flavus* produces AFB_1 and *A. parasiticus* produces B_1, B_2, G_1, G_2 .
Competing microflora	Other microflora may release toxin which may increase or decrease AFB_1 production.
Inoculum size	10^3 spores/ ml is optimum.
Chemical	
Type of substrate	Carbohydrate and fatty acid content in media enhance aflatoxin production
Type of nutrients	Simple sugars are preferred C sources; Gly, Glu, Pro, Asp, Ala and divalent metal ions stimulate aflatoxin production
Environmental	
Temperature	25 – 30 °C
Water activity (a_w)	0.95- 0.99
Atmospheric gases	$CO_2 < 10\%$, $20\% > O_2 > 90\%$
Light intensity	Blue light is favorable
pH	Physiological to acidic range is optimum

2. Control of Aflatoxin

Control of aflatoxins is the need of the hour, since their occurrence in foods and feeds is continuously posing threats to both health and economics all over the world. The increasing number of reports on the toxic nature of aflatoxin suggests a need to prevent contamination of commodities by AFB_1 producing molds or to control the mold growth by manipulation of the microenvironment. Other control methods should be directed at either reducing the concentration of AFB_1 to safe levels or to producing nontoxic degradation products without reducing the nutritional value of the treated commodities. The methods for control of aflatoxin can be broadly categorized into prevention of mold contamination and growth and detoxification of toxic commodities.

2.1 Prevention

Prevention is the best way to control mycotoxin problem. Control at preharvest stage would include improved farm management, use of antifungal agents, genetic engineering approaches, rapid screening techniques, and control of storage environment. Prevention by proper agronomic practices includes using healthy seeds, proper irrigation, rotation of crops; harvesting after full maturity, proper drying and storage. Since aflatoxin contamination is restricted to a fewer number of seeds, the physical segregation of infected seeds and their rejection provides a good method of decontamination (Shantha, 1994). Varietal differences in corn with respect to resistance to *Aspergillus flavus* and aflatoxin production. Vesonder et al. (1978) from study on a series of corn hybrids, concluded that the evaluation and selection would probably be more difficult for plant resistance to infection by *A. flavus* and aflatoxin production than it had been for plant resistance to diseases.

2.2 Detoxification

The detoxification treatments of AFB_1 should be aimed either at removing the double bond of the terminal furan ring or in opening the lactone ring. Once the lactone ring is opened, further reactions can occur to alter the binding properties of terminal furan ring to DNA and proteins. Such structural changes can be induced through supply of energy in a form absorbable by AFB_1 or by using any chemical to block or remove the active sites. Biological or enzymatic treatment also modify the structure such that to result some less toxic or nontoxic compounds. In the following paragraphs physical chemical and biological methods of aflatoxin detoxification have been reviewed.

2.2.1 Physical detoxification

Application of heat energy between 100°C for 120 min to 300°C has been tried since sixties for detoxification of aflatoxin in food and feed with variable success. But quality of the product deteriorated. Microwave treatment at high energy levels showed great potential for aflatoxin degradation. A 95% destruction of toxin was achieved by Staron et al. (1980) at 6 Kwt for 4 min. Other heat detoxification of aflatoxin in food processing are steam flaking, dry-heat roasting, micronizing (infrared heat). Studies by Applebaum et al., (1982) on pasteurization, sterilization and roller drying of milk resulted in partial detoxification.

Radiolysis by gamma radiation of food constituents may initiate free radical reactions leading to degradation of aflatoxin. However, the dose required for foods were higher than the permissible level (Temcharoen and Thilly, 1982). The combined hydrogen peroxide with gamma radiation by enhanced generation of oxygen free radicals and more efficient reaction with aflatoxin molecule. UV irradiation produced a series of AFB_1 breakdown to less toxic products. A few percentage of aflatoxin degradation was observed after the exposure of foods to light from fluorescent and incandescent bulbs. Solar radiation which possesses energy in the range of UV and visible spectra, has been shown to have greater efficiency in degrading aflatoxin in foods. Joseph et al., (1976) concluded that riboflavin quenched aflatoxin photo-degradation perhaps by complexing with the toxin and thus enhancing the process. Solar radiated edible oil was shown to be nontoxic to experimental rats and one day old ducklings (Gamage et al., 1985). Absorption by kaolin and clay loam soil at a level of 1, 1.5 and 3% with contact time 15–30 min; satisfactory adsorption occurred at 80°C for 15 min.

2.2.2 Chemical detoxification

Important chemical reagents used for chemical detoxification are (a) chlorinating agents – sodium hypochlorite, chlorine dioxide and gaseous chlorine (b) oxidising agents – hydrogen peroxide, ozone and sodium bisulfite and (c) hydrolytic agents – acids and alkalis. Chlorination with sodium hypochlorite at concentration of 0.2 % -11 %, with 3 % or with gaseous chlorine at 10 % were shown to have almost totally degraded AFB_1 either in its pure form or in spiked foods. Concerns regarding the safety of chlorinated foods still exist. The presence of residual chlorine in treated foods, the production of modified fats and proteins that may be of unknown toxicity, and the reduction in tryptophan content (Rhee et al., 1976) due to chlorine treatment, need to be investigated.

Peanut protein isolates with H_2O_2 shows better efficiency, low cost and ready availability. Residual H_2O_2 present in treated foods degrades readily to leave no toxic residues of its own. These treatments are claimed to cause only mild effects on food quality as supported by the unaltered protein efficiency values in GNM and the negligible loss of proteins and lipids in corn. Ozone reduced AFB_1 levels by 91% in cottonseed meal containing 22% moisture at 100 °C for 2 h and by 78% detoxification in GNM with 30% moisture for 1h exposure. Longer treatment duration, however, reduces protein efficiency ratio and available lysine. Ozone oxidises 8, 9 – double bond of furan ring through electrophilic attack. Inactivation of aflatoxin by sodium bisulfite involves a free radical degradation mechanism. They also reported that methanol and citric acid retard the degradation of aflatoxin by bisulfite.

Walther et al., (1983) studied the fast atom bombardment mass spectrometry of aflatoxin and reaction products of sodium bisulfite with aflatoxin. The degradation of AFB_1 by ammonia was equally effective in a gaseous or in aqueous phase in decontaminating aflatoxin in feeds. Up to 5% ammonia, 10-20% moisture at 80-120°C or high pressure for 15-30 min could achieve effective decontamination. Alkaline treatments caused hydrolysis of the lactone ring in AFB_1. Samarajeewa (1990) ranked the relative efficiency of different alkali in degrading AFB_1 at 110°C in following order: potassium hydroxide> sodium hydroxide> potassium carbonate> sodium carbonate> dipotassium carbonate> ammonium hydroxide> sodium bicarbonate> ammonium carbonate. Acid treatment leads to hydration of AFB_1 at the 8,9 – olefinic bond of the terminal furan ring to form AFB_{2a}. The AFB_{2a} still possess toxicity, so the use of strong acids to degrade AFB_1 in foods was discouraged. A variety of other chemicals were also screened by Chakrabarty (1981). Of them 75% methanol, 5% dimethyl amine hydrochloride, aldehydes, benzoyl peroxide, osmium tetroxide, iodine, ferrous ammonium sulphate are noteworthy.

2.2.3 Biological detoxification

Many microorganisms including bacteria, actinomycetes, yeasts, molds and algae can degrade aflatoxins. Ciegler et al (1966) at the Northern Regional Research Laboratory had been the most active in searching for microorganisms to test biological detoxification. They screened approximately 1000 microorganisms for their ability to either destroy or transform AFB_1 and AFG_1. Some molds or mold spores partially transformed AFB_1 to new fluorescing compounds. Only one of the bacteria assayed, *Flavobacterium aurantiacum* – NRRL B-184, removed aflatoxin from solution. Both growing and resting cells of this culture took up AFB_1 irreversibly from solution. Toxin contaminated milk, oil, peanut butter, groundnuts and corn were completely detoxified and contaminated soybeans were partially detoxified. Duckling assays showed that detoxification of aflatoxin solution by *F. aurantiacum* – NRRL B-184 was complete with no new toxic compounds being formed. The effect of divalent chelators (EDTA and OPT) was also explored. Karunaratne et al., (1990) showed that *Lactobacillus acidophilus, L. bulgaricus* and *L. planatarum* could be used to either prevent mold growth or to degrade aflatoxin.

Chaurasia and Sinha (1994) published experimental results on kernel infection and aflatoxin production in peanut (*Arachis hypogaea* L) by *A. flavus* in presence of geocarposporic bacteria. They hypothesized that geocarpospheric bacteria would be ideal for protecting the developing groundnut pods against aflatoxigenic fungi. Role of steroid metabolizing fungi was also explored. Aflatoxin degradative activity was demonstrated in 6 –12 day old mycelium and cell free extracts of *A. flavus*. Addition of cycloheximide, SKF 525-A on metyrapone to cultures of *A. flavus* prevented subsequent degradation of the toxin. Conversion of AFB_1 to isomeric hydroxy compounds by *Rhizopus* species produced two new fluorescent compounds. The compounds thus produced were characterized by them using ^{14}C labeled AFB_1. Fungal strains *A. niger, Eurotium herbariorum* a *Rhizopus* sp. and non-aflatoxigenic *A. flavus* was found to convert AFB_1 to aflatoxicol.

Overall, it can be concluded from above studies that very few, especially food grade microorganisms, can effectively degrade aflatoxin. They may release certain more toxic compounds. So the use of these organisms to detoxify aflatoxin in commercial scale needs to be investigated.

3. Enzymatic Control of Aflatoxin

Enzymes have now become a part of a number of new industrial processes and offer a great potential in a wide variety of applications. The market demand for enzymes in India was Rs. 300 crores by the beginning of this century. It is expected that the market demand for enzymes will rise to Rs. 500 crores. Among these the food industry has a major share. Enzymes thus have a good hold in the Indian as well as global market of food industry. This scenario shows an opening to explore the possibility of using enzymes as a means of detoxification process of aflatoxin, a major contaminant of foods and feeds.

3.1 Mechanism of enzyme action

For the last two decades much emphasis has been given to enzymatic detoxification of aflatoxin *in vivo*. This, however, does not solve the problem of bringing down the toxin content of the infected commodity before ingestion. Enzymatic detoxification process is advantageous in many respects. It is mild and fast, selective and can be performed in physiological conditions. As far as the quality and nutritional aspects of the commodity is concerned, enzymatic detoxification is also safe to use. Reports from literature clearly revealed the potency of enzymatic detoxification process and peroxidase group of enzyme could be used efficiently for this purpose. Peroxidase group of enzymes are efficient in detoxifying aflatoxin toxic residues as below.

- Formation of active enzyme-substrate complex (ES) ie. Compound I.

  Peroxidase group of Enzyme + H_2O_2 $\rightarrow$ Compound I.

- Transition of Compound I to active Compound II.

  Compound I + AH $\rightarrow$ Compund II + A.

- Recovery of enzyme by reduction of Compound II

  Compound II + AH $\rightarrow$ Peroxidase enzyme + A + 2 H_2O

  A= donor, AH = donor in reduced form.

3.2 Types of enzyme and their activity

Compared to microbial degradation, enzymatic treatment offered higher rates and was generally less prone to environmental changes. Haeme- containing peroxidases and in particular the enzymes obtained from white rot fungi eg., *Phanerochaete chrysosporium* (lignin peroxidase) had been shown to oxidize a wide variety of toxins, polyhydroxy aromatic hydrocarbons (PAH) and polychlorinated phenols. Barman (1972) revealed that, the soil and waste-water toxins were decontaminated by immobilized peroxidase. Ellis et al., (1991) first concluded that the biological degradation reactions might occur through enzymatic activity and these enzymes produced products or byproducts that react with aflatoxins. Peroxidase was speculated to be one such enzyme since it catalyzes the decomposition of hydroperoxides to produce free radicals, which may then react with aflatoxin. Furthermore, certain peroxidases produce hypochlorite and singlet oxygen in presence of hydrogen peroxide and chloride ion (Allen, 1975). Microsomal peroxidase dependent detoxification of aflatoxins were also studied. Various microbial P-450 enzymes, play a significant role in the detoxification of xenobiotic chemicals.

Janssen and Schanstra (1994) reviewed protein engineering for environmental application. New insight was obtained into the structure and catalytic mechanism of enzymes that convert environmental pollutants and toxins. Recent advances in this field have made it possible to use this information for improving catalytic performance of engineered enzymes to achieve increased stability and expanded substrate range. Some nonsteroidal anti-inflammatory drug also interacts with HRP in *in-vitro* assay system for H_2O_2 scavenging. Class III peroxidases were evaluated in terms of structure function relationship. Adak et al. (1996) revealed the possible role of arginine and tyrosine residues at the active site of HRP in aromatic donor (guaiacol) oxidation. The former played an obligatory role in aromatic donor binding whereas the later residue played a facilitating role presumably by hydrophobic interaction or hydrogen binding.

Guenrich and Parikh (1997) reviewed the expression of drug metabolizing enzymes. They concluded that a principal advance in the production of drug metabolizing enzymes had been the development of catalytically self- sufficient cytochrome P-450 reductase fusion proteins and *E. coli* and baculovirus coexpression constructs. Work on GSTs resulted in the identification of important residues by random mutagenesis screening techniques as well in the engineering of model *S. typhimurium* strains for genotoxicity analysis. Plant gltoathione S transferase was studied for its potential role in the detoxification of both xenobiotics and endogenous compounds, three dimensional structure, molecular diversity and substrate specificity. Das and Mishra (2003) critically reviewed all the existing detoxification methods of aflatoxin, their advantages and disadvantages.

3.3 Factors affecting enzyme activity

3.3.1 Substrate concentration

Any enzyme reaction is dependent on available substrate concentration. AFB_1 was dissolved in methanol to be used as stock solution. In a closed isothermal vessel containing 50 mM phosphate buffer, (adjusted to required pH) measured volume of substrate was added. Substrate concentration initially favours enzyme reaction provided other variable parameters remain constant, but after certain point the reaction reaches at equilibrium.

3.3.2 Enzyme and cofactor concentration

For any enzymatic reaction enzyme concentration also is a vital limiting factor. Order of reaction and concentration of enzyme is interdependent. Peroxidases were used based on amount of substrate followed by fast addition of measured amount of hydrogen peroxide, the cofactor.

3.3.3 Effect of pH

Enzymes have a characteristic optimum pH at which their activity is maximum. The pH activity profile of any enzyme reflects the pH at which important proton donating or proton accepting groups in the enzyme catalytic site are in their required state of ionization. The optimum pH of an enzyme is not necessarily identical with the pH of its normal surroundings, which may be just above or below the optimum pH. The catalytic activity of enzymes in cells may thus be regulated in part by changes in the pH of surrounding medium. HRP is stable in a wide range of pH, but as it plays a key role in controlling the rate of reaction, pH of the reaction mixture was varied from extremely acidic to basic ranges.

3.3.4 Other factors

Effect of temperature and incubation time needs to be studied for each enzyme-substrate reaction as this is unique for each set of experiment. Generally liquid medium favors enzyme reaction. However, conditions need to be optimized in case of solid medium or immobilized enzyme. In the last case enzyme may be immobilized on some surface on which the substrate containing raw material in liquid form may be passed.

4 Technological Aspects of Enzymatic Detoxification of Aflatoxin

4.1 Reaction medium

4.1.1 Reaction in liquid medium

During the reaction in liquid medium, 2 mM AFB_1 is dissolved in methanol to be used as stock solution. In a closed isothermal vessel containing 50 mM phosphate buffer (adjusted to required pH), measured volume of substrate is added. Enzyme is added followed by fast addition of measured amount of hydrogen peroxide. The vessel was kept on gentle stirring up to 65 min. Effect of tempcrature, pII, incubation time, substrate and enzyme concentration is studied individually and finally the optimum levels of these parameters are found out to get maximum detoxification. HCl is used to stop the reaction. Initial and final concentration of AFB_1 is measured.

4.1.2 Reaction in solid medium

Ground nut meals (GNM) are autoclaved to kill all the fungal spores and other contaminating organisms and mixed with AFB_1, a process commonly called spiking. One hundred grams of finely powdered (0.25 mm), defatted GNM containing 7-8% wb moisture (AOAC, 1995), is spiked with known concentrations of AFB_1 in chloroform solution. After the toxin was mixed homogeneously, the solvent was driven off in an inert atmosphere and dried at room temperature. For reaction in GNM, the enzyme dissolved in sterile phosphate buffer at pH 6 is used to detoxify the substrate AFB_1 spiked in 100g of sterile groundnut meal sample and incubated up to 30h. Hydrogen peroxide is always used in the reactions. The reaction mixture is stirred gently. The concentration of AFB_1 in the sample is estimated by TLC of the chloroform extract taking 1g of reaction mixture and later quantified by luminescent spectrofluorometer. Moisture content of the reaction mixture was maintained between 12-15%. Moisture content is determined by the process of AOAC (1995). Calcium propionate was used as the fungistat. Subsequent steps are like reaction in liquid medium as described earlier.

4.2 Process control

4.2.1 Determination of K_m and V_{max}

Kinetic study of an enzyme- substrate reaction is a prerequisite for design and analysis of a reacting system. A typical experiment to measure enzyme kinetics starts at time zero when solution of substrate and an appropriate enzyme are mixed well in a closed isothermal vessel containing a buffer solution to control pH. The substrate/ product concentration is monitored at specific intervals. Similarly in the case of AFB_1-enzyme reaction the experimental conditions are optimized based on individual parameters and an ideal set of data are selected for kinetic study. Specific K_m and V_{max} are determined following Michaelis - Menten equation and Lineweaver-Burk plot.

4.2.2 Determining the role of inhibitor

In an optimum AFB_1- enzyme reaction condition the substrate concentration is varied keeping other parameters (pH, temperature, time, etc) unchanged and the reaction velocity is plotted against substrate concentration; V_{max} and K_m were determined. In the next set of experiments concentration of inhibitor of HRP enzyme (amino pyrine) is kept constant and substrate concentration is varied, V_{max} and K_m are determined. This later experiment is repeated keeping inhibitor concentration at other fixed values. By comparing the velocity of reaction in presence of 7 with the uninhibited reaction, the percent inhibition at different concentration is calculated and plotted. The double reciprocal plot of aflatoxin in presence of aminopyrine at different concentration reveals the mode of inhibition.

4.2.3 Determination of arrhenius parameters

AFB_1, the substrate in presence of 50 mM phosphate buffer and 5mM hydrogen peroxide was taken in a conical flask and enzyme is added quickly to the reaction mixture. The conical flask is precooled or preheated in the range of 0 - 45 °C and kept on a magnetic stirrer equipped with a thermometer; pH 6 was maintained constant during all these experiments. As soon as the reaction temperature is attained, 1 mL aliquot of reaction mixture is withdrawn. This is done at every 15 min interval from 0 - 60 min and cooled to room temperature. Aflatoxin content in the reaction mixture is detected and estimated by TLC and luminescent spectroscopy respectively. From percent conversion data of substrate (AFB_1) at different time intervals at each specific temperature, first order reaction rate constant, k values are calculated. To calculate the Arrhenius parameters, log k values were plotted against 1/T, T being expressed in absolute scale.

4.3 Effect of detoxification process on product quality

Functional properties of the treated product spiked with aflatoxin B_1 needs to be analyzed using standard laboratory procedures.

4.3.1 Protein quality assessment studies

To determine the nitrogen solubility profile, proteins are extracted in 1M NaCl, 20 mM sodium phosphate buffer at a ratio 1:18 (w/v, flour: extraction medium) as described by Basha and Cherry (1976). The pH of the extract is adjusted to 2, 4, 6, 8, 10 using appropriate amounts of 1N - 6N HCl and 1N - 6N NaOH. After 1h of constant stirring on a magnetic stirrer at 25°C the slurry is centrifuged at 15,000 g for 30 min at 4°C and filtered to obtain a clear extract. Twenty five mL sample of each extract is analyzed for nitrogen content by the Kjeldahl method (AOAC, 1995). The percent soluble nitrogen in each sample is calculated and plotted against the corresponding pH values. The powdered sample can be analyzed for the total nitrogen content by Kjeldahl method. Briefly, to know the amount of nonprotein nitrogen in ground nut meal a known quantity of powdered material was dissolved in ice cold 10% trichloroacetic acid. Proteins were precipitated, while non-protein nitrogen gets extracted. After centrifugation the precipitate is washed and analyzed for nitrogen content. Multiplying the nitrogen value with 5.7 (conversion factor for groundnut protein) gives the crude protein content. Deduction of protein nitrogen value from total nitrogen gave the nonprotein nitrogen content of the sample (Pelett and Young, 1980).

4.3.2 SDS- Polyacrylamide gel electrophoresis (PAGE)

SDS- PAGE is performed using the method of Laemmli (1970). Electrophoretic mobility of the groundnut proteins was determined on 12% resolving gel. From each protein samples to be tested, 50 μL was taken and loaded into the wells of the gel containing sodium dodecyl sulfate (SDS) as anionic detergent. The protein-SDS complex carries net negative charges, hence moves towards the anode. The separation is based on the size of the protein. Gels are stained by coomassie brilliant blue R- 250 and destained by keeping the gel overnight in destaining buffer. Coloured bands are checked for protein mobility.

4.3.3 Rat feeding studies

Male Wister rats of almost equal body weight are housed in four individual groups and fed four types of meal for 6 weeks. The control diet consisted of carbohydrate 40%, groundnut oil 7-8%, non-nutritive cellulose 2%, salts 4%, vitamin B complex 1% and GNM 45%. Cages are provided with feeders. To each group of 8 rats 100g diet is provided per day of different formulation under study. Food consumption and body weights of the animals are recorded daily. Mortality, if any, is noted. The surviving animals are sacrificed after certain intervals and organs are weighed. Food efficiency ratio is calculated as gain in body weight per gram of the food consumed by the experimental animals. Results are statistically analyzed and reproducibility was checked.

4.4 Quality analysis & management

4.4.1 Ames test for mutagenicity and carcinogenicity

The mutagenicity and the carcinogenicity are tested using the Ames' *in vitro* microbial detection system (Ames et al, 1975). The compounds are tested on petri plates with specially constructed mutants of *S. typhimurium* TA 100 as tester strain. The sensitivity of the test lies in the fact that in presence of a variety of mutagens the tester strains revert back from a histidine requiring auxotroph to prototrophs.

4.4.2 Toxicity test on Bacillus megaterium

The toxicological test profile of the isolated fractions before and after enzymatic treatment of any commodity is determined using the above organism. On sterile petri dishes containing warm molten nutrient agar medium 5mL chloroform extracts of detoxified samples were added and allowed to dry. The solvent evaporates in warm temperature. A 100 μL of *B. megaterium* suspension culture of 10^5 cells /mL concentration is added onto the petri plates and incubated at

37 °C for 16 hours. A blank set without toxin is also incubated at the same condition. The bacterial colonies are counted under an electronic colony counter and percent inhibition is calculated.

4.4.3 Somatic embryogenesis in presence of toxin

Embryogenic callus maintained on standard MS medium containing 1 mg/L 2, 4-D was used as inoculum. Embryo developed from callus tissue were manually selected under a stereomicroscope and divided into 4 developmental stages viz., globular, heart, torpedo and mature cotyledonary stage (Bela and Shetty, 1999). In the next step isolated cotyledonary embryo were cultured in petri dishes (90x15 mm) with 30 mL agar solidified medium (20 - 30 embryo / dish) sealed with parafilm and incubated. Hormones used were 0.1 mg/L IAA and 0.1 mg/L BAP. To the warm molten agar medium 20% of solutions A, B, and C were added respectively in three separate set of experiments. The data were expressed as germination percentage of the embryo, morphological abnormalities, if any, and root or shoot development (Das and Mishra, 2001).

4.4.4 Germination of seeds

After removal of seed coats the groundnut seeds were immersed in 0.05% gibberellic acid (GA) overnight, washed thoroughly by tap water and 5%Teepol and then surface sterilized. Finally the seeds were incubated in solidified MS agar medium supplemented by Solution A, B, C and with no toxin in test tubes for germination.

5. Developed Process for Enzymatic Detoxification of AFB_1

5.1 Detoxification of aflatoxin in ground nut meal using horse radish peroxidase

5.1.1 Extraction, purification and assay of HRP

Freshly harvested roots of radish were crushed and the juice was extracted through fine mesh. This was purified by 68% and 31.2% ammonium sulfate precipitation, respectively along with subsequent dialysis after each step. Total protein content was measured by Bradford method (1976) and unit was determined by guaiacol assay. One unit of enzyme was defined to form 1 mg of tetraguaiacol from guaiacol in 1 min at pH 6.0 at 20°C. Specific activity of the enzyme has been calculated as the gram mole of substrate converted (AFB_1) per miligram of protein per hour. For reaction in liquid medium enzyme having 200 U/mg protein is referred to as enzyme A, the fraction obtained by 68% ammonium sulfate precipitation having 20 U/mg as enzyme B and fraction obtained by 31.2% ammonium sulfate precipitation having 30 U/mg as enzyme C.

During AFB_1-HRP reaction in GNM enzyme purification was extended one more step. After 68% ammonium sulfate precipitation of the crude juice extract of freshly harvested enzyme the supernatant was subjected to 31.2% (Pool A) and 25% (Pool B) ammonium sulfate precipitation along with subsequent dialysis. Pool A having very low activity was further purified by passing through calcium phosphate gel, 24.2% ammonium sulfate precipitation and dialysis respectively and then mixed with pool B. Total protein content and unit of the purified enzyme at each step was determined as before. For AFB_1-HRP reaction in GNM, the commercially procured enzyme having 200 U/ mg protein is referred to as enzyme P, the fraction obtained by 68 % ammonium sulfate precipitation having 20 U/ mg as enzyme A, and the fraction obtained by 31.2 % ammonium sulfate precipitation having 30 U/mg as enzyme B and the enzyme after third step purification as enzyme C having 50 U/ mg.

5.1.2 Detoxification of AFB_1 by horse radish peroxidase in liquid medium

AFB_1 extracted from *A. flavus* ATCC 15517 was detoxified upto 60% by HRP (200 IU/ mg) *in vitro*. Partially purified peroxidase enzyme from freshly harvested radish roots having 20 IU/ mg and 30 IU/ mg activity showed approximately 30% and 38% conversion *in vitro* respectively. Enzyme A, B, and C showed 0.055×10^{-3}, 0.580×10^{-6}, 6.3×10^{-6} g mol/ mg/h of specific activity, respectively. AFB_1 concentration was varied from 0.2 – 2 mM to observe the velocity of reaction by peroxidase enzyme. Velocity of HRP-AFB_1 reaction was plotted against substrate concentration and a hyperbolic curve was obtained. K_m and V_{max} of the reaction were determined using Michaelis –Menten equation and Lineweaver – Burk plot (Das and Mishra, 2000b).

Enzyme concentration is the limiting factor in cost determination of detoxification process. AFB_1- HRP reaction was set up using optimum substrate concentration ie, 1 mM AFB_1 in presence of a wide range of enzyme concentration. The effect of enzyme concentration on velocity of reaction is presented in Figure 3. The reaction rate was maximum between 2-4 units of enzyme in this specific reaction condition. Experiments were performed using 1 mM substrate, 2 IU of enzyme and percent conversion were noted down starting from 10 seconds to 65 min. No appreciable reaction rate was noticed till 15 min. Incubation period of 40 min was found to be optimum.

Optimum pH for AFB_1 - HRP reaction was found to be 6. Very high rate of degradation of AFB_1 in pH 2 may not be due to enzymatic action only, but also due to instability of the AFB_1 molecule in the extreme pH values. HRP enzyme

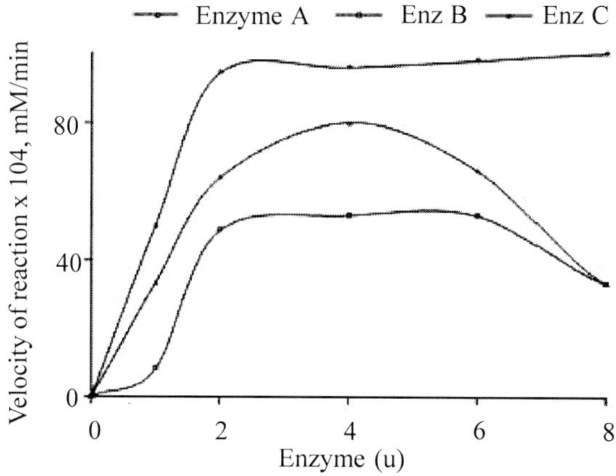

Fig. 3: Effect of enzyme concentration on velocity of reaction

may also get denatured over these pH ranges. Concentration of hydrogen peroxide was optimized and the consumption of the same during the reaction time was noted down. The rate of AFB_1 – HRP reaction increased with increase in temperature within the range in which the enzyme is stable and retains full activity. To see the effect of temperature on AFB_1- HRP reaction in a liquid medium the reaction was performed from 0-45°C. This range was chosen because in India the atmospheric temperature generally reaches 45°C in summer and 5-10 °C in winter. Data shows that the reaction maintain a first order profile till 30 °C. Beyond this temperature the rate of increase in reaction velocity become slow. The temperature coefficient Q_{10} is approximately 2. The percent conversion of aflatoxin B_1 at every 15 min interval at each specific temperature were noted down and first order reaction rate constant k was calculated as,

$k = 2.303/t \log a/(a-x)$

where, a is initial concentration of AFB_1 in mole, x is moles of AFB_1 decomposed at time t and a-x represents moles of undecomposed AFB_1. Using aminopyrine in different concentration as an inhibitor to HRP, Lineweaver – Burk Plot was plotted as Figure 4. The profile reveals the mode of inhibition is noncompetitive type.

To get the Arrehnius plot, log k was plotted against 1/T, and activation energy E_a for the reaction was calculated to be 36.9 K Joule/ mole. Other thermodynamic parameters were calculated using following relationships.

Enthalpy of activation, $\Delta H = E_a - RT$,

Free energy change $\Delta F = RT \ln 2.0842 \times 10^{10} \times T/ k$ (in sec^{-1}),

Change in entropy, $\Delta S = (\Delta H - F)/T$.

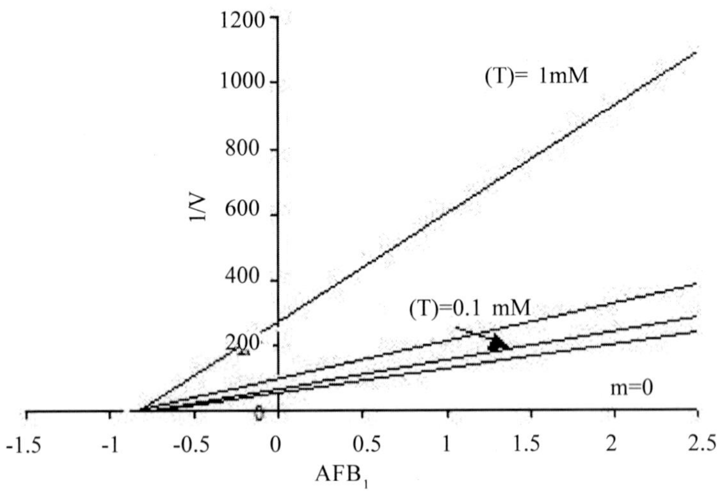

Fig. 4: Double reciprocal plot for AFB_1 – HRP reaction in presence of different concentration of aminopyrine

5.1.3 Detoxification of AFB_1 contaminated in GNM by HRP

The standardisation of reaction parameters for AFB_1- HRP in a buffered liquid reaction mixture helped to optimize the reaction conditions to detoxify AFB_1 in GNM. Reaction was performed at room temperature and normal pressure in 50 mM phosphate buffer having pH 6. During the optimization of substrate concentration AFB_1 was varied from 0.2 to 2 mM. Double reciprocal plot revealed the K_m is 1.25 mM and V_{max} is 50 x 10^{-5} mM/ min (Das and Mishra, 2000a) .

In case of GNM spiked by AFB_1 the reaction was set up using 2 – 16 IU of enzyme for 1.4 mM AFB_1 spiked in 100 g of GNM sample and incubated upto 24 h. A steep increase in activity was evident upto 8 IU/mg of enzyme concentration. Beyond 10 IU the percent conversion achieves a steady level. Data confirmed that 8-10 IU of enzyme may be optimally used to detoxify 1.4 mM of AFB_1 spiked in 100g of GNM.

Horse radish peroxidase does not react fast on GNM matrix. No change in AFB_1 concentration in the sample was noticed upto 5 h of incubation period. Prolonged incubation, however, resulted in a very slow initial reaction which then increased sharply upto 20 h. Beyond that the conversion percent increased slowly for further 4 h and then achieved a steady level without any more conversion (Figure 5). The necessity of H_2O_2 was also more; in present case 20 mM proved to be optimum.

The high moisture content which is a prerequisite for enzyme substrate action in the meal results in better detoxification but at the same time GNM becomes prone to fugal attack as well as qualitative deterioration. To overcome this problem calcium propionate was used as fungistat. Concentration of fungistat was varied in the range of 0.1 to 0.3g/ 100g meal. Good growth inhibition occurred at 0.3g calcium propionate per 100g meal, which was used for all experiments.

5.1.4 Optimization and modelling using response surface methodology

Effect of individual independent variables on reaction velocity of AFB_1 - HRP reaction have been discussed so far. Optimization of reaction parameters was done by response surface methodology (Figure 5 a and b). Some 25 sets of experiments in liquid reaction medium and 15 sets of experiments in GNM were performed varying more than one parameter at a time. The effect was expressed by some 3-D plots and contour graphs. By superimposing the contour graphs, optimum of reaction parameters were found out. A mathematical function *f* was assumed for describing the relationship between response, η and factors, ξ_i as Eq. (1) and (2)

$$\eta = f(\xi_{1,}\ \xi_{2\ldots}\xi_4) \tag{1}$$

$$\eta = \beta_0 + \sum_{i=1}^{4} \beta_i X_i + \sum_{i=1}^{4} \beta_{ii} X_{ii}^{2} + \sum_{i=1}^{3} \sum_{j=i+1}^{4} \beta_{ij} X_i X_j \tag{2}$$

Where β_0 is the regression coefficient at the centre; β_i, $\beta_{ii}\beta_{ij}$ are linear, quadratic and second order coefficients respectively.

The coding of the ξ_i into X_i was performed by the following Eq. (3),

$$X_i = \frac{(\xi_i - \overline{\xi_i})}{d_i} \tag{3}$$

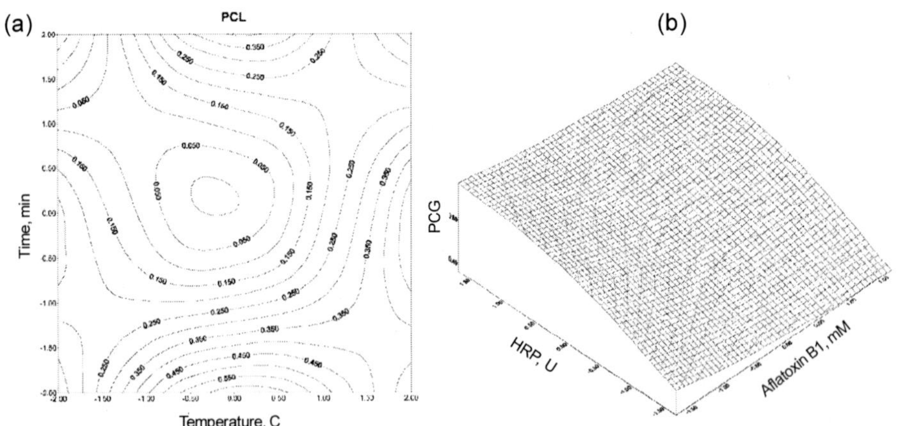

Fig. 5: (a)Isoresponse contour plot of Interaction of time and temperature at 4U HRP and 1 mM AFB_1 in liquid reaction mixture (b). Response surfaces showing interactions of independent factors on percent conversion of AFB_1 by HRP in groundnut meal (PCG), Interaction of AFB_1 and HRP at 4U HRP at 24 h incubation time

5.1.5 Effect of Aflatoxin B_1 detoxification on the physicochemical properties and quality of groundnut meal

A comparative evaluation of the capacity of AFB_1 contaminated GNM and the same after detoxification to initiate changes in the mean body weight, and liver to body weight ratio was made on 2 weeks old rat feeding tests. The mean body weight of the rats was increased when fed with all the 4 types of diet upto 6 th week. But after 6 weeks the mean body weight started decreasing in case of rats fed with diet A, i.e., containing 1.2 mM AFB_1 without any treatment. The mean body weight remained unaltered incase of rats fed with diet B ie, the diet supplemented by HRP treated AFB_1 and in creased for the two other sets. The liver to body weight ratio was measured at 2 weeks interval up to 8 weeks continuously. Data on mortality, body weight, food consumption and food efficiency ratio was recorded. The dose response curves of different organs and body weight with increase in AFB_1 content in diet were drawn (Figure 6 A-E).

The results on toxicity test for the reaction mixture after the reaction of AFB_1 with HRP in liquid medium showed that the inhibitory effect on *Bacillus megaterium* suspension culture started at 4.5 μg/mL of AFB_1. Growth rate remarkably deteriorated beyond 7.37μg /mL of AFB_1 and dropped to 3% growth in case of 10 μg/mL of toxin leading to complete inhibition with further increase in concentration. The inhibitory effects of enzyme treated GNM containing AFB_1 shows that the commercial enzyme having 200 U/ mg protein allows highest bacterial growth with respect to control suggesting that the toxic effect

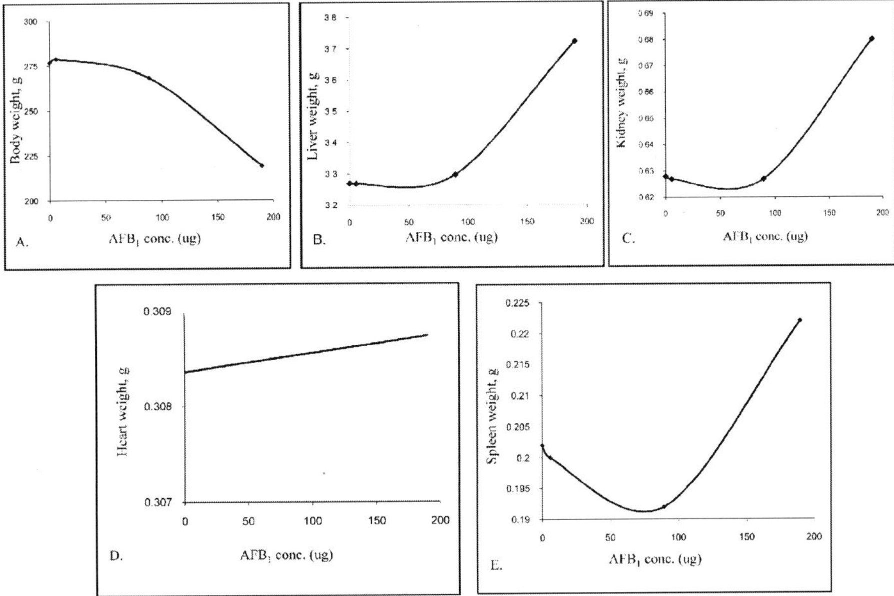

Fig. 6: (A-E) Dose response curves for AFB_1 concentration against body weight and individual organs after eight weeks of feeding experiment on rats

is least in the corresponding sample. The mutagenicity and carcinogenicity of AFB_1- HRP reaction products were tested and compared to authentic AFB_1 using *S. typhimurium* TA 100 as the tester strain. Purified AFB_1 judged mutagenic at a concentration of 0.9 µg/ plate beyond which number of revertants/ plate increases steeply. Other fractions elicited much less mutagenicity as expressed by lesser number revertants/ plate. A dose response curve resulted from Ames' test on *S. typhimurium* TA 100 was drawn. The assessment of physicochemical properties of treated GNM was discussed by Das and Mishra (1998, 2000c, 2000d).

5.2 Detoxification of aflatoxin in red chilli powder using garlic peroxidase

The specific activity of garlic peroxidase in raw juice was fifty four times less than the finally purified peroxidase fraction. The crude enzyme also had short stability; with drop in enzyme activity by two times from the initial in every ten min compared to insignificant change in enzyme activity over several hours at room temperature in the purified fraction. The autodigestion of enzyme in presence of other plant proteases or inhibitors in raw juice may be the reason for low stability (Parisi et al., 2002). The presence of strong pungent aroma of allicin and other flavor compounds present in garlic also made less practicability of the use of unpurified fraction of peroxidase in detoxification process. This states the necessity of enzyme purification. The enzyme was only partially purified by the use of simple biochemical procedures (salting and desalting) in place of advanced procedures like capillary electrophoresis and high pressure liquid chromatography to reduce the complexity of the process. The partially purified enzyme had high specific activity (236.6 U/mg protein) and was free from interfering plant components suggesting that the purification process adopted was adequate.

5.2.1 Extraction and purification of enzyme

An attempt is made to look for locally available alternative cheap source of peroxidase which exhibit properties similar to or better than horseradish enzyme which have aflatoxin detoxifying abilities. Garlic as an enzyme source is advantageous due to its easy availability round the year. The enzyme was partially purified to reduce the cost and complexity of the process. Its application in foods and feeds for aflatoxin degradation can be promising and thus requires proper investigation for finding the best conditions for its activity.The two step purification of enzyme increased its specific activity (Umg^{-1} protein) from 121.9 to 246.6. Peroxidase of specific activity greater than 100 Umg^{-1}proteins are widely used in various bioprocesses. Though extracellular peroxidase activity was seen in the crude enzyme extract but the specific activity was very low due to presence of interfering plant pigments and proteins; also the pungent smell of the principal flavor component of garlic (allicin) was prominent in this fraction making it unfit to be used in red chilli powder for degradation experiments.

Therefore, there was need to purify the enzyme further. The advantage of the purification procedure adopted in present investigation lies in its easy and fast operation compared to advanced purification procedures like capillary electrophoresis and high pressure liquid chromatography, etc. adopted for high purification of proteins. The main aim was to reduce the purification steps, while retaining the enzyme activity. The pH and temperature optima for the partially purified garlic peroxidase were 7 and 40 ºC, respectively. The residual activity of free enzyme at 40 °C after 140 min of incubation was 80% and at 90ºC complete loss in enzyme activity was recorded. The half life of enzyme at 70°C was 55 min and it could retain 37% of its initial enzyme activity at the end of four months of storage at 37 °C in 0.01 phosphate buffer. The K_m *and* V_{max} calculated for free partially purified garlic peroxidase was 14.2 mmol and 0.05 mmol/min respectively (Figure 7).

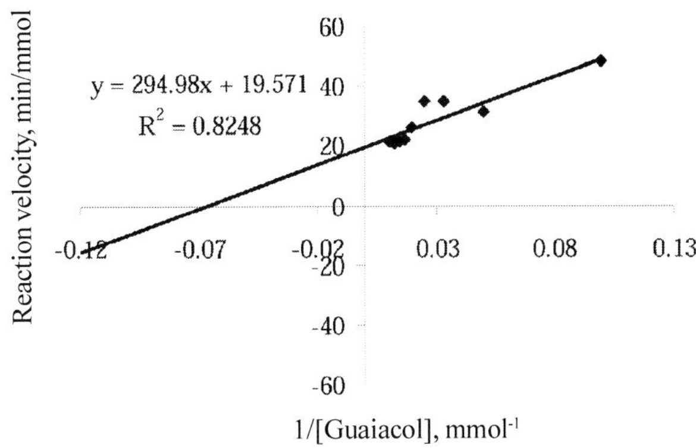

Fig. 7: LB Plot for free garlic peroxidase and guaiacol at 37 °C, pH 7

5.2.2 Enzymatic degradation of aflatoxin B_1 in red chilli powder

All the reactions were performed at 37±1 °C temperature and at neutral pH with varying enzyme and substrate concentrations over different incubation time. Hydrogen peroxide (80 mmol) was used as an oxidant as it is a prerequisite for any peroxidase catalysed reaction. Moisture content of the reaction mixture was recorded as 12±1% wb upon addition of AFB_1 and POD. For AFB_1-POD reaction; one unit of enzyme is defined as the amount of enzyme that is used to degrade one nanomole (nmol) of AFB_1 in 24 h.

5.2.3 Effect of substrate concentration on reaction velocity and AFB_1 conversion

Red chilli powder samples each of 100 g were spiked with 10 to 100 nmol AFB_1. The samples were made to react with 10U of enzyme and incubated for 24 h. When enzyme concentration is kept constant; the reaction velocity increases with increasing substrate concentrations (Figure 8).The maximum AFB_1 conversion (66.6%) was recorded at 60 nmol [AFB_1]. The reaction velocity and AFB_1 conversion (%) decreased further on increasing the AFB_1 concentration beyond 60 nmol when enzyme amount was fixed. At maximum substrate level (100 nmol AFB_1) only 32.6% AFB_1 degradation was recorded.

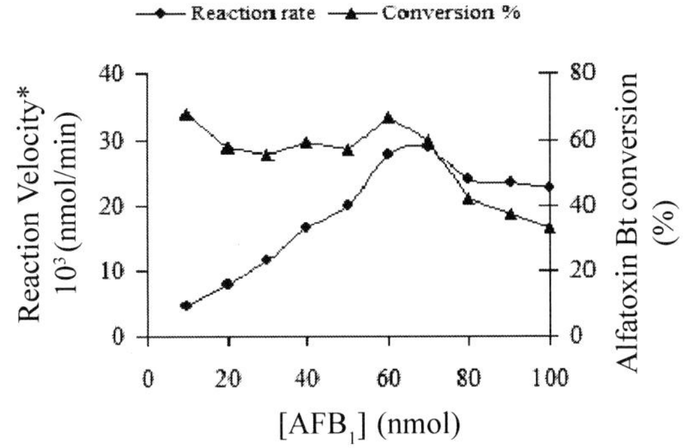

Fig. 8: Effect of POD concentration on AFB_1-POD reaction velocity and AFB_1 conversion % in red chilli powder (60 nmol AFB_1, 37 °C, 24h)

5.2.4 Effect of enzyme concentration on reaction velocity and AFB_1 conversion

For optimizing the enzyme concentration reaction was set up using 2 to 16U of enzyme for 60 nmol of AFB_1 per 100 g chilli powder at 37 °C. From the Figure 9, it is clear that at lower enzyme concentration the reaction velocity is almost negligible. The maximum reaction velocity (22.6x10^{-3} nmol/min) and AFB_1 conversion (54.3%) was achieved at 12U POD/nmol AFB_1. The reaction velocity beyond 12U POD was almost constant and increase in enzyme amount further had no significant effect on AFB_1 conversion % at fixed substrate concentration. Thus it can be inferred that the maximum level of enzyme for degrading 60 nmol of AFB_1/100 g chilli powder is 12U.

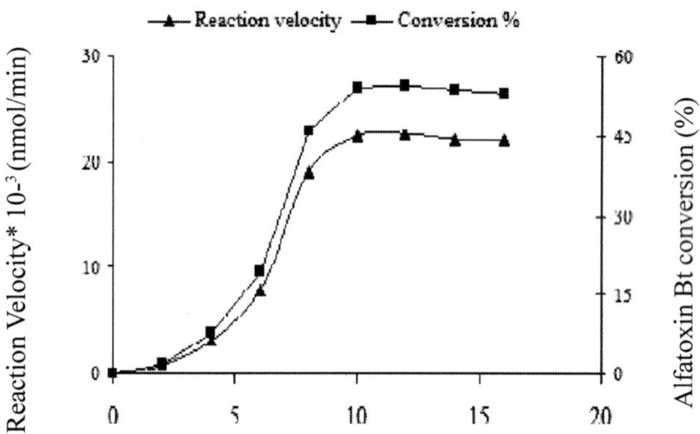

Fig. 9: Effect of POD concentration on AFB_1-POD reaction velocity and AFB_1conversion % in red chilli powder (60 nmol AFB_1, 37 °C, 24h)

5.2.5 Effect of incubation time on reaction velocity and AFB_1 conversion

Figure 10 shows effect of incubation time on AFB_1-POD reaction velocity and AFB_1 conversion % at optimum substrate and enzyme concentrations. As discussed earlier, 60 nmol of AFB_1/100 g chilli powder and 12U of enzyme was considered optimum; the time was varied from 1 to 30 h. The reaction of POD with AFB_1 was very slow in red chilli powder. There was no change in AFB_1 up to 8 h of incubation. Only at the end of 8 h of incubation the AFB_1 degradation started. Appreciable amount of degradation (42.21%) was only observed after 20 h of incubation. Maximum reaction velocity and AFB_1 conversion of 66.4% was achieved at 24 h.

Thus, overall by comparing the results from different parameters it is observed that maximum AFB_1 degradation of 66% was achieved with partially purified garlic POD at optimum reaction parameters (60 nmol AFB_1 per 100 g chilli powder, 12 U of enzyme) after 24 h of incubation.

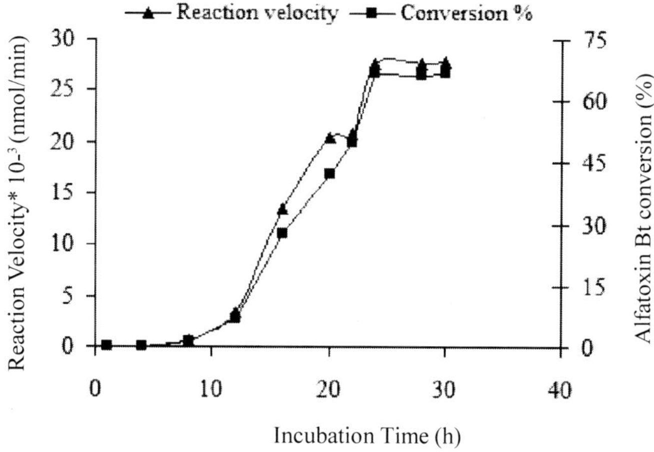

Fig. 10: Effect of incubation time on AFB_1-POD reaction in chilli powder (12 U POD, 60 nmol AFB_1, 37 °C)

5.2.6 Determination of K_m and V_{max}

From the previous experimental data on effect of varied AFB_1 concentration on reaction rate of POD-AFB_1 reaction in red chilli powder, a double reciprocal (LB) plot was drawn. The K_m and V_{max} of all the reactions catalysed by the partially purified garlic peroxidase using guaiacol and AFB_1 as substrates are summarized in Table 2.

Table 2: The K_m and V_{max} values for the partially purified garlic peroxidase (POD) catalysed reactions under different conditions

Enzyme-substrate	$K_{m,}$ mol	$V_{max,}$ mol/min
POD-guaiacol	$1.4*10^{-2}$	$5*10^{-5}$
Immobilised POD-guaiacol	$1.0*10^{-1}$	$2*10^{-4}$
POD-AFB_1 in buffer	$1.0*10^{-7}$	$50*10^{-9}$
POD-AFB_1 spiked in chilli powder	$1.0*10^{-7}$	$20*10^{-10}$

5.2.7 Modeling and optimization of enzymatic degradation of AFB_1 in chilli powder

The enzymatic degradation of AFB_1 in red chilli powder was also optimized using statistical modeling by RSM. The experiments on single factor optimization gave an idea about choosing the central values and levels of factors (variables) for RSM. The reduced quadratic model (Equation 4) was used to fit the experimental data (R^2= 0.92; model F value = 21.61; p <.001). The square and the linear terms of the independent variables in the quadratic model were found significant. The predicted R^2 (0.92) for this model was also in reasonable agreement with the adjusted R^2 (0.86).

$$Y = -101.92 + 0.48X_1 + 11.94X_2 + 7.14X_3 - 0.007X_1^2 - 0.44X_2^2 - 0.144X_3^2 \quad (4)$$

Where, Y = AFB_1 degradation (%); X_1 = substrate (AFB_1) concentration, nmol; X_2 = Enzyme (POD) concentration, U/nmol; X_3 = Time, h

The numerical and graphical optimizations were done for the experimental data. The optimum values predicted for the variables X_1, X_2, X_3 were 32.0 nmol, 13.0 U/nmol, 23.7 h, respectively with predicted response value of 74.5%. The actual percent degradation achieved at these optimum conditions was 71.2% which is highly correlated with the predicted value (R^2 = 0.99). The relative percent difference (RPD %) between actual and predicted was low (4.5%). The optimum conditions and maximum % AFB_1 degradation obtained from the study of individual effects of enzyme, substrate and incubation time on reaction rate and those obtained from the experiments according to CCRD are presented in Table 3. The one-at-a-time technique of optimization generally used in enzyme technology has some major flaws like it is time consuming; does not take into account simultaneous interactions of all the variables. Also, all the possible combinations of different variables to achieve maximum output in the reaction are not well established. Due to these drawbacks the simultaneous optimization using experimental design has become more common and when using experimental design, full factorial, partial factorial or central composite were the techniques of choice. From the present study also it is observed that after RSM optimization there was increase in the AFB_1 degradation % (Table 3). Compared to initial enzyme kinetic studies of AFB_1 degradation in red chilli powder, only one variable was varied at a time and other two were kept constant. It prevented the proper study of the interaction effect of all the variables simultaneously.

5.2.8 Effect of UV exposure on AFB_1 detoxification efficiency in red chilli powder

After the enzymatic detoxification of AFB_1 spiked chilli powder, the residual AFB_1 was still present. The residual AFB_1 in the detoxified red chilli powder under the parameters obtained after RSM optimization (32 nmol AFB_1, 13U POD, 37 °C after 24 h incubation) was 2.8±0.8 µg/kg (AFB_1 degradation = 71.2%). To enhance the degradation % in the enzyme treated sample were exposed to long wave UV (365 nm) light for different time intervals. The sample could be detoxified upto 90.9% (final AFB_1 =0.9±1.3 µg/kg) at the end of 60 min of UV exposure (Table 4). These results indicate that ultraviolet energy could be an effective factor in aflatoxin degradation. The degradation % of AFB_1 by UV exposure was time dependent.

Table 3: AFB_1 Degradation (%) as obtained by classical enzyme-substrate study and after optimization using RSM

Experimental parameters	Substrate (AFB_1), nmol	Enzyme, U/nmol	Incubation time,H	AFB_1 degradation (%)
One factor study (E1)	60	12	24	66.0
Optimization and modeling as per CCRD and RSM (E2) (Experimental)	32	13	26	71.2
Optimization and modeling as per CCRD and RSM (Predicted)	32	13	26	74.5

Table 4: Effect of long wave UV exposure on AFB_1detoxification from the spiked red chilli powder (after RSM optimum parameters)

Sample type / Treatments	Initial AFB_1 (µg/kg)	Residual AFB_1 (µg/kg)	Overall detoxification efficiency(%)*	t_{stat}	t critical (two tail)	Probability (tcritical<=tstat)^
Control (Originally spiked)	9.9±0.4	-	-	-	-	-
Enzyme treated (E2)	9.9±0.4	2.8±0.8	71.2	13.34	6.31	0.04*
UV(15min)	9.9±0.4	2.8±0.3	71.7	16.65	6.31	0.03*
UV(30min)	9.9±0.4	1.6±0.9	83.8	14.3	6.31	0.04*
UV(45min)	9.9±0.4	1.4±1.1	85.8	19.4	6.31	0.03*
UV(60min)	9.9±0.4	0.9±1.3	90.9	19	6.31	0.04*

Readings are mean ± S.D. for triplicate determinations. *Detoxification efficiency w.r.t control sample, *Significant change w.r.t. control using student's t test

5.2.9 Quality of the detoxified red chilli powder

The quality parameters which were highly affected by the AFB_1 detoxification were total carotenoids, ascorbic acid, total proteins, ASTA color index and antioxidant activity (RSA). Effect of all the detoxification treatments were insignificant on the capsaicin, sugar, fat, fiber, ash and phenol contents of the untreated and treated samples. Effect of various detoxification processes on quality parameters of red chilli powder is summarized in Table 5. Overall from the results of the quality parameters and effect of UV exposure on AFB_1 degradation % it can be concluded that incubation of infected material at 60 nmol initial AFB_1 (with 12U POD) and 32 nmol initial AFB_1 (with 13U POD) for 24 h followed by UV exposure (365 nm) for 30 min leads to AFB_1 degradation up to 70.3% (final AFB_1=5.8 μg/kg) and 83.8% (final AFB_1 = 1.6±0.9 μg/kg), respectively. These optimum detoxification parameters produced minimal nutritional loss in the detoxified chilli powder.

5.2.10 Safety of the detoxification process

Toxicity studies done with the detoxified red chilli powder extract showed that the untreated sample (originally spiked, AFB_1 = 20 μg/kg) caused maximum bacterial inhibition (92%).The detoxification treatment lowered the bacterial growth inhibition. The inhibition (%) of 42.8 and 33.2 was recorded in enzymatically detoxified and enzyme-UV detoxified sample, respectively. The rat feeding studies revealed that the rats fed on detoxified and control diet utilized the food effectively with high FER values (0.04 and 0.07 g gain/g feed consumed) respectively. The presence of the high concentration of the toxin in AFB_1 fed group resulted in poor metabolism in the rats, high mortality, low FER and weight gain. Rats fed with diet containing standard AFB_1 showed marked increase in all the serum enzymes (specific to liver) studied when compared to control. The levels of the serum enzymes in the experimental group fed with the detoxified reaction mix and the control group differed insignificantly ($p>.05$). Though, the amount of ALP in the group fed on diet containing detoxified mix showed some differences from the control sample but the increase was not as significant as in the case of group fed on AFB_1 spiked diet ($F=20020.0$; $p<.001$). The elevated levels of the serum hepatic enzymes in aflatoxin spiked diet are the positive sign of AFB_1 induced liver damage. The increase in liver to weight ratio and total bilirubin also suggests the proliferation of liver cells due to some toxicity in this group. Though the slight increase in the total bilirubin and protein content was also observed in the treated group (detoxified diet) the increase was not as significant as in the case of untreated group. The triglycerides levels were also elevated in group fed on AFB_1 spiked diet (Table 6).

Table 5: Quality attributes of AFB_1 detoxified red chilli powder

Quality Parameter	Control (AFB_1 free sample)	Enzym etreated [*1]	Enzyme-UV treated sample[*2] Duration of UV exposure (time in min)			
			15	30	45	60
Capsaicin (mg/100g)	0.99 ± 0.09	$0.99\pm1.6^{n.s}$	$0.99\pm0.3^{n.s}$	$0.97\pm0.3^{n.s}$	$0.98\pm0.7^{n.s}$	$0.98\pm1.1^{n.s}$
Carotene (mg/100g)	128.5 ± 0.4	123.3 ± 0.1	123.6 ± 0.1	$98.9\pm0.7^{*}$	$62.9\pm1.2^{***}$	$62.0\pm1.3^{***}$
ASTA ColorIndex	110 ± 1.0	98.3 ± 0.5	98.6 ± 1.15	97.7 ± 0.4	$75.7\pm0.2^{***}$	$70.5\pm0.1^{***}$
Ascorbic acid (mg/100g)	88.1 ± 0.5	86.3 ± 1.1^{ns}	85.3 ± 0.5^{ns}	$71.5\pm1.1^{*}$	$28.62\pm0.5^{***}$	$24.94\pm1.1^{***}$
RSA (%)	95.8 ± 1.6	90.7 ± 1.1	$87.3\pm0.3^{*}$	$80.0\pm0.5^{*}$	$27.4\pm0.5^{***}$	$25.61\pm1.1^{***}$
Total phenol (mg/100g)	494.9 ± 0.5	$548.67\pm0.5^{***}$	$535.5\pm2.8^{***}$	$487.3\pm0.1^{*}$	495.2 ± 0.0^{ns}	$462.5\pm0.5^{**}$
Total protein (%)	18.0 ± 0.1	$32.1\pm0.1^{***}$	$28.1\pm0.0^{***}$	$24.3\pm0.2^{***}$	$24.0\pm0.0^{***}$	$24.08\pm0.7^{***}$
Total sugars (%)	25.0 ± 0.0	25.0 ± 0.4^{ns}	25.3 ± 0.7^{ns}	25.4 ± 0.2^{ns}	25.0 ± 0.0^{ns}	25.2 ± 0.4^{ns}
Total Fat (%)	13.2 ± 0.3	13.3 ± 0.5^{ns}	13.4 ± 0.2^{ns}	13.2 ± 0.5^{ns}	13.3 ± 0.1^{ns}	13.3 ± 0.7^{ns}
Crude Fiber (%)	19.2 ± 0.4	18.4 ± 0.5^{ns}	18.6 ± 0.5^{ns}	18.4 ± 0.6^{ns}	18.4 ± 0.6^{ns}	18.2 ± 0.6^{ns}
Ash (%)	9.3 ± 0.1	9.7 ± 0.0^{ns}	9.3 ± 0.2^{ns}	9.3 ± 0.8^{ns}	9.6 ± 0.4^{ns}	9.5 ± 0.0^{ns}

*1: 13 U POD/nmol AFB_1 incubated for 24 h at 37 °C

*2: Enzyme treated followed by UV exposure (365 nm) for different durations

Significant change w.r.t. control using one way ANOVA with post-hoc comparison by tukeys test $^{***}p<.001$; $^{**}p<.01$; $^{*}p<.05$; $^{n.s.}$ not significant ; Readings are mean ± S.D. for triplicate determinations

Table 6: Liver function test data for the three experimental groups of male wistar rats at the end of 21 days experimental period

Parameter	Control*1	Untreated*2	Treated*3
AFB_1 (ìg / kg)	0.00	40.00	5.00
Mortality	0.00	3.00	1.00
Liver to body weight ratio	0.02	0.05***	0.03n.s.
Total bilirubin (mg/dl)	0.42	1.78***	0.59*
Total Protein (mg/dl)	7.20	15.24***	8.20*
SGOT (U/L)	32.66	154.45***	33.64 n.s
SGPT (U/L)	30.50	129.01***	34.06 n.s.
ALP (U/L)	112.10	192.20***	116.40*
ACP (U/L)	116.40	180.90***	112.60 n.s.
GGT (U/L)	42.72	70.48***	42.24 n.s.
HTG (mg /g)	10.04	22.20***	10.41 n.s.

*1: Control aflatoxin free standard diet; *2: Group fed on AFB_1 spiked diet; *3: Group fed on enzyme-UV detoxified diet. One way ANOVA followed by post- hoc test to see the LSD at 5%, ***p<.001; **p<.01; *p<.05

6. Summary

To conclude the overall process it can be stated that, peroxidase group of enzymes can be successfully used to detoxify aflatoxin B_1 in liquid medium as well as contaminated ground nut meal. The cost of detoxification is cheap and affordable. The enzyme performed better in the liquid medium than in GNM. Detoxified meal had no adverse effect on its organoleptic qualities and nutritional value. Treated meal was highly acceptable by animals (mice), which showed least or no deleterious effects after 8 months of feeding experiments. Enzymatic detoxification decreased toxicity and mutagenicity of the treated product.

The reaction velocity in red chilli powder at constant enzyme and substrate concentration was nearly twenty five times slower than the liquid buffer system. This may be due to lack of optimum enzyme-substrate contact in a heterogeneous reaction mixture. The increasing substrate concentration increased the reaction rate or in other words increased AFB_1 degradation in red chilli powder. Number of effective collisions between enzyme and substrate grows with increased substrate concentration.

Microwave/ UV exposure was only given in post enzyme treated sample when level of aflatoxin is already reduced (low amount of aflatoxin present which needs to be degraded further) so that less exposure is needed. The longer UV exposure resulted in significant change in carotenoids, ascorbic acid and color of the treated red chilli powder. Some loss in color and carotenoids was also recorded in enzyme treated sample alone but it was comparatively less significant than the enzyme-UV treated sample. The reduction in final AFB_1 content in post enzyme-UV treated was three times more than the enzyme treated sample alone.

The safety of the detoxification process was tested in terms of decrease in carcinogenic, mutagenic and hepatotoxic potential of AFB_1 molecule after the detoxification process. The bacteria cultured in media having detoxified reaction mix showed lower growth inhibition or higher colony count, while the media containing standard AFB_1 resulted in poor bacterial survival. This suggests that the toxic manifestation of the AFB_1 molecule was lowered considerably after detoxification process. Similar results were observed with mutagenic studies on *S. typhimurium*. The liver function test of rats fed with diet spiked with detoxified reaction mix was also normal and comparable to the normal healthy control group. The toxicity, carcinogenicity and mutagenicity of the AFB_1 molecule dropped significantly followed by enzymatic and UV degradation. The detoxification treatment might have produced some change in structure of the AFB_1 molecule making it less toxic. This needs further investigations. The proper pathway of the degradation is still not understood completely. From the overall results discussed so far reduction in AFB_1 content is confirmed by the partially purified garlic peroxidase and UV light. Overall from the nutritional and safety studies it is confirmed that process is safe to be used in foods. Genetically engineered enzymes and their mechanism of action is yet to be investigated. Usability in large scale application and optimization of post treatment storage parameters should also be considered.

References

Adak, S., Mazumdar, A. and Banerjee, R. K. 1996. Probing the active site residues in aromtic donor oxidation in Horse radish peroxidase: involvement of an arginine and tyrosine residue in aromatic donor binding, *Biochemical Journal*, 314(3): 985- 991.

Ames, B.N., Mc. Cann J. and Yamasaki E. 1975 b. Methods of detecting carcinogens and mutagens with the Salmonella / mammalian microsome mutagenicity test. *Mutation Research*, 31, 347- 364.

Allen, R. C. 1975. The role of pH in the chemiluminescent response of the myeloperoxidase-halide- HOOH antimicrobial system. *Biochem. Biophys. Res. Commun,* 63: 684- 685.

AOAC, 1995. *Official Methods of the Association of Official Analytical Chemists* (15th edn.), Washington DC.

Applebaum, R. S., Brackette R. E., Wiseman D.W. and Marth E. H. 1982. Aflatoxin: Toxicity to dairy cattle and occurrence in milk and milk product- A review. *Journal of Food Protection,* 45: 752- 777.

Aydin, A., Erkan ME, Baokaya R and Ciftcioglu G 2007. Determination of Aflatoxin B_1 levels in powdered red pepper. *Food Control,* 18:1015–1018

Barman, T. 1972. Valuable summary of the major properties of the known enzymes classified according to the International rules, *Enzyme Handbook.* Vol –1, Springer- Verlag, New York.

Basha, S.M.M. and Cherry J. 1976. Composition, solubility and gel electrophoretic properties of proteins isolated from Florunner (*Arachis hypogaea* L.) peanut seeds. *Journal of Agricultural and Food Chemistry,* 24: 359-365.

Bela ,J. and Shetty, K. 1999. Somatic embryogenesis in Anise (*Pimpenella anisum* L): The effect of proline on embryogenic callus formation and ABA on advanced embryo development. *Journal of Food Biochemistry,* 23:17-32.

Bradford, M. M. 1976. A rapid and sensitive method for the quantitation of microgram quantities of protein utilizing the principle of protein die-binding. *Analytical Biochemistry*, 72:248-254.

Bankole, S. A. and Mabekoje, O.O 2004. Occurrence of aflatoxins and fumonisins in preharvest maize from south-western Nigeria. *Food Additives and Contaminants,* 21(3): 251-5.

Bhat, R.V. 1989. Alatoxin contamination of groundnuts: Proceedings of the International Workshop. 6-9 Oct (1987), ICRISAT, India.

Brown, R. L. andChen Z.Y, 2001. Resistance to aflatoxin accumulation in kernels of maize inbreds selected for ear rot resistance in West and Central Africa. *Journal of Food Protection,* 64(3): 396-400.

Chakrabarty A. B. 1981. Detoxification of aflatoxin in Corn. *Journal of Food Protection,* 44:173-176.

Chaurasia, H.K. and Sinha, R. K. 1994. Potential of biological control of aflatoxin contamination in developing peanut (*Arachis hypogaea* L) by atoxigenic strains of *A. Flavus*. *Journal of Food Science and Technology,* 31: (5):62-366.

Chu, F. S. 1984. Immunoassays for analysis of mycotoxins. *J. of Food Protection*, 47(7):562-569.

Ciegler, A., Lillehoj E.B., Peterson R. E. and Hall, H. H. 1966. Microbial detoxification of aflatoxin. *Applied Microbiology*, 14:(6): 934-939.

Cullen, J.M. & Newberne, P.M. 1994. Acute hepatotoxicity of aflatoxins. In: Eaton DL, Groopman JD, eds. The toxicology of aflatoxins: human health, veterinary, and agricultural significance, London, Academic Press.1993:1-26.

Das, C. and Mishra, H. N. 2003. A review on Biological control and metabolism of aflatoxin. *CRC Critical Review on Food Science and Nutrition,* 43(3):245-264.

Das, C. and Mishra, H.N. 2001. Effect of Aflatoxin B_1 detoxification on some *in vivo* and *in vitro* plant physiological systems.*Acta Physiologaea Plantarum*, 23(3): 311-317.

Das, C. and Mishra H. N. 2000a. *In vitro* degradation of Aflatoxin B_1 by horse radish peroxidase. *Food Chemistry*, 68:309-313.

Das, C. and Mishra H.N. 2000b. *In vitro* degradation of Aflatoxin B_1 in ground nut (*Arachis hypogaea*) meal by horse radish peroxidase. *LWT-Food Science and Technology,* 33(4):308-312.

Das, C. and Mishra H. N. 2000c. Effect of Aflatoxin B_1 detoxification on the physicochemical properties and quality of Ground nut meal. *Food Chemistry,* 70(4):483- 487.

Das, C. and Mishra H.N. 1998. Biodegradation of Aflatoxin. Poster abstracts of 'Fourth International Food Convention', Mysore, 23-27 Nov.

Das, C. and Mishra H.N. 2000d. A novel approach to enzymatic detoxification of Aflatoxin B_1. Proceedings of 'The International Conference on Processed Foods for 21st Century', Calcutta during14- 16 Jan.

Ellis, W. O., Smith, J. P., Simpson, B. K., Oldham, J. H., & Scott, P. M. 1991. Aflatoxins in food: occurrence, biosynthesis, effects on organisms, detection, and methods of control. *Critical Reviews in Food Science & Nutrition*, 30(4), 403-439.

Fandohan, P., B. and Gnonlonfin B 2005. Natural occurrence of Fusarium and subsequent fumonisin contamination in preharvest and stored maize in Benin, West Africa. *International Journal of Food Microbiology,* 99(2):173-183.

Fung, F, Clark RF 2004. Health effects of mycotoxins: a toxicological overview. *Journal of Toxicology: Clinical Toxicology*, 42(2):217-34.

Gamage, T. V., Samarajeewa U., De S. G., Wettimuny S. and Arseculeratne S. N. 1985. Nontoxicity to ducklings of solar irradiated coconut oil contaminated with aflatoxin. *Applied Microbiology and Biotechnology,* 1:239-245.

Gong, Y. Y., K. Cardwell, et al. 2002. Dietary aflatoxin exposure and impaired growth in young children from Benin and Togo: cross sectional study. *Bmj,* 325(7354): 20-21.

Guenrich F. P. and Parikh A. 1997. Expression of drug metabolizing enzymes. *Current Opinion in Biotechnology,* 8: 623- 628.

Hawkins, L. K. and Windham, G. L. 2005. Effect of different postharvest drying temperatures on Aspergillus flavus survival and aflatoxin content in five maize hybrids. *Journal of Food Protection,* 68(7): 1521-1524.

Hamid, A. B., Smith J.E. 1987. Degradation of aflatoxin by *Aspergillus flavus. J. genl. Microbiology,* 133, 2023- 2029.

Heathcote, J. G. and Hibbert J. R. 1978. Biochemical effects structure activity relationship. In Goldblatt L. A. ed. Aflatoxin: Chemical and Biological aspects. Elsevier Scientific Publishing Co. Amsterdam, 112-130.

Janssen, D. B. and Schanstra 1994. Engineering proteins for environmental applications. *Current Opinion in Biotechnology,* 5: 253- 259.

Jiang, Y., P. E. Jolly, et al. 2005. Aflatoxin B1 albumin adduct levels and cellular immune status in Ghanaians. *International Immunology,* 17(6): 807-14.

Joseph, Bravo P. I., Findley M., Newbern P.M. 1976. Some interactions of light, riboflavin and aflatoxin B_1 *in vivo* and *in vitro. Journal of Toxicology and Environmental Health,* 1(3): 353- 376.

Karunaratne, A. Wezenberg E. and Bullerman L. B. 1990. Inhibition of mold growth and aflatoxin production by *Lactobacillus* sp. *Journal of Food Protection,* 53: 230-234.

Laemmli, U.K. 1970. Cleavage of structural proteins during the assembly of the head of bacteriophage T_4. *Nature* (London). 227:680-685.

Lee, L. S., Dunn J. J., Delucca A. J. and Ciegler A. 1981. Role of lactone ring of aflatoxin B_1 in toxicity and mutagenicity. *Experientia,* 37:16- 17.

Monteiro, P. V., Rao K. S., Prakash V. 1966. *In vitro* interaction of Groundnut proteins with AFB_1. *Journal of Food Science and Technology,* 33(1):27 -31.

Moss M. O., and Smith J. E. 1985. Mycotoxins formation, Analysis and significance. John Wiley and Sons, Chichester.

Parisi, M., Moreno, S. and Fernández, C. 2002. Characterization of a novel cysteine peptidase from tissue culture of garlic (Allium *sativum* L.). *In Vitro Cellular & Developmental Biology – Plant,* 38: 608-612.

Patten, R.C. 1981. Aflatoxins and disease. *The American Journal of Tropical Medicine and Hygiene,* 30(2):422-425.

Pellet, L.P. and Young, V.R. 1980. Nutritional evaluation of Protein food. UN University publication.

Rhee, K. C., Natarajan K. C., Cater C. M., Mattiel K. F. 1976. Processing edible peanut protein concentrates and isolates to inactivate aflatoxin. *Journal of the American Oil Chemists' Society,* 42:467-471.

Samarajeewa, U., Sen A. C., CohenM. D. and Wei CI. 1990. Detoxification of aflatoxins in Foods and Feeds by physical and chemical methods. *Journal of Food Protection,* 53(6):489- 501.

Sharma, R.S., Trivedi K. R., Wadodkar U. R., Murthy T. N. and Punjarath J. S. 1994. Aflatoxin B_1 content in deoiled cakes, cattle feeds and damaged grains during different seasons in India. *Journal of Food Science and Technology,* 31(3):244-246.

Shantha, T. 1994. Methods of estimation of aflatoxins: A critical appraisal. *Journal of Food Science and Technology,* 31(2): 91-103.

Staron, T. D., Thirouin, Perrin, L. & Frere, G. 1980. Microwave treatment of biological food production. *Indust. Aliment. Agricol,* 12:1305-1312.

Temcharoen, P., Thilly, W.G. 1982. removal of aflatoxin B_1 toxicity but not mutagenicity by 1 Mrad gamma radiation on peanut meal. *Journal of Food Safety,* 4: 199-205.

Turner, P. C. and Sylla, A. 2005. Reduction in exposure to carcinogenic aflatoxins by postharvest intervention measures in West Africa: a community-based intervention study. Lancet 365(9475): 1950-6.

Vesonder, R.F., Ciegler A., Rogers R.F., Burbridge K.A., Botharst R.J. and Jensen A.H. 1978. Survey of 1777 crop year preharvest corn for vomitoxin. *Applied and Environmental Microbiology,* 36:885-888.

Walther, H. J., Parker, C. E., Harvan D. J., Voyksner R. D., Hernandez O., Hagler W. M., Hamilton P. B., Hass J. R. 1983. Fast atom bombardment Mass spectrometry of aflatoxins and reaction products of sodium bisulfite with aflatoxins. *Journal of Agricultural and Food Chemistry*, 31:1.

Wantabe, C. M. H., David, W., Linz J. E., Townsend C. A. 1996. Demonstration of catalytic roles and evidence for the physical association of type I fatty acid synthases and polyketide synthase in biosynthesis of AFB_1. *Chemistry and Biology*, 3:463-469.

Williams, J.H., Phillips, T.D., Jolly, P.E., Stiles, J.K., Jolly, C.M. and Aggarwal, D. 2004. Human aflatoxicosis in developing countries: a review of toxicology, exposure, potential health consequences, and interventions. *American Journal of Clinical Nutrition,* 80: 1106-1122.

Wilson, D. M. and G. A. Payne 1994. Factors Affecting *Aspergillus flavus* Group Infection and Aflatoxin Contamination of the Crops. The Toxicology of Aflatoxins: Human Health, Veterinary and Agricultural Significance. D.L. Eaton and J.D. Groopman. San Diego, CA, Academic Press, Inc.: 309-325.

Index

A

B

C

D

E

G

H

I

J

K

L

M

N

O

P

Q

R

S

T